ENVIRONMENTAL MANAGEMENT IN PRACTICE: VOLUME 3

Methods of environmental management, and especially the 'tools' of environmental management, are increasingly being relied upon world-wide to deliver a degree of sustainability in all human activities. A thorough understanding of the nature, capabilities and limitations of these 'tools', as well as the conditions under which they can be best applied, is essential for students, researchers and practitioners within the field of environmental management.

Environmental Management in Practice presents three comprehensive volumes containing the most up-to-date research and practical applications in the field. Spanning the four main aspects of environmental management: instruments, compartments, sectors and ecosystems, this three-volume work contains over sixty contributions from leading specialists in each field and offers the first major source of contemporary international research and application within environmental management in practice.

Volume 1: Instruments for Environmental Management, focuses on the instruments and tools currently available to the environmental manager. A theoretical background to the instruments is given together with an overview of those instruments that are in common use today, with particular attention to the physical, economic, legislative and communication instruments.

Volume 2: Compartments, Stressors and Sectors, deals with the problems that occur in the three 'compartments' of the environment – namely, air, water and soil. The contributors also address the socio-economic sectors of industry, traffic, energy, agriculture and tourism.

Volume 3: Managing the Ecosystem, focuses on those ecosystems in which human intervention has been or continues to be predominant, specifically within cities and rural areas.

Packed with accessible and up-to-date information, these three volumes provide a comprehensive overview of environmental management for those studying, researching and practising in the field.

Bhaskar Nath is Director of the European Centre for Pollution Research, London. **Luc Hens** is Professor and Head of the Human Ecology Department at the Free University of Brussels. **Paul Compton** is an environmental and demographic consultant. **Dimitri Devuyst** is Co-ordinator of the programme of Environmental Impact Assessment in the Department of Human Ecology, Free University of Brussels.

ENVIRONMENTAL MANAGEMENT IN PRACTICE: VOLUME 3

Managing the Ecosystem

EDITED BY
B. NATH, L. HENS,
P. COMPTON AND D. DEVUYST

London and New York

First published 1999
by Routledge
11 New Fetter Lane, London EC4P 4EE

Simultaneously published in the USA and Canada
by Routledge
29 West 35th Street, New York, NY 10001

Typeset in Galliard by The Florence Group, Stoodleigh, Devon
Printed and bound in Great Britain by TJ International, Padstow, Cornwall

British Library Cataloguing in Publication Data
A catalogue record for this book is available from the British Library

Library of Congress Cataloguing in Publication Data
Environmental management in practice / edited by B. Nath . . . [et al.].
p. cm.
Includes bibliographical references and index.
Contents: v. 1. Instruments for environmental management
– v. 2. Compartments, stressors and sectors – v. 3. Managing
the ecosystem.
1. Environmental management.   I. Nath, Bhaskar.
GE300.E577 1999   97–52071
363.7–dc21   CIP

ISBN 0-415-18791-5

CONTENTS

FIGURES

TABLES

BOXES

CONTRIBUTORS

Bostjan Anko, Ph.D., is a full professor in the Department of Forestry at the University of Ljubljana, Slovenia. He is co-chairman of the IUFRO Working Party on Landscape Ecology and a member of the IUCN Commission on Environmental Education and Communication.

Fernando Dias de Avila-Pires holds a D.Sc. in zoology from São Paulo State University and is a member of the Brazilian Academy of Sciences. He has been a professor in the Department of Tropical Medicine in the Oswaldo Cruz Institute in Rio de Janeiro since 1993. Professor Avila-Pires is a past Fellow of the John Simon Guggenheim Foundation and has served as Co-Associate Director of the US National Committee for the International Biological Program. He is a member of several professional associations including the American Society of Mammalogists and the Association for Tropical Biology.

Jozef Buys is a Chargé de Mission at the Belgian Agency for Development Co-operation. He is a 'licentiaat' in the history of art and archaeology and has spent the bulk of his professional career in South America where he has been involved in different excavation projects of pre-Columbian cultures.

Paul A. Compton, M.Sc., Ph.D., formerly Professor in the School of Geosciences at the Queen's University of Belfast, Northern Ireland, is an environmental and demographic consultant. His academic interests lie at the interface between the natural and social sciences.

Dimitri Devuyst received his Ph.D. degree in human ecology from the Free University of Brussels (Belgium), as well as a master's degree in human ecology and a bachelor's degree in botany from the same university. His main interests are in the field of environmental impact assessment. Currently he is Postdoctoral fellow of the Fund for Scientific Research – Flanders in the Department of Human Ecology at the Free University of Brussels.

Thomas A. Eddy holds a Ph.D. in agriculture from Kansas State University, Manhattan, Kansas, USA. His graduate studies were in plant ecology, entomology and wildlife biology. Currently he is an associate professor in the Division of Biological Sciences at Emporia State University, Kansas.

Susan Gubbay received a B.Sc. in ecology and a D.Phil. in marine ecology from York University, UK, and is a Fellow of the Royal Institution of Chartered Surveyors. She was a senior conservation officer with the Marine Conservation Society and is currently a self-employed consultant in the field of coastal management. She has acted as a specialist adviser to the House of Commons Environment Select Committee on coastal zone management and protection.

Pham Hoang Hai is a Chief Researcher at the Institute of Geography in the National Centre for Natural Science and Technology, Hanoi, Vietnam. His research focuses on the ecological approach to landscapes and the environmental aspects of country planning. In this work special emphasis is given to rural areas. Dr Hai is a specialist in environmental impact assessment of dams.

Richard J. Huggett was educated at the University of London where he received his BA and Ph.D. degrees. Currently, he is a senior lecturer in geography at the University of Manchester.

Bruno Kawasange is a wildlife management officer in the Department of Wildlife, Tanzania. He holds a diploma in wildlife management from the College of African Wildlife Management, Mweka, Moshi, Tanzania and a B.Sc. in zoology and wildlife ecology from the University of Dar Es Salaam. He holds a master's degree in human ecology from the Free University of Brussels.

Nguyen Ngoc Khanh is a geographer and researcher in the Department of Environmental Impact Assessment at the Institute of Geography in the National Centre for Natural Science and Technology, Hanoi, Vietnam. He is a close collaborator of Dr Pham Hoang Hai, with whom he shares interest in landscape ecology, urban and country planning and environmental effects of dams. Dr Khanh also specialises in impacts of tourism and industrial developments.

Philippe Lefèvre-Witier is a medical doctor and haematologist with a Ph.D. in human biology and genetics. He has turned to anthropological genetics. As such he directed interdisciplinary research projects in Algeria, Nepal, Mali, Niger, France and Mexico. In 1966 he joined the French National Centre for Scientific Research where he is currently research director. At the University of Toulouse he manages the International Certificate in Human Ecology.

Julian D. Orford, BA, M.Sc., Ph.D., is Professor of Physical Geography in the School of Geosciences at the Queen's University of Belfast. For over twenty-five years he has been conducting research on coastal geomorphology and sedimentology and is a noted author on the subject of beach and coast development. Recently he has been examining the response of coastal morphology both to sea level change and to increased storminess related to climatic change.

R. Laurie Robbins holds a Ph.D. in botany from Texas Tech University, Lubbock, Texas, USA. She had a National Science Foundation post-doctoral fellowship at Missouri Botanical Garden, where she is currently a research associate. She has held positions at Allegheny College, Pennsylvania, at Texas Tech University and is presently an associate professor in the Division of Biological Sciences at Emporia State University, Kansas.

Paul P. Schot is a geohydrologist in the Department of Environmental Studies at the University of Utrecht. His interests lie in the relationship between hydrology and wetland ecosystems, and the impact of human activities. He has worked in the field of integrated water management and eco-hydrology in the Netherlands and has acted as a consultant on water resources management for the Ministry of Water in Burkina Faso. Dr Schot is an independent consultant for the Dutch Commission on Environmental Impact Assessment.

Bernard J. Smith, B.Sc., Ph.D., is Professor of Physical Geography at the Queen's University of Belfast. Prior to joining Queen's University he was a soil surveyor in the Cote D'Ivoire and lecturer in geography at Ahmadu Bello University, Zaria, Nigeria. Professor Smith specialises in environmental issues and tropical geomorphology. Since 1990 he has led an interdisciplinary study of erosion hazard and landscape change in south-eastern Brazil.

Roy W. Tomlinson, B.Sc., Ph.D., was a lecturer in geography at the University of Rhodesia from 1970 to 1974. Since 1970 he has been a lecturer and senior lecturer in geography at the Queen's University of Belfast. His research interest lies in upland land cover, conservation and remote sensing.

Brian Whalley, B.Sc., Ph.D, is Professor at the Queen's University of Belfast. His work is focused on aspects of mountain geomorphology and glaciology with current field work in Norway, Iceland and the Alps. Other current research is the analysis and simulation of building stone weathering.

David N. Wilcock earned a BA in geography at the University of London and a Ph.D. in geography at the University of Liverpool. He was Fulbright-Hays scholar and visiting lecturer in geography at the Western Washington State University. He has been a university teacher in Northern Ireland since 1964 and was appointed Professor of Environmental Studies at the University of Ulster in 1987. He is also a member of the Northern Ireland Council for Nature Conservation and the Countryside.

PREFACE AND ACKNOWLEDGEMENTS

Environmental management draws its knowledge base from across the spectrum of disciplines – the natural, social and medical sciences, the humanities and engineering. It aims to maintain a harmonious relationship between the environment and human society, and in its approach to this adopts a holistic, interdisciplinary stance. Since value judgements are an integral part of environmental management, it is as much an art as a science in its methodology and application.

The growth of interest in environmental management is relatively recent. It reflects a widely held perception of accelerating environmental deterioration caused by the pressure of human activities, as evidenced by worsening problems of pollution and the destruction of natural landscapes and habitats. These concerns can be traced back to the 1960s, when the interconnectedness of nature was vividly demonstrated by the way in which seemingly benign activities such as the chemical control of pests could, by diffusing through the food chain, produce adverse environmental effects in regions ostensibly untouched by man's activities.

As our knowledge of the global environment has grown, other worrying effects have come to light. The emission of greenhouse gases is linked to global warming and climate change. Although we do not fully understand the probable effects of this, it may well result in greater temperate aridity and so jeopardise the world grain supply, with potentially disastrous consequences. Moreover, resultant changes in sea level could submerge major coastal sites of population.

There is also the well-established connection between CFC emissions, the depletion of upper atmosphere ozone, and increased ultra-violet radiation at the planet's surface. This has negative implications not only for human health but also for the well-being of other species. Likewise, the destruction of the tropical rain forest is seen as a grave threat to biodiversity and the world's gene pool. The fact that these hazards are the subject of internationally agreed measures of amelioration (albeit implemented with variable commitment by individual countries) testifies to the potential gravity of global warming, ozone depletion and loss of biodiversity.

These global issues also raise concerns at the level of ecosystems. The effects of modern agricultural practices on environmental quality are a case in point. Pesticide and fertiliser residues pollute the groundwater; animal and plant habitats are destroyed as hedgerows are removed and wetlands drained in the interest of intensive cultivation; soil structure is broken down, creating problems of soil erosion. Now, in addition, intensive rearing of plants and animals is even causing concern for the safety and wholesomeness of the food produced. Populations are no longer willing to accept assurances from experts that genetically engineered crops are safe, or that it is right to feed natural herbivores, such as cattle, protein supplements derived from the rendered remains of other animals.

Of course it is not only agriculture that is problematic. Urban living and its associated activities can be just as destructive of the environment; not least, the creation of built environments where residential, commercial and industrial areas and communications infrastructures either obliterate or radically change pre-existing landscapes and ecosystems. Moreover, urban systems depend upon the mobility of people and goods for their effective functioning, and so create the traffic problems associated with further detrimental effects on the environment. The problems caused by excessive use of energy and natural resources in production and consumption, and their implications for future generations, also have to be tackled. Measures to ensure effective waste disposal and the curbing of air and water pollution are vital for the maintenance of environmental quality.

When viewed over a longer time-scale, however, the environmental picture is somewhat less gloomy. For example, popular coverage might lead one to suppose that human activity is the only cause of climatic change: the evidence does, after all, appear compelling, with the atmospheric content of the major greenhouse gas, carbon dioxide, having risen progressively since the start of industrialisation in the seventeenth century. But the record also shows that the world's climate has fluctuated markedly both in the recent geological and even historical past, and scientists still disagree on whether the rise now observed in global temperature should be attributed solely to greenhouse emissions. That this rise may be part of a natural progression readily absorbed by global systems cannot be ruled out at this stage.

It is also worth bearing in mind that ever since the domestication of plants and animals and the discovery of fire, human beings have moulded their natural surroundings to suit their purposes. During the medieval period, for instance, much of the forest that once covered the continent of Europe was destroyed (a process analogous with the present destruction of the tropical rain forest), without any apparent harmful consequences in the long term. Moreover, it was the agricultural and industrial revolutions that created those environmental conditions we consider 'natural' and with which we are comfortable. It is debatable, in other words, whether any truly natural landscapes and ecosystems remain – they have all to a greater or lesser extent come under human influence.

For most of human history our attitude to the environment has been purely exploitative: nature was there to be conquered, and the resource endowment to be used in the furtherance of human development. Little or no attention was paid to the possibility of detrimental environmental impacts – indeed, in most instances these were simply not appreciated because the complex relationships and linkages of environmental systems were not understood. It is only in this century that this attitude – of man as the conqueror of nature – has changed (indeed in Eastern Europe it persisted, with disastrous consequences, right up until the demise of communism). We now think more in terms of stewardship, whereby humans owe a duty of care to the environment, and in terms of sustainability. However, it is still invariably the case that when choices have to be made economic self-interest wins the day.

The broad scope of environmental management creates its own particular problems. The information on which it relies is scattered across disciplines isolated from one another by the traditional boundaries that demarcate major branches of academic endeavour. It follows from this that relevant advances in the natural sciences may not be appreciated by those working from a social science perspective, and so on. It is therefore a major objective of this book to bring together the expertise found within the diverse fields of environmental management, with the aim of providing an accessible overview of its content and methods. The treatment is biased towards environmental management as practised at the regional level – the so-called meso-scale. Global issues such as climate change and loss of biodiversity lie outside the scope of this book and so receive only incidental mention.

The idea for this book came initially from the involvement of the four editors in environmental training programmes in Eastern Europe, and a book was duly published by the Free University of Brussels Press in 1993. This publication is a revised, improved and extended edition of that earlier version and is presented in three volumes. The theoretical principles of environmental management are illustrated with the use of up-to-date examples and case studies, and self-assessment questions are included to aid students who may wish to use it as a textbook. It should also be of interest to policy-makers and researchers seeking information about the management of today's environmental problems.

Volume 1 considers the instruments for environmental management under four main headings – predictive and scientific instruments; economic instruments; legal instruments; and instruments for environmental communication and education. It not only covers relatively long-established instruments of management, such as environmental impact assessment and risk assessment, but also introduces more recently developed approaches such as material flow analysis and life-cycle assessment. Volume 1 aims at up-to-date, comprehensive coverage and includes discussion of such important topics as the concept of sustainable development, environmental legislation in the European Union and the USA, and the management of environmental conflict.

Volume 2 is devoted to the environmental management of compartments, stressors and sectors. It covers the impact of population and the way environmental information is processed and interpreted through the filter of human culture. Soil, air and water are the environmental compartments referred to in this volume. They are subject to stress through overuse and pollution, and the manner in which these stressors should be managed constitutes a major strand of enquiry. The sectors referred to are agriculture, forestry, industry, transport and tourism, and how these should be managed in the interests of preserving environmental quality.

The theoretical and practical considerations involved in the appropriate management of major natural ecosystem types – wetlands, tropical forests, desert areas, the coastal margin, river and inland water environments – are discussed in Volume 3. These are, of course, 'natural' ecosystems only in a relative sense; the question of management arises precisely because of human impacts. Of equal importance are those environments created entirely by human activities, and in recognition of this Volume 3 also deals with rural and urban environments, as well as human ecosystems under threat, and the management of archaeological sites.

The three volumes of this textbook contain over sixty chapters written by more than eighty authors. This large project would have been impossible without the support and active contributions of many colleagues whose names are not mentioned in the individual chapters. We would like to thank especially the secretarial staff of the Human Ecology Department, Free University of Brussels (VUB) for their excellent work; especially Mr Glenn Ronsse, who was responsible for the final formatting and overall secretarial management of the project. Sincere thanks are due to the peer reviewers of the chapters.

For Volume 3 these were: A.K. Agyare, Department of Game and Wild Life, Accra, Ghana; F. Bartaletti, University of Genoa, Italy; A. Begossi, Universidade Estadual de Campinas, SP, Brazil; R.B. Bening, University of Development Studies, Tamale, Ghana; E. Bernus, Paris, France; P.B. Bridgewater, Australian National Parks and Wildlife Service, Canberra, Australia; S.E. Carter, Tropical Soil Biology and Fertility Programme, Nairobi, Kenya; F. Dahdouh-Guebas, Free University Brussels, Belgium; J. de Smidt, Universiteit Utrecht, Netherlands; V. Demoulin, University of Liège, Sart Tilman, Belgium; W.R. Erdelen, Universität Würzburg, Rauhenebrach, Germany; G. Folke, University of Stockholm, Sweden; M. Gast, University of Aix-Marseille, France; A. Grant, University of East Anglia, Norwich, Norfolk, United Kingdom; I.W. Heathcote, University of Guelph, Ontario, Canada; G. Hendrickx, Redacteur Natuur en Techniek, Beek, Netherlands; R. Johnstone, Stockholm University, Sweden; P. Ketner, Bennekom, Netherlands; Ph. Lefèvre-Witier, Université Paul Sabatier – Toulouse III, France; W.V. Luneburg, University of Pittsburg, USA.; J.A. McNeely, IUCN, Gland, Switzerland; R. Massa, Universita degli Studi di Milano, Italy; M. Mugica, Universidad Complutense Madrid, Spain; H.K. Murwira, Chemistry and Soil Research Institute, Harare, Zimbabwe; M.G. Paoletti, University of Padova, Italy; M. Perttilä, Finnish Institute for Marine Research, Finland; R. Pesci, Fundacion CEPA, La Plata, Argentina; L. Rahm, Linköping University, Sweden; J. Rammeloo, National Botanic Garden, Belgium; P.P. Schot, Utrecht University, Netherlands; C. Susanne, Free University Brussels, Belgium; A.Y. Troumbis, Biodiversity Conservation Laboratory, Lesbos, Greece; K. Van Balen, Catholic University Leuven, Belgium; W. Van Cottem, Zaffelare, Belgium; P. Van Damme, University of Ghent, Belgium; J.M. van Groenendael, University of Nijmegen, Netherlands; B. Van Puijenbroeck, Antwerp Zoo, Belgium; F. Verhaeghe, Free University Brussels, Belgium; J. Verneaux Jean, Institut des Sciences et Techniques de l'Environnement, Besançon, France; H. Verschure, Catholic University Leuven, Belgium.

We are deeply indebted to Mr K. Van Scharen, executive director of the VUB University Press, publishers of the first edition of this textbook in 1993, who was most supportive in the publication of this second edition. We thank Routledge for guidance during the establishment of this project and in its final production.

The publishers have made every effort to contact copyright holders of material reprinted in *Environmental Management in Practice: Volume 3.* However, this has not been possible in every case and we would welcome correspondence from those individuals/companies we have been unable to trace.

LIST OF ABBREVIATIONS

ADMADE	Administrative Management Design for Game Management Areas, Zambia
AIA	Archaeological Impact Assessment
AIDS	Acquired Immune Deficiency Syndrome
BOD	Biological Oxygen Demand
CAMPFIRE	Communal Areas Management Programme For Indigenous Resources, Zimbabwe
CAP	Common Agricultural Policy
CBNRM	Community Based Natural Resources Management, Namibia
CDRS	Charles Darwin Research Station
CE	Council of Europe
CEC	Commission of the European Communities
CZM	Coastal Zone Management
EA	Environment Agency
ECNC	European Centre for Nature Conservation
EIA	Environmental Impact Assessment
EN	English Nature
ESA	Environmentally Sensitive Area
EU	European Union
FAO	Food and Agricultural Organization (of the UN)
GIA	Geomorphological Impact Assessment
GIS	Geographic Information System
GNP	Gross National Product
GNPS	Galapagos National Park Service
HYRROM	UK Institute of Hydrology's lumped catchment model
IBP	International Biological Program
ICZM	Integrated Coastal Zone Management
IFIM	Instream Flow Incremental Methodology
ILO	International Labour Organisation
INIA	Illinois, USA Natural Areas Survey
IRBM	Integrated River Basin Management
ITCZ	Inter Tropical Convergence Zone
IUCN	International Union for Conservation of Nature and Natural Resources
IUCN-CNPPA	(IUCN) Commission on Natural Parks and Protected Areas
KNP/FR	Kilimanjaro National Park and Forest Reserve
LAs	Local Authorities
LFA	Less Favoured Area
LPAC	London Planning Advisory Committee
MAFF	Ministry of Agriculture, Fisheries and Food, UK
NBPP	North Branch Prairie Project, Illinois, USA
NGO	Non-Governmental Organisation
NOAA	National Oceanographic and Atmospheric Administration
NRMP	Natural Resources Management Programme, Botswana

OEA	Organization of the American States	UNCED	United Nations Conference on Environment and Development (Rio de Janeiro, Brazil 1992)
OECD	Organization for Economic Co-operation and Development	UNCOD	United Nations International Conference on Desertification
OPP	Orangi Pilot Project, Karachi, Pakistan	UNDP	United Nations Development Programme
P	Phosphorus	UNEP	United Nations Environment Programme
PACD	Plan of Action to Combat Desertification	UNESCO	United Nations Educational, Scientific and Cultural Organization
PLA	Port of London Authority		
PP	Particulate Phosphorus	UNSO	United Nations Sudano-Sahelian Office
RIVPACS	River Invertebrate Prediction and Classification System	USFWS	United States Fish and Wildlife Service
RSPM	Resource Systems Planning Model	WCED	World Commission on Environment and Development
SARD	Sustainable Agriculture and Rural Development	WCMC	World Conservation Monitoring Centre
SCP	Selous Conservation Programme, Tanzania	WHO	World Health Organization
SEA	Strategic Environmental Impact Assessment	WRI	World Resources Institute (New York)
SRCP	Serengeti Regional Conservation Programme, Tanzania	WWF	World Wide Fund for Nature (formerly World Wildlife Fund)
SRP	Soluble Reactive Phosphorus		
TGLP	Tribal Grazing Land Policy		
TP	Total Phosphorus		

LIST OF UNITS

Prefixes to the names of units

G	giga (10^9)
M	mega (10^6)
k	kilo (10^3)
d	deci (10^{-1})
c	centi (10^{-2})
m	milli (10^{-3})
μ	micro (10^{-6})
n	nano (10^{-9})
p	pico (10^{-12})
f	femto (10^{-15})

Units

a	annum
Å	ångstrom (0.1 nm)
atm	atmosphere
bbl	billions of barrel
boe	barrels of oil equivalent
Bq	becquerel
°C	degree Celsius or centigrade
cal	calorie
d	day
dB	decibel
g	gram
Gtce	gigatons of coal equivalent
Gwe	gigawatt electricity
h	hour
ha	hectare
hrs	hours
Hz	hertz
J	joule
K	degree absolute (kelvin)
l	litre
m	metre
M	molar (mol/litre)
min	minute
P	phon
Pa	pascal (unit of pressure; 100 kPa = 1 bar)
pe	percentage
PM_{10}	fraction of particulates in air of very small size ($\leq$ 10 μm)
S	sone
s	second
t	tons
Tcf	tons of carbon fuel
TW	terawatt
Twyr	terawatt per year
V	volt
W	watt
yr	year

Other abbreviations

kg_{bw}	kilogram body weight
ln	logarithm (natural, base e)
log	logarithm (common, base 10)
n or N	total number of individuals or variates
ppb	parts per billion
ppm	parts per million
ppmv	parts per million (volume)
s^2	sample variance

INTRODUCTION

Richard J. Huggett and Paul A. Compton

Human civilisations have always 'managed' eco-systems to a greater or lesser degree. Before the eighteenth century, humankind tended to take a responsible attitude towards Nature (see Merchant, 1982; Sheldrake, 1990). The rise of a mechanistic world-view bred an exploitative attitude, at least in the western world and its fast-growing colonies. Ecosystem riches were plundered profligately, with little heed to conservation or sustainability. The result was an unprecedented transformation of the bio-sphere, a radical shift in land cover. Dissenting voices against the rape of the Earth first cried out during the second half of the nineteenth century. However, only since 1945, with the rise of modern environmentalism, has a large body of people spoken with one voice to demand the judicious use of ecosys-tems.

Modern views on ecosystem management evolved largely in response to the current biodiversity crisis. Biodiversity provided a new rallying point in an ecological world that late twentieth-century theorists had shown to be largely chaotic and unpredict-able. Nature may have disturbing, perverse and unpredictable ways, and an abiding ability to evade our understanding, but it is gloriously diverse and still needs our love, our respect, and our help (cf. Worster, 1994: 420). By the late 1980s, an eco-system approach to land management was advocated by many scientists and other people interested in the environment (e.g. Agee and Johnson, 1988). Its ultimate aim is to enhance and to ensure the diversity of species, communities, ecosystems and landscapes.

WHAT IS ECOSYSTEM MANAGEMENT?

Ecosystem management is now a much used term. But what does it mean?

This question is surprisingly difficult to answer. About all that may be said with confidence is that ecosystem management is not the traditional model of resource management. Traditional resource man-agement lays emphasis on maximising production of goods and services through sustained yield from balanced ecosystems. It gives much credence to util-itarian values that regard human consumption as the best use of resources, and that hold a continuous supply of goods for human markets as the purpose of resource management (Cortner and Moote, 1994). This blatant 'resourcism' is patently flawed. It fails to recognise limits to exploitation and, in consequence, a growing number of species, and even entire eco-systems, are currently endangered. But, flawed or not, it persists: even now, ecosystems are viewed by some as long-lived, multi-product factories (Gottfried, 1992), or, if you prefer, as Nature's superstores.

Ecosystem management, though not universally welcomed, is a new and emerging model of resource management. Some of its advocates see their endeav-ours as an extension of multiple use, sustained yield policies (e.g. Kessler *et al.*, 1992). They prosecute a stewardship approach, in which the ecosystem is seen merely as a human life-support system. In this view, public demands for habitat protection, recreation and wildlife uses are simply seen as constraints to maximising resource output (Cortner and Moote, 1994). A more radical approach, which seems to be making headway in discussions of ecosystem manage-ment, is to accept Nature on its own terms, even where doing so means controlling incompatible human uses (e.g. Keiter and Boyce, 1991). This extreme, but eminently sensible, form of ecosystem management reflects a willingness to place environmental values, such as biodiversity, and social and cultural values, such as the upholding of human rights, on an equal footing.

Ecosystem management is defined in several ways, it embraces many and various approaches, and it has several dimensions. It is intimately linked to radical ideas about humanity's place in Nature. Here is a working definition: 'Ecosystem management integrates scientific knowledge of ecological relationships within a complex socio-political and values framework toward the general goal of protecting native ecosystem integrity over the long term' (Grumbine, 1994: 31); it is, at root, 'an invitation, a call to restorative action that promises a healthy future for the entire biotic enterprise' and promises a means of bridging the growing gap between people and Nature in 'a world of damaged but recoverable ecological integrity' (Grumbine, 1994: 35).

Ecosystem management involves several dominant themes (Grumbine, 1994). These fall into three categories – ecosystem considerations, management practices and human concerns.

ECOSYSTEM CONSIDERATIONS

Studying all connections in the ecological hierarchy

A tenet of modern ecology is that everything is connected to everything else, so that nothing in Nature can be understood in isolation. Such ecological interdependence has profound management implications: an ecosystem component cannot be altered without effects being felt throughout the system. This is why a solution to one problem often creates new, unforeseen problems. A 'single problem–single solution' approach is ecologically unsound, and may create more problems than it solves. Ecosystem managers should instead opt for solutions that attempt to embrace the gamut of problems (Gerlach and Bengston, 1992). This will normally mean considering connections between all hierarchical levels of the ecosphere – genes, species, populations, communities,

ecosystems and landscapes (cf. Kaufmann et al., 1994: 6; Huggett, 1995: 13–14).

Including the true ecological boundaries

Ecosystems function on their own scales of space and time. Many are larger than a single administrative or management institution. This was recognised by the first exponents of modern ecosystem management, Frank and John Craighead, whose work showed that the needs of the grizzly bear (*Ursus arctos horribilis*) population could not be met within the confines of Yellowstone National Park (e.g. Craighead, 1979). The management of most ecosystems requires co-operation of managers from several different administrative and political units (Gerlach and Bengston, 1992). It also requires an integrated ecological approach that often demands international action and co-operation. This is true for managing coastal resources (see Alexander, 1993; Ngoile and Horrill, 1993), as well as terrestrial resources. A case in point concerns human impacts acting through terrestrial runoff on coastal marine ecosystems. In managing these impacts, it is useful to adopt the 'marine catchment basin' as a unit of study (Caddy and Bakun, 1994): such catchments may cross regional or even state boundaries.

Managing ecological integrity

This is the overall goal of ecosystem management. It means protecting the total native diversity (of species, populations, ecosystems and landscapes) and the ecological patterns and processes that maintain them (Norton, 1992). In practice, this normally includes one or more specific goals (Grumbine, 1994: 31):

1   Maintaining viable populations of all native species *in situ*, and reintroducing extirpated species where necessary.
2   Ensuring that, within protected areas, all native ecosystem types are represented. In other words, protecting the full natural range of ecosystem variation.
3   Maintaining ecological and evolutionary processes. This means guaranteeing the continuance of

natural disturbance regimes, hydrological process and ecological processes.

4 Managing ecosystems for long enough periods to maintain the evolutionary potential of species, communities and ecosystems.

5 Accommodating human use and occupancy in the light of the above. This acknowledges the vital, if unsure, role that humans must play in all aspects of ecosystem management.

Ecosystem management should be informed by sound conceptual models of ecosystem structure and function (e.g. Le Roux, 1993; Straskraba, 1993). It is essential to understand natural ecosystem processes before attempting to manage an ecosystem. This means understanding the full gamut of ecological processes and patterns, from trophic relationships between species (e.g. Whitlatch and Osman, 1994) to the principles of landscape ecology (e.g. Agardy, 1994) and the question of disturbance. The ecological role of disturbance is now widely recognised and complicates ecosystem management practice. For instance, an ecological framework of natural disturbance and a knowledge of its component processes and effects provide a sound basis for managing forests as a renewable resource (Attiwill, 1994). The practice of ecosystem management has evolved alongside ecological theory. Traditional ecosystem management draws heavily on the ecological notions of balance and equilibrium. Newer management practices tend to incorporate new ecological ideas that stress imbalance, disharmony, disequilibrium, disturbance and unpredictability (chaotic behaviour). Any new paradigm of ecosystem management must somehow incorporate the principles of ecosystem disequilibrium and chaotic behaviour. A sticky problem is that disequilibrium ecologists are divided among themselves over the advice they should give to society on how to act over the environment. One group, reflecting some of the new disequilibrium thinking, began to challenge the public perception that ecology and environmentalism were the same thing. Some ecologists became disenchanted with trying to conserve a healthy planet. Nature is characterised by highly individualistic associations, they argued, so why attempt to constrain it? This anarchic argument, if taken to the extreme, could have revived social Darwinism and stood in antithesis to the conservation ethic of co-operation and collective action suggested by the older notion of balanced ecosystems. In the event, a mildly anarchic view of Nature was taken by some ecologists. Another group of ecologists, in stark contrast, drew different conclusions from the disequilibrium trends of their discipline. They favoured an environmentalism that was more friendly towards manipulating and dominating Nature, but that tolerated modern technology and progress, and human desires for greater wealth and power (e.g. Botkin, 1990).

Even more problematic is the question of chaotic ecosystem dynamics. The implications of ecological chaos for conservation and ecosystem management are far from clear. How are conservationists and managers to use a chaotic ecological design? Answers to this question have to confront a paradox: a chaotic view of Nature is at once exhilarating and threatening. Chaotic Nature, so irregular and individualistic in character, appears almost impossibly difficult to admire or to respect, to understand or to predict. It seems to be a world in which the security of stable, permanent rules is gone forever, a dangerous and uncertain world that inspires no confidence (Prigogine and Stengers, 1984: 212–213). This dark and direful aspect of chaos might promote a feeling of alienation from the natural world, and cause people to withdraw into doubt and self-absorption (cf. Worster, 1994: 413). It might also set people wondering how they should behave in a world where chaos reigns. With natural disturbance found everywhere, why should humans worry about doing their own bit of disturbing? Why not join individuals of all other species and do their own thing without feeling guilty about it? As Worster (1994: 413) put it, what does the phrase 'environmental damage' mean in a world so full of natural upheaval and unpredictability? However, chaos does not have to be portrayed in such gloomy and doom-laden terms. Chaotic Nature has a bright and edifying aspect, too. In a chaotic world, communities, ecosystems and societies are sensitive to disturbance. Small disturbances can, sometimes, grow and cause the communities, ecosystems or societies to change. Consequently, it is a world in which individual activity may have major significance (cf. Prigogine and Stengers, 1984: 313). It is a post-modern world in which increased individuality and diversity encourage a great overall harmony. Moreover, the new-fangled theory of chaotic dynamics is leading to the discovery of hidden regularities in natural processes, the application of which is proving most salutary (e.g. Stewart, 1995). For instance, some forests and fisheries are being better managed through an understanding of their chaotic behaviour.

Environmental monitoring and data collection

Ecosystem management requires continuing research and data collection, as well as better management and use of existing data. This includes data on modern processes and historical data. An example of the use of modern data is the collection of resource inventory data on abiotic, biotic and cultural components of the Great Lakes shoreline environment, which has enabled key issues and areas of concern to be identified (Lawrence *et al.*, 1993). Historical data on environment, climate and culture in the fragile high Andean puna ecosystems have contributed to their management and preservation (Baied and Wheeler, 1993).

The behaviour of managed ecosystems must be carefully monitored so that the relative success or failure of actions can be tested and adjusted if necessary. Sometimes the remedial measures are more damaging than the problem. This is sometimes the case with marine pollution, and notably with oil slicks (e.g. Mearns, 1993). It is also desirable to know whether environmental assessment objectives have been met. This requires long-term monitoring programmes, as instigated, for example, on streams in the Pacific north-west forests of North America (Wissmar, 1993).

MANAGEMENT PRACTICES

Three facets of management practice are highly pertinent to ecosystem management – adaptive management, interagency co-operation, and organisational change.

Ecosystem management is adaptive in the sense that scientific knowledge is regarded as provisional. New findings, especially from monitoring programmes,

may alter advice to managers. Managers must thus remain flexible and adapt to the changeability of 'expert' scientific opinion. Likewise, scientists must adapt to social changes and demands. Ecosystem management calls for all parties to be adaptable. It is claimed, for instance, that the eventual restoration of depleted fish stocks will require adaptive management procedures (Alexander, 1993).

Ecosystem boundaries transgress administrative and political boundaries. National, regional and local management agencies, as well as private parties, must therefore learn to work together and integrate conflicting legal mandates and management goals. An example of this, from the United States, is the National Research Council's *Mandate for Change* report (1990) that, to develop a new paradigm of forestry research, urged close collaboration between forest managers and forest-user groups. Another example is the remedial action plans for Great Lakes water management, where the need for interdisciplinary and intergovernmental participation, and for political and public support, was recognised (MacKenzie, 1993).

Ecosystem management demands that land management agencies change the way in which they operate. This may be done by forming interagency committees or changing profession norms and altering power relationships. There is little doubt that the collaborative decision-making called for by ecosystem management is likely to require extensive revision of traditional management practices and institutions (Cortner and Moote, 1994). Adaptability and flexibility by all parties are the twin keys to success in ecosystem management.

HUMAN CONCERNS

Humans in nature

Ecosystem management accepts that people are part of Nature. Humans exert a profound influence on ecological patterns and processes, and in turn ecological patterns and process affect humans. The connection between the two is largely uncharted territory: interactions between ecosystems and social systems require far more research (Gerlach and Bengston, 1992). However, policies are tending to move away from the 'administrator-as-neutral-expert' approach to policies that engender public deliberation and the discovery of shared values (Reich, 1985). Naturally, such extension of ecological matters to the social and political arena presents difficulties, though these may not be insuperable (e.g. Irland, 1994).

Human values

Ecosystem management accepts that human values must play a leading role in policy decision-making. Conservation strategies must take account of human needs and aspirations (Le Roux, 1993); and they must integrate ecosystem, economic and social needs (Kaufmann *et al.*, 1994). The key players in ecosystem management are scientists, policy-makers, managers and the public. The public, many of whom have a keen interest in environmental matters, are becoming more involved in ecosystem management as professionals recognise the legitimacy of claims that various groups make on natural resources (Cortner and Moote, 1994). In Jervis Bay, Australia, the marine ecosystem is used by many existing and proposed conflicting interests (national park, tourism, urbanisation, military training) (Ward and Jacoby, 1992). Similarly, in the forests of the south-western United States, ecosystem, economic and social needs are considered in policy decision-making concerning ecology-based, multiple-use forest management (Kaufmann *et al.*, 1994) (Figure 0.1).

The human dimension of ecosystem management is encompassed by the notion of sustainable development. This urges that diverse, functioning ecosystems should be preserved without damaging the economy; and that economies and human welfare need preserving as much as the integrity of ecosystems (Gerlach and Bengston, 1992). Integrating human and biophysical factors is a daunting task. Humans interact with Nature through culture, often in symbolic ways that are not comprehended by biological or physical ecosystem models (Gerlach and Bengston, 1992). In other words, humans can generate wants and capabilities of meeting these wants that lie outside the natural ecological order. This is often difficult for scientists to understand, and leads to their focusing on environmental protection and restoration, rather than facing the greater

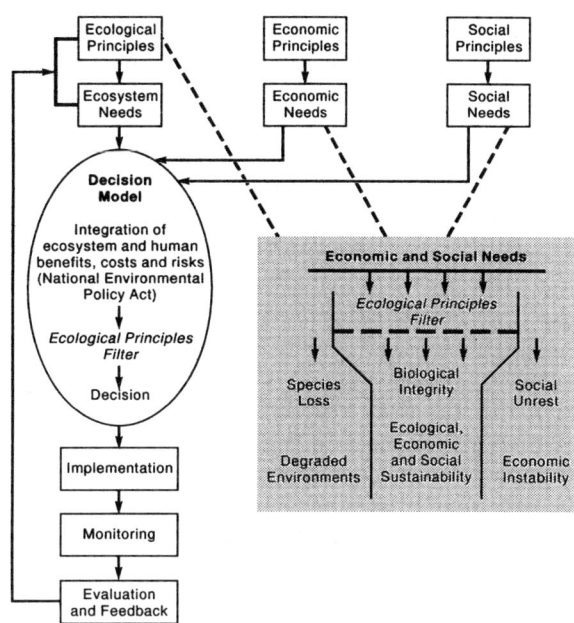

Figure 0.1 The integration of ecological, economic and social needs in a decision-analysis model

Economic and social needs are tested against an 'ecological filter', which is shown in the shaded box. The aim is to determine economic and social actions that will produce the most desirable balance between biological integrity and ecological, economic and social sustainability. Bowing fully to economic and social needs would lead to species loss and environmental degradation. Bowing fully to ecological needs would lead to social unrest and economic instability. A compromise position allows the maintenance of biological integrity while catering for economic and social needs. The resulting decision model leads to the implementation of an environmental policy. The effects of the policy are carefully monitored and evaluated. If the policy should fail to work as desired, then amendments can be made and the process started anew, until a satisfactory outcome is achieved.

Source: After Kaufmann et al., 1994: 11.

challenge of understanding social and cultural interactions with the biophysical world. The very idea of sustainability itself is a curious human construct. Laudable though the idea of sustainability might be, there are problems with it, not the least of which is the lack of a definition. It was originally taken to mean meeting the needs of the present generation without compromising the ability of future generations to satisfy their needs (World Commission on

Environment and Development, 1987), but at least eight interpretations are available (Gale and Cordray, 1991). It is therefore a buzzword that lacks bite. Another problem is its inadvertent arrogance – should the present generation be so presumptuous and foolhardy as to anticipate the needs of future generations? These pithy questions should not be ducked in ecosystem management initiatives.

This introduction has emphasised general philosophical principles and broad management practices connected with ecosystems and is meant to complement the individual chapters in the volume. These are more concerned with the importance of comprehending different natural ecosystems and processes as the only sound basis for making informed decisions about practice. It is only through understanding the complexity and inter-connectedness of the natural world that we may hope to avoid the mistakes of the past and make progress towards protecting the environment for future generations.

Of necessity, the volume adopts a fairly broad-brush approach. Myriad natural ecosystems exist at the micro-level but here it is the continental and even world scales that are emphasised. Hence, individual chapters deal with broad generic types – wetlands, uplands and mountain environments, savannas, river and inland water environments, tropical forest ecosystems, coastal environments and desertification. Although topics were selected for inclusion on the basis of the perceived threat posed by human activities it is recognised that the list is not comprehensive. For instance, high latitude ecosystems like the tundra and the unique environment of Antarctica are not covered, while the specific threats in temperate regions posed, for instance, by agriculture are included in Volumes 1 and 2. Each chapter follows a similar format; the general issues are first identified and then illustrated with material drawn from the individual experience of each author. Management strategies to combat environmental threats are described.

The treatment of natural ecosystems is complemented with a discussion of the human-created environments that exist as cities and rural areas. Cities generate complex energy and material flows with pollution of the air and water, land contamination and the creation of waste being the principal by-products. The traditional means of managing the environmental problems that arise in cities has been through planning regulations but it is now becoming clear that these are inadequate for the task and new approaches involving the concepts of healthy cities and sustainable communities must be seriously considered. Rural environments, by contrast, are different in being the partial creations of human activity and the question is one of finding a more harmonious, less exploitative accommodation with nature. This matter is explored via the interesting but little known case of Vietnam.

The concluding chapters deal with three specific examples of environmental management in human ecosystems. The first of these concerns our archaeological heritage. Although this should, of course, be studied and managed for its own sake, it is also becoming a matter of utilitarian significance in the sense that there may well be lessons to be learned from pre-industrial societies about sustainability, particularly sustainable agriculture. In the second example, a case is made for managing landscapes as a whole, whereby natural ecosystems are integrated with the cultural aspects of human society. The growing sophistication of information technology and the ability to manipulate large amounts of data, within a GIS for instance, now make this a feasible proposition. Lastly, given our concern about the natural environment, there may be a tendency to overlook the human ecosystems that are under pressure. These are discussed in the final chapter 'Disappearing human ecosystems'.

REFERENCES

Agardy, M.T. (1994) 'Advances in marine conservation: the role of marine protected areas', *Trends in Ecology and Evolution* 9: 267–270.

Agee, J.K. and Johnson, D.R. (1988) *Ecosystem Management for Parks and Wilderness*, Seattle, Wash., USA: University of Washington Press.

Alexander, L.M. (1993) 'Large marine ecosystems: a new focus for marine resources management', *Marine Policy* 17: 186–198.

Attiwill, P.M. (1994) 'The disturbance of forest ecosystems: the ecological basis for conservative management', *Forest Ecology and Management* 63: 247–300.

Baied, C.A. and Wheeler, J.C. (1993) 'Evolution of high

Andean puna ecosystems: environment, climate, and culture change over the last 12,000 years in the central Andes', *Mountain Research and Development* 13: 145–156.

Botkin, D.B. (1990) *Discordant Harmonies: A New Ecology for the Twenty-First Century*, New York, USA: Oxford University Press.

Caddy, J.F. and Bakun, A. (1994) 'A tentative classification of coastal marine ecosystems based on dominant processes of nutrient supply', *Ocean and Coastal Management* 23: 201–211.

Cortner, H.J. and Moote, M.A. (1994) 'Trends and issues in land and water resources management: setting the agenda for change', *Environmental Management* 18: 167–173.

Craighead, F. (1979) *Track of the Grizzly*, San Francisco, Calif., USA: Sierra Club Books.

Gale, R.P. and Cordray, S.M. (1991) 'What should forests sustain? Eight answers', *Journal of Forestry* 89: 31–36.

Gerlach, L.P. and Bengston, D.N. (1992) 'If ecosystem management is the solution, what is the problem?' *Journal of Forestry* 92: 18–21.

Gottfried, R.R. (1992) 'The value of a watershed as a series of linked multiproduct assets', *Ecological Economics* 5: 145–161.

Grumbine, R.E. (1994) 'What is ecosystem management?' *Conservation Biology* 8: 27–38.

Huggett, R.J. (1995) *Geoecology: An Evolutionary Approach*, London, UK: Routledge.

Irland, L.C. (1994) 'Getting from here to there: implementing ecosystem management on the ground', *Journal of Forestry* 92: 12–17.

Kaufmann, M.R., Graham, R.T., Boyce, D.A. Jr., Moir, W.H., Perry, L., Reynolds, R.T., Bassett, R.L., Mehlhop, P., Edminster, C.B., Block, W.M. and Corn, P.S. (1994) *An Ecological Basis for Ecosystem Management* (General Technical Report RM–246, United States Department of Agriculture, Forest Service, Rocky Mountain Forest and Range Experiment Station, Fort Collins, Colorado), Washington DC, USA: United States Government Printing Office.

Keiter, R. and Boyce, M. (1991) *The Greater Yellowstone Ecosystem*, New Haven, Conn., USA: Yale University Press.

Kessler, W.B., Salwasser, H., Cartwright, C. Jr. and Caplan, J. (1992) 'New perspective for sustainable natural resources management', *Ecological Applications* 2: 221–225.

Lawrence, P.L., Nelson, J.G. and Peach, G. (1993) 'Great Lakes shoreline management plan for the Saugeen Valley Conservation Authority', *Operational Geographer* 11: 26–33.

Le Roux, Z. (1993) 'Conservation at landscape level: a strategy for survival: the role of research', *Environmentalist* 13: 105–110.

MacKenzie, S.H. (1993) 'Ecosystem management in the Great Lakes: some observations from three RAP sites', *Journal of Great Lakes Management* 19: 136–144.

Mearns, A.J. (1993) '"Appropriate" technologies for marine pollution control', *Sea Technology* 34: 31–37.

Merchant, C. (1982) *The Death of Nature: Women, Ecology, and the Scientific Revolution*, London, UK: Wildwood House.

National Research Council (1990) *Forestry Research: A Mandate for Change*, Washington DC, USA: National Academy Press.

Ngoile, M.A.K. and Horrill, C.J. (1993) 'Coastal ecosystems, productivity and ecosystem protection: coastal ecosystem management', *Ambio* 22: 461–467.

Norton, B.G. (1992) 'A new paradigm for environmental management' in R. Costanza, B.G. Norton and B.D. Haskell (eds) *Ecosystem Health*, Washington DC, USA: Island Press, pp. 23–41.

Prigogine, I. and Stengers, I. (1984) *Order out of Chaos: Man's New Dialogue with Nature*, London, UK: William Heinemann.

Reich, R.B. (1985) 'Public administration and public deliberation: an interpretative essay', *The Yale Law Journal* 94: 1617–1641.

Sheldrake, R. (1990) *The Rebirth of Nature: The Greening of God and Science*, London, UK: Century.

Stewart, I. (1995) *Nature's Numbers: Discovering Order and Pattern in the Universe*, London, UK: Weidenfeld & Nicolson.

Straskraba, M. (1993) 'Ecotechnology as a new means for environmental management', *Ecological Engineering* 2: 311–331.

Ward, T.J. and Jacoby, C.A. (1992) 'A strategy for assessment and management of marine ecosystems: baseline and monitoring studies in Jervis Bay, a temperate Australian embayment', *Marine Pollution Bulletin* 25: 163–171.

Whitlatch, R.B. and Osman, R.W. (1994) 'A qualitative approach to managing shellfish populations: assessing the relative importance of trophic relationships between species', *Journal of Shellfish Research* 13: 229–242.

Wissmar, R.C. (1993) 'The need for long-term stream monitoring programs in forest ecosystems of the Pacific northwest', *Environmental Monitoring and Assessment* 26: 219–234.

World Commission on Environment and Development (1987) *Our Common Future*, New York and Oxford: Oxford University Press.

Worster, D. (1994) *Nature's Economy: A History of Ecological Ideas*, 2nd edition, Cambridge, UK: Cambridge University Press.

I
COASTAL ENVIRONMENTS

Julian D. Orford

SUMMARY

The basic ecological principle embodied in the study of physical environment systems is that of continuity of energy and mass among constituent parts of the system. In a coastal context this principle is recognised in the sediment transport linkages that exist between coastal environments. Changes in beach sediment transport rate control the rate at which coasts may alter. The spatial control on sediment continuity defines the structure of coastal cells which offer the ecologically sustainable basis for coastal management programmes. Coastal erosion is only a problem where human activity impinges upon the normal erosional processes of coastal evolution. Coastal protection as a response to erosion is the historical basis for coastal management programmes and is still the mainstay of modern coastal zone management. Most human coastal intervention makes it difficult to base management approaches on ecological principles without considerable adaptation or cost. In this chapter some examples of management policies are examined for evidence of ecological principles at work. Not surprisingly, economic and cultural principles appear still to be the controlling aspects of coastal management policy.

ACADEMIC OBJECTIVES

This chapter focuses on the ecological principles of management of the physical development of coastal environments, since the ecological management of coastal biological realms has already received detailed attention elsewhere (Clarke, 1974; Carter, 1988).

On completion of this chapter you should be able to:

- recognise the nature and value of the ecological principles that are important to the dynamics and maintenance of physical coastal environments;
- understand how these ecological principles can be easily broken when humans intervene in coastal environments;
- observe how these principles conflict with the economic and cultural principles that structure human activities at the shoreline;
- understand how coastal managers find it difficult to integrate all of these principles into an effective management programme.

THE NEED FOR COASTAL ZONE MANAGEMENT (CZM)

There has been substantial concern expressed, action undertaken and money spent on issues related to the world's coastlines. One can legitimately question why coastal zone issues and their management have come to concern so many agencies. The management policy of the coastal zone is complex in that practitioners can hold a number of aims as their *raison* *d'être* (Barrett, 1989). Coastal management can be about (a) protecting man's investment at the terrestrial edge; (b) protecting a series of productive and diverse physical and biological domains worthy in their own right of being protected (conservation); and (c) integrating and adjusting competing demands on those domains (management of resources). Whatever the ethos of coastal management, it can be argued that it would be pointless to examine its nature without considering the role of people and their interaction with the natural systems of the coastal environment (Clark, 1978).

Historically, coastal erosion was, and is still, regarded as an affront to people's superiority over nature, such that losses must be resisted. However, there are other contemporary reasons for CZM. Coastlines' unique resources are coming under new competing pressures which can be loosely split into economic and cultural categories. Economic pressures relate to economic development that uses coasts and cannot be easily (i.e. cheaply) located elsewhere, for example, energy generation dependent on trans-oceanic sources of fuel. Cultural pressures involve elements of aesthetics and social choice, exemplified by recreation and retirement. These are usually issues of the advanced economically developed world, although a modern phase of coastal management requirement is being generated by the transplanting of recreation demands on to those less economically advanced countries that supply the 'S-requirements' of modern tourism (sun, sand and surf). All these pressures produced a further stimulus for late twentieth-century CZM, that of environmental concern about the coastal zone as a system of physical and biological domains threatened by economic and cultural forces, and in need of conservation for their value *per se* at the interface between land and ocean.

The increasing need for CZM also casts light on a different ethos, consensual rather than confrontational, by which coastal zone problems can be approached. This change, reflecting an ecological perspective on the understanding of coasts, is not a change due solely to the virtues of ecological thinking *per se*. Rather the change is due to the inability of most governments to meet the inexhaustible financial demands of an ever-increasing coastal protection problem (confrontational). Those countries that took centrally directed action over their coastline in the late nineteenth and early twentieth century (e.g. the UK) tended to be economically advanced with a history of coastal investment built on a technological and fiscal base sufficient to confront the perceived coastal protection requirements. However, at the turn of the twenty-first century, a crisis of ability is occurring in these countries, where technology and finances are proving uncertain in the face of continuing coastal

pressures (e.g. the USA). Central governments are seeking new approaches to coastal management that will be low cost as well as cost-effective. Such low-cost measures shift the emphasis of people's response to coasts from protection to management, and to manage coasts effectively one has to achieve a better understanding of the processes and elements that define the coastal system.

THE BEACH AS A SYSTEM

Understanding the evolution and variation of beaches is the precursor to coastal management. The mobile nature of a beach acts as a buffer to breaking waves. Without a beach, waves erode land and property damage ensues. Coastal erosion is a natural part of coastal development; only when erosion affects human activity does coastal defence emerge as the central problem of coastal management.

Beaches are formed from material generally 0.1–1 mm in diameter. Sediment greater than 1 mm can form beaches that are known by the dominant particle size, for example, gravel, pebble, boulder. However, the 'protection' definition of a beach can be extended to any unconsolidated material found as an absorbing buffer to ocean forcing. The term 'forcing' covers the sea's power in terms of wave action and tidal action. The action of waves in developing a sand or gravel beach is more obvious than the action of tidally induced currents that can form an energy buffer through the deposition of very fine sediments which in the intertidal zone are often associated with salt-tolerant marsh vegetation. The coarser the natural beach sediment, the more exposed the beach is to wave energy. All sediment regardless of size has value as beach fill material, but coastal managers should remember that the equilibrium between erosion and beach stability is both sediment-size and energy-exposure dependent.

Beach retention should be a major goal of coastal management. In this respect any beach should be viewed as the central element of a coastal system. The keyword to describe beaches and their behaviour is 'continuity'. Continuity has temporal and spatial dimensions that are at the heart of management needs. Spatial continuity reflects the nature of a beach as a conveyor by which sediment is moved alongshore, onshore and offshore. The direction and rate of sediment movement are variable depending on energy forcing that occurs at a range of temporal scales.

Sediment sources, that is the origin of the sediment, are historically dependent on sea level change. In general, sea level change will force fresh sediment from the nearshore into the intertidal zone. Unfortunately the same change in sea level may promote seaward loss as well. Sediment motion alongshore is generally irrespective of sea level change, though any sea level change that induces further sediment into the beach face is likely to add to longshore supply. Landward sediment comes from two main sources: rivers and cliffs. The former is particularly important in the lower latitudes where large catchments support significant discharges of sand and mud to the coast zone. When such sources are impeded by dams or water abstraction, severe depletion on down-drift coasts is inevitable. Sea level change may be associated with climate shift that directly affects river sediment and water discharges and hence beach volume. Cliff sources are available in all latitudes but become more important in the mid- and upper latitudes with the conjunction of storm wave climates and increased incidence of unlithified terrestrial material. The inci-dence of major sources of heterogeneous sized material in the upper latitudes is a function of the Pleistocene 'inheritance' by which glacial activity generated extensive and easily eroded terrestrial cover. This has been rapidly reworked by rising seas into extensive beaches. The volume of sediment from any cliff source is not necessarily a direct indicator of the beach volume that the cliff supports, as the beach volume only retains that portion of the cliff sediment which is coarser than the minimum size that can be held on the beach depending on wave exposure.

Coastal sediment must go somewhere, given the idea of a coastal conveyor (Figure 1.1). It is rare for the beach to be a permanent feature. A beach can lose fine sediment to dunes by wind action. Coarser sediment can move onshore under storm wave action to form back-beach deposits that are only reincorporated into the beach if the beach rolls onshore under rising sea level. Beach sediment can be lost to offshore sinks in storms, due to offshore flow asymmetry in the surf zone. If the coast is indented due to estuary occurrence (crenellate or closed coast) then sediment can be lost from the beaches into estuary shoals, while up-estuary river transported sediment may bypass beaches entirely and sink directly offshore. Beaches are rarely

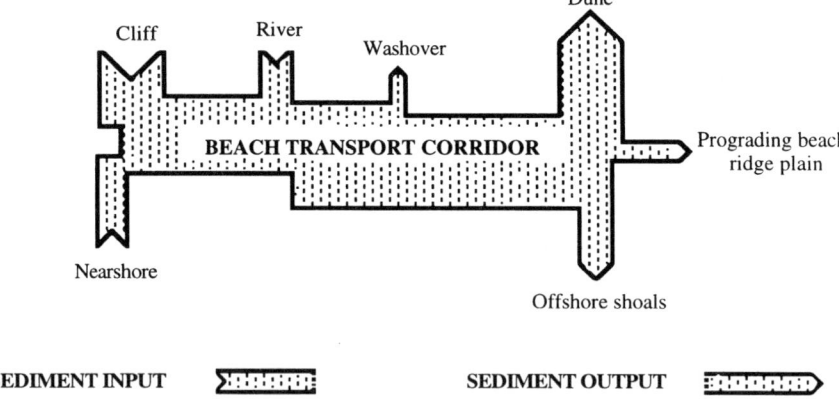

Figure 1.1 The beach sediment conveyor

The beach can be regarded as a coastal sediment conveyor in which there is sediment input (sources, for example, cliffs and rivers) at the beginning of the conveyor and sediment output (sinks, for example, sand dunes, offshore shoals) to both land and sea at the end of the conveyor. The beginning and end of the conveyor are set by the boundaries of the coastal cell. The beach volume at any point alongshore is a function of the initial sediment supply minus up-drift losses. Sediment is rarely static on a beach unless the beach itself is a sink. Changes in beach supply due to source sediment depletion will be seen in reduced beach volume along the conveyor.

regarded as sinks *per se* in that most are transport corridors, although if transport stops then the beach is technically a sink. Beaches with no longshore transport tend to be found on irregular coasts where transport is physically stopped by headlands.

Longshore transport is organised through two basic mechanisms: (a) the kinetic energy of the breaking wave (wave thrust) which carries sediment along in the swash zone; and (b) longshore currents in the surf zone which dissipate the momentum energy caused by wave presence (radiation stress). Both of these components of transport are strongly controlled by the height of the incident breaking wave and the angle of breaker approach. These two factors are a function of wave refraction by which offshore waves are transformed as they adjust to a continuously reducing depth of water as they approach the shoreline. Theoretical knowledge of this process enables coastal managers to model this process, and thus translate offshore wave climate into shoreline transport potential. Considerable research has been undertaken into developing and testing predictive equations for longshore transport using wave thrust only. Comparison of the numerous versions of predictive equations shows a great variation in outcome (Horikawa, 1988). The objective of establishing joint transport potential for longshore currents and wave thrust is a developing science though not yet one with great predictive ability.

Beach variation is a reflection of variable longshore transport which controls the conveyor between the source sediment (Qs) entering the beach and the sink sediment leaving the beach (Qk). If Qs > Qk, then the beach expands seawards. If Qs < Qk, then the beach volume declines and causes plan-view shoreline changes which are inevitably inconsistent along the shoreline. The shoreline will develop interlaced erosional stretches and depositional nodes leading to a cuspate shoreline that reflects the beach system's capacity to reduce longshore transport (deposition) at a cost of increasing transport elsewhere (erosion). It is this situation that causes many protection problems given the interdependency of erosion and deposition zones. Alteration to one will cause disturbance to the next one. If Qs = Qk, beaches appear never to change, a rare occurrence!

COASTAL CELLS AS THE BASIS FOR CZM

The recognition of longshore variation in beach continuity is of strategic importance in coastal management. Emphasis has been placed on the relative areas of sediment source and sink connected by a transport corridor that is the beach. These three zones, spatially linked, form a longshore *wave-sediment cell* that needs to be used as the basis for physical coastal management. The basic mechanism for cell formation is found in breaking wave height and direction variation at the shoreline. Such breaker differences are irrespective of offshore incident wave similarity and are caused during the shoaling transformation of incoming waves by wave refraction as they move over decreasing depths of water. Wave refraction on an irregular offshore bathymetry leads to shoreline breaking waves both parallel and non-parallel to the shoreline. Positions of parallel wave breaking (no longshore transport) represent the boundaries of wave-sediment cells. These can be one of either of two types; (a) where two longshore transport paths meet; or (b) the boundary between two transport paths moving off in different longshore directions. Cells are defined by (a) the source zone where breaker power is more than required to carry the available sediment and surplus power provides further sediment by erosion; (b) the sink zone where breaker power is less than the transport requirement and deposition of sediment occurs; and (c) the transport corridor that connects the source and the sink where power and transport load are balanced. By modelling wave refraction for any shoreline we can obtain a theoretical perspective on wave cell development that can be monitored and calibrated against the real world.

Where a coast shows an indented plan-form and has a variable wave climate, sediment supply is intermittent in time and space. Sediment size and availability will condition the relative lengths of the cell elements. The position and presence of high and wide beaches capable of absorbing wave energy will be dictated by the development and structure of the cell. Wide and high beaches are more likely where two transport paths meet at a cell boundary. Two divergent longshore paths will leave an eroding zone with no beach development. The inherent variability of cells in terms of wave climate and sediment supply needs further exploration as cell modelling is generally based on a unit year of averaged wave climate conditions (Bowden and Orford, 1984) (Figure 1.2). Studies should also focus on extreme events (i.e. storms) whose impact on coastal sediment lasts far longer than the event's duration. The effect of rising sea level on sediment pathways can also be studied in this way.

The physical structures of wave-sediment cells need to be mapped as a basis for effective coastal management. Identification of coastal cells is only a first stage in the planning process as cells often show integration into hierarchies of different sized cells which show a regional integration with drift direction and macroscale spatial morphology (Bray *et al.*, 1995). Identification of hierarchies is central to a regional perspective of coastal planning (Hall, 1989). This integration should also allow for the spatial and temporal connections between apparently disparate units of the coast as few cells are self-contained, with fine sediment liable to move considerable distances beyond the generating cell.

THE PROBLEM POSED BY HUMAN INTERVENTION ON THE BEACH

What then are the ecological principles by which coastal managers should be guided? The overriding principle is that human activity should not be allowed to deflect the continuity of the sediment budget along the shoreline. Human intervention in coastal systems is much more likely to prevent source sediment from entering the beach and hence into decline, than prevent beach sediment from entering sinks. The greater the distance the source area is from a position on the beach the more likely it is that human intervention will not be recognised as the cause of beach loss despite the linear spatial nature of the beach system which ensures that any disturbance to the beach supply will inevitably create further problems alongshore. All beach systems should have terrestrial sediment inputs to compensate for transport losses. This may mean that areas of coastline have to be zoned to allow coast recession to continue. The spatial integrity

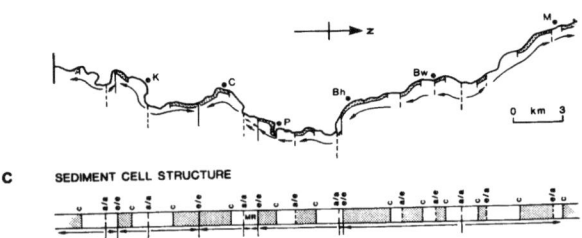

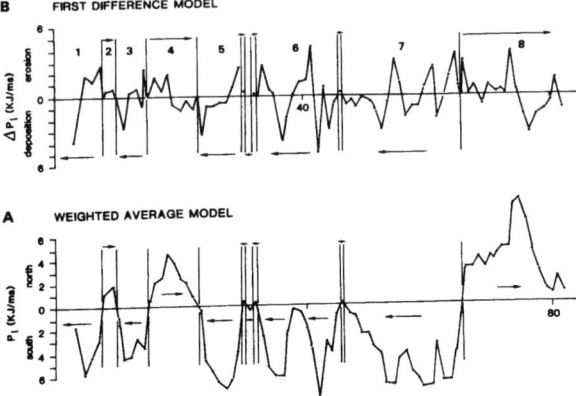

Figure 1.2 The identification of coastal cells based on numerical modelling of breaking wave energy received along a section of the Northern Ireland coastline
(A) Residual braking wave power based on summation of wave power computed at the coast from all principal wave directions weighted by time of occurrence in an average year. Note arrows indicate direction of wave transport potential alongshore.
(B) By subtracting adjacent values of breaking wave power at the shoreline the zones of potential erosion (increasing wave energy alongshore) and deposition (decreasing wave energy alongshore) can be established. Reversals in wave power gradients identify cell boundary positions.
(C) Reversals in drift direction, erosion and deposition zones can be used to identify wave-sediment cells which comprise a source (net erosion) and sink (net deposition) linked by a beach conveyor.
Source: after Bowden and Orford, 1984; copyright 1984 The Royal Irish Academy

of the coastal transport system has to be formally recognised in any planning programme. This means that obstructions to beach transport at all scales should be resisted. This is a difficult rule to enforce as it is a principle that goes against engineering protection design. A property crisis caused by coastal cliff recession is scarcely likely to be ignored simply because the erosion provides sediment to the down-drift beaches. In an effort to reduce the transport rate and build up the beach buffer volume, cross-beach structures (groynes) have been readily introduced by engineers on to sand and gravel shores. Such features (Figure 1.3) are counter-productive to the central ecological principle of beaches, that is sediment continuity in space and time. They invariably induce seaward (walls) and down-drift (groynes) beach erosion (Komar, 1983; Krause and Pilkey, 1988). Intervention is often found when the areal remit of coastal authorities artificially segment the beach system and the intervening authority is not responsible for alleviating the down-drift effects.

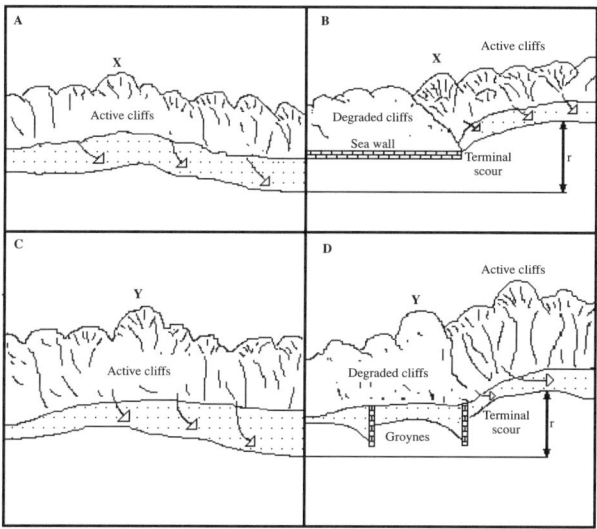

Figure 1.3 Examples of how man-made beach impedance can cause down-coast erosion

Erosion of cliffs (A) is often reduced by building a seawall at the cliff base (B). This reduces the source sediment supply rate to the beach and thus depletes the down-coast beaches. This allows higher breaking waves to attack the undefended sections of cliff line, accelerating the recession rate. Another favourite engineering tool is the groyne, used to reduce the beach transport rate and retain a sediment buffer to protect the land (D). This reduction in sediment has the same effect as in (B) with accelerating coast erosion down-drift of the last groyne. Both groynes and seawalls induce 'terminal scour', that is increased erosion down-drift of the structure defined by 'r'. X and Y are reference markers.

If human activity has to be allowed on the coastal zone then it should be concentrated in those areas where sediment budgets are positive, otherwise development should be controlled on an ephemeral-life basis, that is short-lived and without massive infrastructure costs that cannot be recovered in the short term. This does not mean that areas of sediment surplus can be sealed off under concrete and separated from the beach system. Many sediment sinks are only temporary and may be an integral part of the system to be called upon during extreme conditions. Dune areas are identifiable in this category yet should be recognised as a reserve for the beach system that can be called upon under storm activity to replenish the beach system. Even the idea of defending dunes as a coastal buffer by building bulkhead walls should be resisted, as this may prevent movement of the dune sediment back on to the beach at times of storms. Other areas of coastal deposition often regarded as of low value and available for exploitation are coastal wetlands. To some extent recognition of the biodiversity and ecological importance of wetlands is now a barrier to mass exploitation but for generations such areas were regarded as natural sites for new-ground schemes (termed incorrectly 'reclamation'). Such new-ground schemes take out a natural buffer to extreme wave conditions in estuaries, and encourage reflection of wave energy rather than dissipation. Reflected wave energy tends to mobilise nearshore sediment thus steepening the offshore profile and allowing bigger waves to break closer to the shoreline causing further problems for the reclaimed land.

The underlying ecological principle to coastal management is the essential coherence and integrity of the coastal system. It has to be viewed and managed in a holistic way. Piecemeal attack may not be seriously rated as a major problem but in summation even piecemeal changes can create major disturbances due to the linear nature of coastal systems. The degree to which these principles can be employed depends on the problems presented by the human impact on coasts. Political and economic expediency will force managers to breach these guidelines virtually every time human interference in coasts causes a problem. In general managers are reactive rather than proactive and therefore are likely only to use these guidelines in an attempt to minimise damage of intrusion. How does one then successfully intervene in the shoreline? The short answer is that it is impossible unless economic and physical prices are paid which government and private concerns are increasingly reluctant to cover. Intervention may be possible if it is planned in conjunction with the known physical limits of the coastal system. Once the spatial attributes of the beach system are mapped then natural breaks for intervention may be definable and economically sustainable on a cost-minimisation basis. One of the prices of intervention should be the cost of artificially nourishing or replenishing lost beach volume due to intervention. Such a price would not be a one-off affair as the consensus on beach nourishment is that it needs an ongoing maintenance operation to cover the annual losses of sediment.

Some observers argue that beach intervention on any scale is in the long term unsuccessful and that there is a limited future for humans on some shorelines (Pilkey, 1981). Doyle *et al.* (1984) cogently combine these various principles into a series of statements (Box 1.1) that would bar most engineering on the coast. The cost of engineering an *in situ* solution outweighs the cost of retreating from the shoreline. Clearly this cannot be enunciated as a general principle without detailed cost/benefit analysis of local situations, as in many cases people and property are already substantially embedded on the coast and cost is only part of the evaluation procedure. Recent thinking (Titus *et al.*, 1991; Frankhauser, 1995; Turner *et al.*, 1996) suggests that a more flexible approach is required in considering whether people should hold the coastal line or retreat, but recognises that the cost of staying put is high in that solutions have to be engineered and total. Physical coastal environments will never be the same once an engineered solution is adopted and any ecological principles of management may as well be jettisoned once such a decision is taken.

COMING TO TERMS WITH PROCESSES STRUCTURING THE COAST

Although scientists understand much about contemporary or instantaneous coastal and beach processes, there are still problems in understanding the way coastal systems operate over a year or decade. This is often because the sum of instantaneous processes does not appear to add up to the coastal changes that we see in the longer term. It is also difficult to understand some of the temporal lags found with coastal change. Processes' response times and lags may be of such a scale that human intervention on the coast can be divorced from its results by many decades, if not centuries (Orford, 1988). The coastal manager should be prepared for the spatial uncertainty of coastal change in which the focus of change may spatially jump offshore or even onshore, as well as longshore.

The lack of recognition and understanding of long-term process-response control on the coast is a major handicap to effective coastal management. Public perception can be perverse in that sometimes images of beach status are enshrined beyond reality. Jennings and Smyth (1990) offered an illuminating example of mistaken human perception from the English south coast. They showed that the fight to retain the gravel beaches between Seaford and Dungeness (southern England) by generations of groynes is built on a misapprehension that such gravel beaches that were there had a permanence of thousands of years. They indicated that many of these features only arrived within the last 500 years and that their current

---

BOX 1.1   TRUTHS ABOUT THE SHORELINE

- There is no erosion problem until a structure is built at the shoreline.
- Construction by man on the shoreline causes shoreline change.
- Shoreline engineering protects property on the beach, not the beach itself.
- Shoreline engineering destroys the beach it was intended to save.
- The cost of saving beach property through shoreline engineering can often be greater than the value of the property to be saved.
- Once shoreline engineering begins, it cannot be stopped.

Source: Doyle *et al.*, 1984

depleted state is simply recognition of a near terminal point in a withering-away process caused by long-shore transport of the gravel. Recent major gravel recharging (= beach nourishment) is doomed to be ineffectual in the face of consistent longshore drift, unless substantial extra defences can be provided that would contain the drift rate.

Human memory is a major source of uncertainty for coastal managers. Memory of severe events (hurricanes and storms) is often short-lived in a coastal community due to high population turnover (migration and death). Yet there is always a willingness for new incomers to take the vulnerable positions from older and chastened victims of coastal excesses. The understanding of natural hazards associated with living on the coastline appears to have to be relearnt with succeeding generations.

Most recent investigators have concentrated on mean sea level rise as the major coastal problem over the next century (e.g. Gornitz, 1991). There is a considerable literature on the long-term effect of sea level rise relative to coastal morphology, in particular the likelihood of shoreline recession as a function of sea level rise (Bruun, 1962; Orford, 1987; SCOR, 1991). This issue probably more than any other has driven the concern that underlies the need for effective coastal management policy. To date there has been considerable debate as to the predicted amount of projected sea level rise by the end of the next century (15 to 95 cm, Houghton *et al.*, 1996). Such debate centres over the actual components that will influence sea level. Most damaging and least damaging scenarios of global estimates of eustatic sea level change are often given dependent on the degree to which future emissions of greenhouse gases can be regulated. Clearly the utility of such values *per se*, or ability to interpret such ranges, is low given that coastal managers have little feel for the translation of such values into effective coastal response. It is easier to excite interest in the context of a water rise of 0.5 m, rather than in a rate of 2 mm per year over 25 years, yet the rate of rise is of more importance to the activity of the coastal system. A low rising sea level rate may lead to the preservation of a coastal morphology while a high rate may generate its erosion.

The issue of sea level rise has been generated in the wake of the proposition of accelerated greenhouse warming, yet the real coastal forcing consequences of this global warming must be in the notion of increased storminess (Pugh, 1990). The predicted spatial variation in this increased storminess is imprecise but low latitudes are likely to experience increased extreme tropical storm activities (hurricanes, cyclones and typhoons), while mid-latitudes are likely to experience an increase in cyclonic activity associated with depressions. These weather system changes imply an increase in wave energy and storm-surge levels experienced at the shoreline. New sediment is unlikely to be produced at a rate sufficient to match the enhanced transport potential. Increased peak energies are likely to accelerate the losses from the transport corridor, reduce the beach volume and expose the terrestrial basement to increased erosion.

COASTAL PROTECTION: THE PROBLEMATIC FRONT LINE OF CZM

Most coastlines show the effects of human activity in their recent physical development. The pressure for CZM policies in recent decades originated more due to the need for coastal protection as a consequence of coastal expansion of human populations than as a brake to exploitation of a natural environmental resource. In some cases coastal management is still regarded solely as coastal protection under another name. It is worthwhile considering why coastal management and coastal engineering have a relationship that is uneasy, yet difficult to break.

In western societies people lived in a sympathetic relationship with the coast up to the industrial revolution. This was due to their passive acceptance of the power of nature given their inability to intervene in the coast. There are accounts of positive intervention by the use of protection methods (the Netherlands), but rarely did such methods lead to long-term success. The switch to aggressive intervention came in the nineteenth century when some societies felt that they had a engineering basis for the active exploitation of coastal areas. This engineering approach was essential for the infrastructural support of an expanding trade and industry based on a trans-oceanic perspective as witnessed by the transformation of estuaries into port and harbour facilities and the rapid growth of coastal towns with ports as a focus of this expansion.

At the same time, major urban growth occurred along shorelines indirectly related to industrial expansion. The rise of the seaside resort as a leisure and recreational facility stems directly from the enhanced mobility of industrial masses through railway expansion and from the demand for one-day, and eventually week-long holidays (Lawton, 1968; Walvin, 1978; Funnell, 1983). Whereas the industrial development sought coastal sites that were protected and had good terrestrial access (estuaries), the resort towns were often located in more exposed sites on open coasts where bracing scenic qualities and bathing access via beaches were the important factors (Manning-Sanders, 1951). Distance from inland urban centres was also important in terms of minimising rail link costs, but the initial purpose of resorts was that people should actively participate for health reasons, in sea-bathing and partaking of coastal air by walking or promenading. The first activity inevitably demands a wide beach, the second some form of walkway structure for easy access and comfort.

Inevitably those coastal sites with any kind of beach near to urban centres expanded regardless of the long-term status of the beach. The growth of the southern English seaside town in the nineteenth century is a classic example of the way in which the demands of coastal users for a 'fixed' beach were defeated by the mobility of their actual beaches. The development of static tourist-resort infrastructures meant that the retention of beaches was essential for tourist and trading sustainability. Any beach losses had to be resisted by active intervention. It was inevitable that this would be by means of engineering structures (Bruun, 1972). The failure to understand beach behaviour in nineteenth-century engineering is evident in the rash of structures to control eroding cliffs and hold beach sediment in place that spread around the coastlines of the economically advanced world. Unfortunately coastal engineering proved to be limited, in that symptoms rather than causes of coastal problems were addressed, and then in a piecemeal and aspatial fashion. This meant that many schemes further compounded the original problem or created new ones. After 150 years of defending coasts many original coastal cells have now been changed by engineered solutions (Figure 1.4). Shoreline protection was supplied more as an adjunct of political pressure associated with the population on the coastline.

Schemes were often defined by cultural aspects of people living at the coast, rather than by any serious regard for the coherence of the coastal system. It comes as no surprise that attempts at coastal modification by engineering have in the long term only at best reduced erosion rates, while at worst have accelerated erosion at other points along the coast. This type of management approach has continually failed to recognise the nature of the problem: beach engineering solutions are self-defeating. This basic paradox has been repeatedly encountered along the advanced world's shorelines and is now being exported as the solution to less economically advanced countries with coastal resorts plus attendant problems.

A crisis of approach to coastal problems has been evident over the last 20 years. The increasing relative affluence of some sectors of western societies is shown in the growing pressure for retirement and second homes (leisure) on the coast. This growth, due to planning processes, is felt principally around the old coastal sites where services demanded by this new influx can best be met. The spatial growth of coastal towns causes an increasing demand for protection measures adjacent to the old schemes. However, the extension of such schemes has been increasingly questioned by both coastal scientists and engineers on the grounds of cost, benefits and effectiveness. Engineered structures are proving too expensive, the benefits appear to be limited to a few shoreline dwellers at the expense of the national tax-paying population, while the effectiveness of features in resisting erosion and limiting any impact elsewhere appears to be decreasing.

Early forms of coastal zone management were initiated as a response to the failure to meet expectations of coastal populations for cheap and effective protection. CZM aimed to broaden the management perspective moving from a reactive one dominated by protection to a proactive one in which the remit of management was broadened to accommodate a view of the coast as a coherent system with its own demands where intervention was to be avoided. Human occupancy of the coasts is viewed as a major user of coastal resources and as such the mitigation of its demands and the sustainability of its requirements by planning are key elements in modern management. Unfortunately, this approach does not always answer the specific protection problems of individuals or

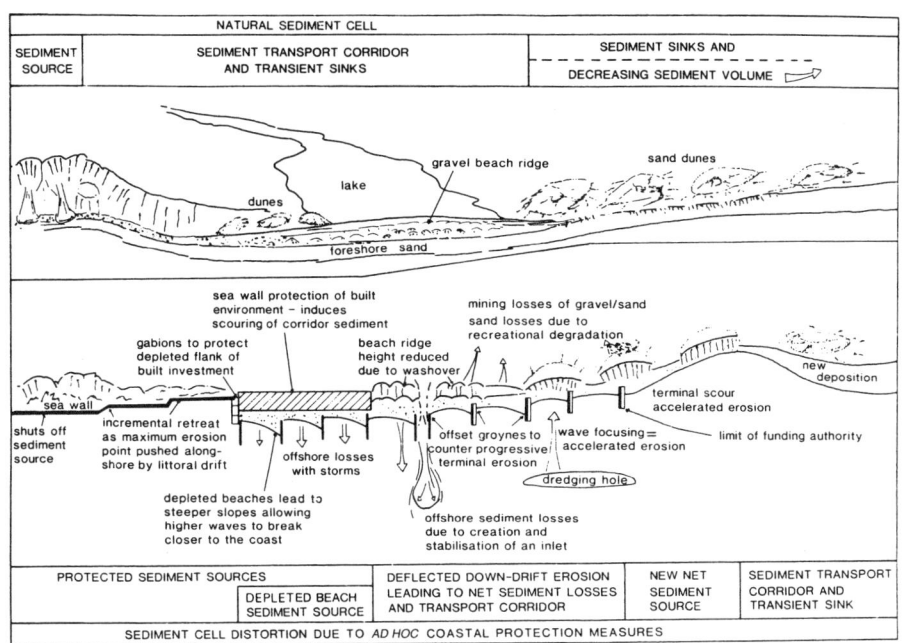

Figure 1.4 A schematic view on how a natural coastline with simple cell structure has evolved into a complex one of deflected and depleted cells due to a variety of man-made obstacles to sediment movement

In many cases early attempts to defend the coast necessitate later structures to help prop up the position generated by the earlier attempts. Hence sea walls that defend coastal properties seal off sediment sources, thus necessitating groynes to reduce the already diminished longshore sediment transport rate in order to create a sediment bulkhead in front of shoreline structures. The last groyne in any series (often at the end of the 'paymaster' authority) causes new erosion downbeach. Coastal mining and offshore dredging for mineral extraction and navigation passage further disturb the original sediment pattern.

Source: after Orford, 1986; copyright: Surrey University Press, 1986 and Blackie and Sons

communities which are still the major problems for CZM. As the demand on the world's shorelines grows the issue of protection for coastal inhabitants is also growing. However, if economically advanced countries cannot afford management based on engineered protection, there is little chance that the less economically developing countries will be able to afford engineering protection management as their front-line management strategy. In the long term, this may prove of benefit to these countries as they will be forced to look at the causes of their coastal problems rather than only treating the symptoms.

COASTAL MANAGEMENT APPROACHES

If protection through intervention is not ecologically realistic what should be considered in an ecological approach? The sustainable essence should be that the coastal system is to be treated as an integrated entity. The movement of sediment should take priority, such that any attempts to control sediment inputs and throughputs need to be considered on a wider front than just at the point of intervention. CZM needs to produce a basic statement of sediment budget for the coast, recognising the inherent spatial and temporal variability and then plan the nature of intervention that has sufficient cultural and economic priority, on a disruption-minimisation basis.

It is important to appreciate that the nature of coastal management in which this ethos is contained is not invariant. There needs to be a flexibility of approach that is dependent on the nature of the process, morpho-sedimentary and cultural domains of the coast. The nature of the sediment pathways in a coastal system will define the morphological and sedimentological structure of the units that make up the coast. World-wide there are diverse coastal environments structured by energy, morphology, sediment type and supply. Management strategies for mid-latitude barrier islands will differ in detail from high latitude fringing-gravel beaches and low latitude mud-dominated coasts in consideration of the connectivity of the system, yet all should consider the production and control of sediment distribution as an essential element of the programme.

The recognition of these morphological domains requires an understanding of the process domains that drive the morphology. Management needs to recognise the nature and distribution of energy that controls such domains. Although it is unlikely that human activity on the coast is liable to shift the initial macro-distribution of such processes on an open coast, human activity may adjust the morphological structure of closed coasts sufficient to alter the distribution of energy. Land fills and estuary barriers are examples of methods by which processes may be changed in a macro sense. All interventionist structures would alter the micro-distribution of process energy around the point of intervention, but are unlikely to alter the macro-distribution. CZM also needs to be adjusted with respect to the nature of extreme events dominating the coastal system. One can draw a distinction between the scale of surge generated by mid-latitude cyclonic activity and surge generated by an extra-tropical storm of hurricane dimensions, in terms of the likely impact on the coastal zone. The management response of societies to the threat of these two extreme situations is clearly different. In the case of depressions, emphasis has been placed on building up coastline defences to withstand the peak flood, while managers in areas threatened by hurricanes place emphasis on evacuation, recognising the impracticality of total surge defence. Therefore management policies in the light of extreme events can range between reactive and proactive. Reactive management is where the emphasis is on containment of the coastal threat by relying on defences that are regarded probabilistically to be adequate. Proactive management assumes that coastal defences will not be sufficient; therefore emphasis is placed on minimising damage either by educating the coastal user into the danger

of living in the threatened zone, or by establishing evacuation procedures or safety refuges (Salmon, 1984).

The choice of management mode is also determined by the mode of cultural development of the people on the coast. Much of this chapter has concentrated on the experience of economically advanced countries whose coastal development has been based around a property-owning population such that coastal policy emphasises property protection. Yet this model is inappropriate for societies where the mass population is agrarian, subsistence based and poor. For example, the use of insurance as a policy instrument to control coastal development would not be effective in Bangladesh whereas it is of considerable importance in regulating barrier island development in the USA (Dawson, 1987). The aesthetic emphasis of management planning is also unlikely to be at the fore of the coastal policies in less economically advanced countries whereas it is probably a major implicit force in western societies. Differences in both policy approach and instruments of action are also likely between countries who see the shoreline as an unfortunate break to the spatial spread of land-based cultural imperatives (e.g. subsistence-based, developing societies) and societies who see the shoreline as an important unit in its own right that has intrinsic economically justifiable imperatives that require separate planning and management (Platt et al., 1987).

To a large extent the ecological principles of continuity and transport in coastal management were first established in the USA as a response to the perceived difficulties that traditional engineering approaches had engendered along the US coastline (Ducsik, 1974; Kaufman and Pilkey, 1979; Nordstrom et al., 1986). Geomorphological development of barrier islands requires a mobility of sediment to ensure the links between beach, dune, back-barrier and lagoon (Nummedal, 1983). Unfortunately barrier stabilisation by protection effectively breaks those sediment links. The lag effect of these broken links sometimes took decades before the changes in sediment budgets became apparent, thus generating a perspective of apparent success of the engineered structures to control the erosion, and allowing time for the innovation of engineering to spread. In 1972 the USA passed coherent federal coastal zone management legislation which demanded an integrationist approach to coastal development which relied on ecological principles. This legislation offered financial incentives to coastal states to develop coherent development plans by which local authority and private coastal proposals could be integrated with state intentions such that coastal development would follow guidelines of minimising impact. The need for CZM was promulgated as much by the demands of the 'eco-concerned' over the continuing destruction of the diverse coastal habitats (in particular the wetlands and their fisheries-related role), as it was by the recognition of the post-1945 explosion of private and public investment along the coast that was vulnerable to coastal erosion and could not be protected without either the substantial fiscal demands of coastal engineering or the expensive commitment of a federal insurance policy on storm damage. The policy core of CZM at the local level is to avoid siting property development that will interfere with the existing processes of coastal development. This sounds reasonable but was and is difficult to enforce in sites where prior shoreline development has taken place (Heikoff, 1976). By 1984 under the pressure of the federal budget deficit, the US Congress passed the Coastal Barriers Resources Act that on face value appears to preserve coastal barrier islands from development by prohibiting federal support for island infrastructure on designated islands that had at that date less development than a specified level. However, this stage of CZM was passed more for the fiscal savings than for saving barrier islands (Godschalk, 1987). Coastal management policy is overtly integrationist and ecologically led at a legislative level; however, protectionist perspectives still effectively drive much of the human response to coastal issues (e.g. Mitchell, 1974; Titus, 1990; Titus *et al.*, 1991).

Further perspectives on types of coastal management are identified for the United Kingdom (Box 1.2) and Bangladesh (Box 1.3).

CONCLUSIONS

Protection of property and lives from effects of coastal processes form essential elements of most coastal management programmes. The principal means of achieving this is by either removing people from the coast or conserving the presence of the beach as a barrier or buffer to wave action. The latter course of action appears to be the essential guiding principle of CZM. However, it is a principle that is hard to adhere to as the slightest intervention by people into the coastal zone tends to disrupt the coherence and continuity of beaches and thus creates the opportunity

---

BOX 1.2   APPROACH TO COASTAL ZONE MANAGEMENT: UNITED KINGDOM

There is no comprehensive coastal zone management policy in the UK although there is considerable pressure for a unified agency based around planning along regional coastal cells (House of Commons, 1992). The physical nature of the coastline is somewhat more simplified than that of the multi-path barrier islands of the USA in that most of the UK coast shows a narrow beach corridor although with multiple cells at all levels of a regional hierarchy (Bray et al., 1995). Coastal policy appears to revolve overtly around the concept of protection. There is a two-strand approach whereby legislation supports fiscal support of engineering structures for the twofold problems of coastal erosion and coastal flooding (Ricketts, 1986; Barrett, 1989). Different combinations of central governmental agencies and local authority structures are used to deal with these two problems. Historically, both approaches have been protection-led through engineering design, mostly taking a symptoms rather than causes approach, and both constrained by the spatial remit of the local authority power, thus coastal developments appear as an *ad hoc* approach on the ground. The concept of sediment continuity is well recognised but the constraints of the legislative and planning process have always been cited as reasons why ecological principles have not been foremost in the coastal policy. In recent years, cost/benefit analysis of engineering approaches has led to formal recognition by engineers that the problems of coastal erosion cannot be approached on this old basis (Penning-Rowsell et al., 1989). Clearly, softer versions of coastal defence should be used including the possibility of a retreat philosophy by property buy-out, although the economic priority of engineered structural intervention is still regarded as the principal weapon of control policy. The current mood of UK coastal scientists and engineers is such that they would welcome legislative changes that recognise the importance of sediment continuity, but until such a time, coastal policy is protectionist and fiscal based.

> **BOX 1.3   APPROACH TO COASTAL ZONE MANAGEMENT: BANGLADESH**
>
> Bangladesh is now regarded as the crucial test of coastal management given that it exemplifies the extremes of conditions that affect coastal zone management: massive extreme energy due to tropical cyclones (Murty et al., 1986); highest mortality from coastal hazards; a highly unstable shoreline due to fluvial and marine disturbances; a substantial area of the country under the limits of projected sea level rise; massive population pressure on the shoreline; a high degree of irrationality and fatalism, by western standards, in the population over coastal hazards; and one of the world's lowest per capita income and GNP (Aminul Islam, 1977; Burton et al., 1978). Coastal zone management is a top-down process that is essentially concerned with how to protect lives from the deadly surges that accompany cyclones moving up the Bay of Bengal (Frank and Husain, 1971). The possibilities of rapidly evacuating the densely populated rural poor are limited, given the cultural imperatives of an Islamic society that may prevent women from leaving their homes and appearing in public and induce a high degree of fatalism about one's chances given surge presence. Culture also dictates that the rapidly prograding coastline in the form of new islands (chars) is open to occupancy on a first-claim basis, thus forcing landless poor to chance their existence in surges by staying put rather than seeking shelter in the limited number of purpose-built government refuges and risking a loss of land claim after the surge is over. Coastal management is not about ecological principles at the macro-scale, instead it is limited to defining and supporting socio-economic infrastructures that can minimise societal destabilisation as well as the physical destruction caused by tropical storms. Bangladesh remains as probably the greatest problem for coastal management in the twenty-first century.

for reduced beaches and hence increased erosion. Although most CZM policies would recognise the importance and primacy of this ecological principle, its implementation is generally more a case of wishful thinking than reality. CZM still seems to vary between policies of protection based on engineering support to policies in which engineering protection is still a high priority but one that is incapable of being delivered because of accelerating cost or because of a refusal to accept the reality of alternative softer methods with their political and legislative implications for more proactive management of the coastal zone.

REFERENCES

Aminul Islam, M. (1977) 'Tropical cyclones: coastal Bangladesh', in F.W. Gilbert (ed.) *Natural Hazards: Local, National, Global*, Ch. 2, New York, USA: Oxford University Press, pp. 19–24.

Barrett, M.G. (1989) 'What is coastal management?', in *Coastal Management*, Ch. 1, Telford, London: Institute of Civil Engineers, pp. 1–10.

Bowden, R. and Orford, J.D. (1984) 'Residual sediment cells on the morphologically irregular coastline of the Ards peninsula, Northern Ireland', *Proceedings of the Royal Irish Academy* 84B: 2–27.

Bray, M.J., Carter, D.J. and Hooke, J. (1995) 'Littoral cell definition and budgets for central southern England', *J. Coastal Res.*, 11: 381–400.

Bruun, P. (1962) 'Sea level rise as a cause of shore erosion', *J. Waterways, Harb. Div.* American Society of Civil Engineers, 88 (WW1): 117–130.

Bruun, P. (1972) 'The history and philosophy of coastal protection', *Proceedings of the 13th. Conference of Coastal Engineers.* 33–74.

Burton, I., Kates, R.W. and White, G.F. (1978) *The Environment as Hazard*, New York, USA: Oxford University Press.

Carter, R.W.G. (1988) *Coastal Environments*, London, UK: Academic Press.

Clark, M.J. (1978) 'Geomorphology in coastal zone environmental management', *Geography* LXIII: 273–282.

Clarke, J. (1974) *Coastal Ecosystems*, Washington DC, USA: The Conservation Foundation.

Dawson, A.D. (1987) 'The NFIP and developed coastal barriers', in R.H. Platt, S.G. Pelczarski and B.K.R. Burbank (eds) *Cities on the Beach*, Research Paper No. 224, Chicago, Ill., USA: University of Chicago, Dept of Geography, pp. 245–260.

Doyle, L., Sharma, D.C., Hine, A.C., Pilkey, O.H. Jr., Neal, W.J., Pilkey, O.H. Sr., Martin, D. and Belknap, D.F. (1984) *Living with the West Florida Shore*, Durham, NC, USA: Duke University Press.

Ducsik, D.W. (1974) *Shoreline for the Public*, Michigan, USA: MIT Press.

Frank, N.L. and Husain, S.A. (1971) 'The deadliest cyclone in history', *Bull. Amer. Met. Soc.* 52: 438–444.

Frankhauser, S. (1995) 'Protection vs. retreat. The economic cost of sea-level rise', *Environment and Planning* A 27: 299–319.

Funnell, C. (1983) *By the Beautiful Sea*, New Jersey, USA: Rutgers University Press.

Godschalk, D.R. (1987) 'The 1982 Coastal Barrier Resources Act: a new Federal policy tack', in R.H. Platt, S.G. Pelczarski and B.K.R. Burbank (eds) *Cities on the Beach*, Research Paper No. 224, Chicago, Ill., USA: University of Chicago, Dept of Geography, pp. 17–27.

Gornitz, V. (1991) 'Global coastal hazards from future sea level rise', *Palaeogeogr., Palaeoclim. and Palaeoecol.* 89: 379–398.

Hall, B. (1989) 'A regional strategy', in *Coastal Management*, Ch. 6, Telford, London, UK: Institute of Civil Engineers, pp. 51–57.

Heikoff, J.M. (1976) *Politics of Shore Erosion: Westhampton Beach*, Ann Arbour Mich., USA: Ann Arbor Scientific Publishers Inc.

Horikawa, K. (ed.) (1988) *Nearshore Dynamics and Coastal Processes*, Tokyo, Japan: University of Tokyo Press.

Houghton, J.T., Meira Filho, L.G., Collander, B.A., Harris, N., Kattenberg, A. and Maskell, K. (1996) *Climate Change 1995: The Science of Climate Change*, Cambridge, UK: Cambridge University Press.

House of Commons (1992) *Report of the House of Commons Select Comittee on the Environment: Coastal Zone Protection and Planning*, 17-I, March 1992, Vol 1.

Jennings, S. and Smyth, C. (1990) 'Holocene evolution of the gravel coastline of East Sussex', *Proc. Geol. Assoc.* 101: 213–224.

Kaufman, W. and Pilkey, O. (1979) *The Beaches are Moving*, New York, USA: Anchor Press/Doubleday.

Komar, P.D. (1983) 'Coastal erosion in response to the construction of jetties and breakwaters', in P.D. Komar, *Handbook of Coastal Processes and Erosion*, Ch. 9, Boca Raton, Fl., USA: CRC Press, pp. 191–204.

Krause, N.C. and Pilkey, O.H. (1988) 'The effects of sea-walls on the beach', *J. Coastal Res.*, Special Issue, No. 4.

Lawton, R. (1968) 'Population changes in England and Wales in the later nineteenth century: an analysis of trends by Registration districts', *Trans. Inst. Brit. Geogr.* 44: 55–74.

Manning-Sanders, R. (1951) *Seaside England*, London, UK: B.T. Batsford Ltd.

Mitchell, J.K. (1974) *Community Response to Coastal Erosion*, Research Paper No. 156, Chicago, Ill., USA: Dept of Geography, University of Chicago.

Murty, T.S., Flather, R.A. and Henry, R.F. (1986) 'The storm surge problem in the Bay of Bengal', *Progr. Oceanography* 16: 195–233.

Nordstrom, K.F, Gares, P.A., Psuty, N.P., Pilkey, O.H. Jr., Neal, W.J. and Pilkey, O.H. Sr. (1986) *Living with the New Jersey Shore*, Durham, NC, USA: Duke University Press. (This is one of a series that deals with the coastal management problems of the USA, state by state.)

Nummedal, D. (1983) 'Barrier islands', in P.D. Komar (ed.) *Handbook of Coastal Processes and Erosion*, Ch. 5, Boca Raton, Fl., USA: CRC Press, pp. 77–121.

Orford, J.D. (1986) 'Coasts: environments and landforms', in P.G. Fookes and P.R. Vaughan (eds) *Handbook of Engineering Geomorphology*, Guildford, UK: Surrey University Press, pp. 193–210.

Orford, J.D. (1987) 'Coastal processes: the coastal response to sea level variation', in R.J.N. Devoy (ed.) *Sea Surface Studies*, London, UK: Croom Helm, pp. 415–463.

Orford, J.D. (1988) 'Alternative interpretations of man-induced shoreline changes in Rosslare Bay, southeast Ireland', *Trans. Institut of Brit. Geogr.* 13: 65–78.

Penning-Rowsell, E.C., Coker, A., N'jai, A., Parker, D.J. and Tunstall, S.M. (1989) 'Scheme worthwhileness', in *Coastal Management*, Ch. 19, Telford, London, UK: Institute of Civil Engineers, pp. 227–241.

Pilkey, O. (1981) 'Geologists, engineers and a rising sea level', *Northeastern Geol.* 3: 150–158.

Platt, R.H., Pelczarski, S.G. and Burbank, B.K.R. (eds) (1987) *Cities on the Beach: Management Issues of Developed Coastal Barriers*, Research Paper No. 224, Chicago, Ill., USA: Dept of Geography, University of Chicago.

Pugh, D.T. (1990) 'Is there a sea level problem?', *Proc. Inst. Civ. Engr.* 88: 347–366.

Ricketts, P.S. (1986) 'National policy and management responses to the hazard of coastal erosion in Britain and the United States', *Applied Geogr.* 6: 197–221.

Salmon, J.D. (1984) 'Veretical evacuation in hurricanes: an urgent policy problem for coastal managers', *Coastal Zone Management J.* 12: 287–300.

SCOR (1991) 'The response of beaches to sea level changes: A review of predictive models', *J. Coastal Res.* 7: 895–921.

Titus, J. (1990) 'Greenhouse effect, sea level rise and barrier island: case study of Long Beach Island New Jersey', *Coastal Management* 18: 65–90.

Titus, J., Park, R.A., Leatherman, S.P., Weggel, J.R., Greene, M.S., Mausel, P.W., Brown, S., Gaunt, C., Trehan, M. and Yohe, G. (1991) 'Greenhouse effect and sea level rise: the cost of holding back the sea', *Coastal Management* 19: 171–204.

Turner, R.K., Subak, S.E. and Adger, W.N. (1996) 'Pressures, trends and impacts in coastal zones: interactions between socio-economic and natural systems', *Environmental Management* 20: 159–173.

Walvin, J. (1978) *Besides the Seaside*, London, UK: Allen Lane.

SUGGESTED READING

Bird, E.C. (1985) *Coastline Changes*, Chichester, UK: J. Wiley.

Carter, R.W.G. (1988) *Coastal Environments*, London, UK: Academic Press.

Devoy, R.J. (1987) *Sea Surface Studies*, London, UK: Croom Helm.

Edwards, S.F. (1987) *Coastal Zone Economics*, London, UK: Taylor and Francis.

Komar, P.D. (1983) *Handbook of Coastal Processes and Erosion*, Boca Raton, Fl., USA: CRC Press.

Viles, H. and Spencer, T. (1995) *Coastal Problems*, London, UK: Edward Arnold.

SELF-ASSESSMENT QUESTIONS

1 Complete the following sentence using a word or phrase from the list supplied, and justify your answer with a brief statement. 'Knowledge of the processes by which sediment is transported on beaches is . . . for successful coastal zone management.'
   Choose from:
   (a) unnecessary;
   (b) unimportant;
   (c) important;
   (d) essential.

2 Select the most appropriate statement from choices given to complete the following sentence: 'Sea level rise over the next 20 years can be regarded in the context of coastal management as. . . .'
   Choose from:
   (a) of no importance to ongoing coastal management;
   (b) of importance but only when evidence is available that sea level has risen;
   (c) of primary importance but only to low-lying areas liable to flooding;
   (d) of sufficient importance to dictate the basic policy of coastal management.

3 A coastal sediment cell must by definition include which items from the following list:
   (a) Source area.
   (b) Cliffs.
   (c) Beaches.
   (d) Transport corridor.
   (e) Dunes.
   (f) Sediment sink.
   (g) Boundaries.

4 A key principle of ecological beach management is promoting 'beach sediment continuity'. Assuming the role of a coastal manager, which of the following activities would you allow/disallow in the context of beach continuity? Do you think there are any mitigating circumstances that you should consider in your decision-taking?
   (a) Cliff protection of an isolated house.
   (b) Cliff protection of a planned row of private houses.
   (c) Cliff protection of a row of existing private houses.
   (d) A sea wall along part of the sediment corridor.
   (e) A sea wall as a promenade along a dune field.
   (f) Removal of pebbles from a beach to be sold as tourist souvenirs.
   (g) Removal of sand from a beach to help lighten the soil texture of gardens adjacent to the coast.
   (h) Removal of pebbles from a beach for road-building purposes.

5 Decide the best match between the type of person and the reason why they live in the coast zone. Be brief in justifying your decision.
   Type of person:
   (a) A retired New York dentist at Miami Beach.
   (b) A hotelier on the south coast of France.
   (c) An Australian oil refinery worker.
   (d) A Bangladeshi landless family.
   (e) An urban squatter in Bombay.
   (f) A foreign exchange dealer in London.
   (g) A Hollywood film star on Malibu beach.
   Reason for living in the coastal zone:
   1   Choice.
   2   Necessity.
   3   Inertia.

6 Indicate which of the following statements are compatible with an ecologically-led coastal management programme:
   (a) Recognition of valued and non-valued coastal environments.
   (b) Continuity of form and dynamics of coastal environments.
   (c) Costed and benefited in terms of human activity alone.
   (d) Sustainability.

(e) Protection by intervention.
7 Describe the existing coastal management approaches of the following countries in terms or whether they are proactive or reactive approaches.
(a) United States of America.
(b) United Kingdom.
(c) Bangladesh.

2

COASTAL ENVIRONMENTS

Integrated coastal zone management

Susan Gubbay

SUMMARY

In recent years an important development concerning the management of coastal environments has been the introduction of integrated coastal zone management (ICZM). The process of ICZM is recognised as providing the necessary overview, direction and environmental basis for sustainable use of the coast and it is being implemented by many coastal nations.

The reasons for embarking on ICZM centre around the pressures of use on a rich natural environment. The key components of the process such as integration, linking planning and management of marine and terrestrial systems, and underpinning programmes with environmental principles are described in this chapter. The main issues that need to be tackled to put the idea into practice are discussed and illustrated with reference to the situation in a variety of countries as well as through the two case studies of the situation in the UK and in China.

There are many consistent themes to ICZM programmes but no single 'right way'. Instead it is important to review and then discard, adapt or adopt the ideas that have been used elsewhere to suit the particular circumstances. Much remains to be learnt about implementing ICZM but considerable progress has also been made on this subject in the last decade.

ACADEMIC OBJECTIVES

Integrated coastal zone management (ICZM) is a process that is being used increasingly to guide the sustainable use of coastal areas. It relies on environmental considerations underpinning decision-making for planning and management at the coast, to achieve goals such as reducing vulnerability of coastal development to natural hazards or ensuring that the natural diversity and essential ecological processes at the coast are not damaged or lost. The aim of this chapter is to illustrate the importance of ICZM to the management of coastal environments and explain the key elements of putting the idea into practice.

On completion, readers should be able to:

- describe and understand the reasons for the interest in ICZM;
- identify the key elements of an environmentally based ICZM programme;
- recognise the constraints on implementing ICZM programmes.

BACKGROUND

Coastal environments are some of the most diverse and productive natural areas on the planet. They support a remarkable variety of habitats which have been, and continue to be, the result of changes in sea level and the influence of exposure to waves and currents, latitude, tidal range and geology. Physiographic features such as islands, fjords, estuaries, deltas and cliffs make up the broad sweep of coastal scenery and within them are habitats such as reefs, seagrass beds, mangroves and lagoons. A particularly spectacular example of coastal habitat in Europe is the Wadden Sea which fringes the coastline of Denmark, Germany and part of the Netherlands. It extends over 900,000 ha and forms the largest unbroken stretch of intertidal mudflats in the world. The area is one of the main European staging points for migratory birds, and is used by around 10 million birds each year as they move from Greenland and the far Arctic Siberian tundras to overwinter along the Atlantic coast of Europe and Africa. The shallow waters are a critical nursery area for North Sea fish stocks with an estimated 80 per cent of the plaice, 50 per cent of the sole and 40 per cent of the herring in the North Sea using the region as a nursery ground (Nijkamp and Peet, 1994).

The diversity of the coastal environment is not limited to its habitats but can also be seen in its wildlife. Species from most phyla occur in coastal waters so it could be viewed as one of the most diverse realms of the planet. This is particularly well illustrated for fish as 50 per cent of fish species occur exclusively on continental shelf areas even though this represents less than 10 per cent of the total ocean area (Nelson, 1984).

Another feature of coastal environments is that most are under considerable pressure from human activity. Urbanisation and population growth are at the root of many coastal problems. Coasts have been of major economic importance for centuries with cities developing where maritime trade routes made their landfall and where easy access to fisheries could bring food and wealth to nearby communities. Today coastal locations attract industry, recreation, trade and other activities as well as housing large numbers of people. In Europe this is particularly marked in the Mediterranean, where the population living in coastal areas is predicted to reach more than 200 million by the year 2025 (Grenon and Batisse, 1989).

The coast is a dynamic environment and the way it is used also changes with new activities and pressures developing alongside more traditional uses for coastal land and inshore waters. Tourism is an example of a more recent trend. This is one of the greatest growth industries in the world, and coastal localities are among the most popular destinations for visitors. In Greece, for example, 90 per cent of the tourist investment is directed to the coast, in Ireland 70 per cent of tourism activity is accommodated on the coast and in Portugal the population density at the coast more than trebles during the tourist season (CEC, 1994).

The economic value of these and other coastal activities is undoubtedly billions of dollars and much of the value is linked to location. Many industries can reduce their costs if sited on the coast, recreational users are attracted to the coast to enjoy water sports, and marine aquaculture is thriving. In 1994 the total production of fin fish, shellfish and aquatic plants was valued at US\$ 39.83 billion and amounted to 25.5 million tonnes. Environmental problems associated with the aquaculture industry are, however, becoming a matter of increasing concern. Clearance of mangroves and other natural coastal vegetation to create shrimp ponds has destroyed key nursery areas for wild fish (AMBIO, 1995). The FAO reports that more than 17 per cent of Thailand's mangrove forests were cleared between 1987 and 1993 to create shrimp ponds. In India the loss was nearly 40 per cent for a similar period (Holmes, 1996). Chemicals used to control disease in fin fish such as salmon have had toxic effects in surrounding waters. The industrial fisheries, which supply the fish meal which is used for feed, add pressures on wild stocks.

Around the world, coastal nations face similar issues. There are problems with supply and access to coastal resources, environmental degradation and conflicts of use. The traditional approach has been to tackle these problems on an individual basis but, in recent years, the emphasis has started to change. Planners and managers are taking a more strategic look at the way in which the coast is used and are more alert in assessing the cumulative effect of activities on the environment. This was given limited consideration in the past despite being an important factor in the deterioration of coastal environments. Newer programmes and proposals are therefore seeking administrative and management arrangements that recognise the linked nature of coastal land and inshore water and acknowledge that environmental considerations should underlie the decision-making process in all sectors. The expression used to describe this different approach is integrated coastal zone management (ICZM).

KEY ELEMENTS OF ICZM

Interest in ICZM has been particularly marked in the last decade. Coastal nations facing pressures of use at the coast, deterioration of their coastal environments and recognising the potential of their coastal zones in economic terms have been looking at ways of improving management of activities taking place in this environment. The prospect of sea level rise, linked to global warming, has also focused attention on the coastal zone, making it imperative that countries work together to reduce the threat, and for coastal nations to have a considered and effective response to the consequences at the coast. Apart from sea level rise (current estimates between 12–92 cm by 2100 (Houghton, 1996)) these may be higher frequency of storms, temperature fluctuations, disturbance in hydrological cycles, changing wave characteristics, salt intrusion and changes in ocean circulation patterns to name a few. These issues raise the question of whether current coastal management arrangements are appropriate and adequate. One outcome has been considerable interest in improving management of the coastal environment and ICZM is seen as a key part of the way forward. The main principles of this approach are summarised in Table 2.1.

A review of international initiatives notes that 177 coastal states have some kind of programme or project on ICZM (Sorensen, 1993). The majority of these are concerned with resolving conflicts of use and balancing demand for coastal resources but other important elements are also being taken forward to put ICZM into practice.

Specific policies and programmes for the coastal zone
The coastal zone is generally taken to include maritime land, the foreshore and inshore waters. This presents unique difficulties in planning and management because elements from both terrestrial and marine planning systems apply and there are few, if any, arrangements to link the different provisions. The need to provide this, combined with the recognised importance of the coastal zone and the pressure it is under, makes it essential to have policies and programmes specific to the coast as well as dealing with coastal problems through planning arrangements which might apply to a wider region or the country as a whole.

Taking an integrated and strategic approach
ICZM programmes look at the whole spectrum of activity in the coastal zone, and are therefore concerned with planning *and* management. One of the most useful aspects of such programmes is that they

Table 2.1 Key features of integrated coastal zone management

A definition	A dynamic process in which a co-ordinated strategy is developed and implemented for the allocation of environmental, socio-cultural and institutional resources to achieve the conservation and sustainable multiple use of the coastal zone (CAMPNET, 1989).
ICZM aims to:	promote sustainable use; balance demand for coastal zone resources; resolve conflicts of use; promote environmentally sensitive use of the coastal zone; promote strategic planning for coasts.
ICZM recognises:	the 'coastal zone' as a unit for planning purposes; that planning and management of coastal land and waters cannot be dealt with separately; the coastal zone as an area that requires special attention for planning and management.
ICZM requires:	a national perspective; a long-term view; an integrated approach to planning and management; communication, collaboration and co-ordination between planners, managers and users; public involvement; a flexible approach; a specific agency to deal with coastal zone matters.

Source: Gubbay, 1990

provide an overview. This should facilitate better integration of plans, policies and management arrangements at a regional and local level, and complement any sectoral planning and management that might be necessary. At the same time a strategic view of coastal matters such as urbanisation can highlight issues such as the threat from small-scale but cumulative loss of coastal wetland areas to development, or an overall picture of decline or improvement in water quality at the coast.

Underpinning the planning and management with environmental principles
Sustainable use is widely quoted as an ideal. ICZM aims to put this into practice at the coast by ensuring

that environmental principles underlie decision-making rather than being an extra consideration. Environmental impact assessments and environmental impact statements help to achieve this but they have their limitations as they are generally tied to particular projects, whereas in ICZM it is recognised that there is also a need for environmental principles to inform policy formulation and decision-making at a strategic level.

Linking the planning and management of terrestrial and marine systems

Activities taking place on coastal land can affect the offshore environment and vice versa. Sedimentation from forestry and agricultural activities, persistent organic toxins such as pesticides entering the coastal zone in runoff, and modified freshwater inputs to estuaries due to extensive damming are some examples. ICZM programmes seek to address this through planning and management systems that span the whole of the coastal zone, as well as by linking coastal policies to those developed for inland areas.

Planning ahead

Taking a long-term view is an important part of ICZM programmes. Such programmes seek to address existing problems but also plan to avoid future difficulties. ICZM measures will also need to operate on various time-scales, some as short-term projects and others running over long periods in order to achieve sustainable management.

A number of international guidelines have been developed to assist with the implementation of ICZM – principally the Noordwijk Guidelines prepared by the World Bank (1993), UNEP (1995) and IUCN (Pernetta and Elder, 1993). These have been reviewed by Cicin-Sain *et al.* (1996) and summarised as a consensus set of ICZM guidelines (Table 2.2).

A number of stages in the evolution of ICZM programmes have been described by Sorensen (1993) and are illustrated in Figure 2.1. The results of a survey of ICZM programmes in 29 countries show that the reasons for embarking on such a programme are varied (Table 2.3) but once a programme is

Table 2.2 A consensus set of ICZM guidelines

Purpose of ICZM	The aim of ICZM is to guide coastal area development in an ecologically sustainable fashion.
Principles	ICZM is guided by the Rio Principles with special emphasis on the principle of intergenerational equity, the precautionary principle, and the polluter-pays principle. ICZM is holistic and interdisciplinary in nature, especially with regard to science and policy.
Functions	ICZM strengthens and harmonises sectoral management in the coastal zone. It preserves and protects the productivity and biological diversity of coastal ecosystems and maintains amenity values. ICZM promotes the rational economic development and sustainable utilisation of coastal and ocean resources and facilitates conflict resolution in the coastal zone.
Spatial integration	An ICZM programme embraces all of the coastal and upland areas, the uses of which can affect the coastal waters and the resources therein and extends seaward to include that part of the coastal ocean that can affect the land of the coastal zone. The ICZM programme may also include the ocean area under national jurisdiction (EEZ) over which national governments have stewardship responsibilities both under the Law of the Sea Convention and UNCED.
Horizontal and vertical integration	Overcoming the sectoral and intergovernmental fragmentation that exists in today's coastal management efforts is a prime goal of ICZM. Institutional mechanisms for effective co-ordination among various sectors active in the coastal zone and between the various levels of government operating in the coastal zone is fundamental to the strengthening and rationalisation of the coastal management process. From the variety of available options, the co-ordination and harmonisation mechanism must be tailored to fit the unique aspects of each particular national government setting.
The use of science	Given the complexities and uncertainties that exist in the coastal zone, ICZM must be built upon the best science (natural and social) available. Techniques such as risk assessment, economic valuation, vulnerability assessments, resource accounting, cost–benefit analysis, and outcome-based monitoring should all be built into the ICZM process

Source: Cicin-Sain *et al.,* 1996

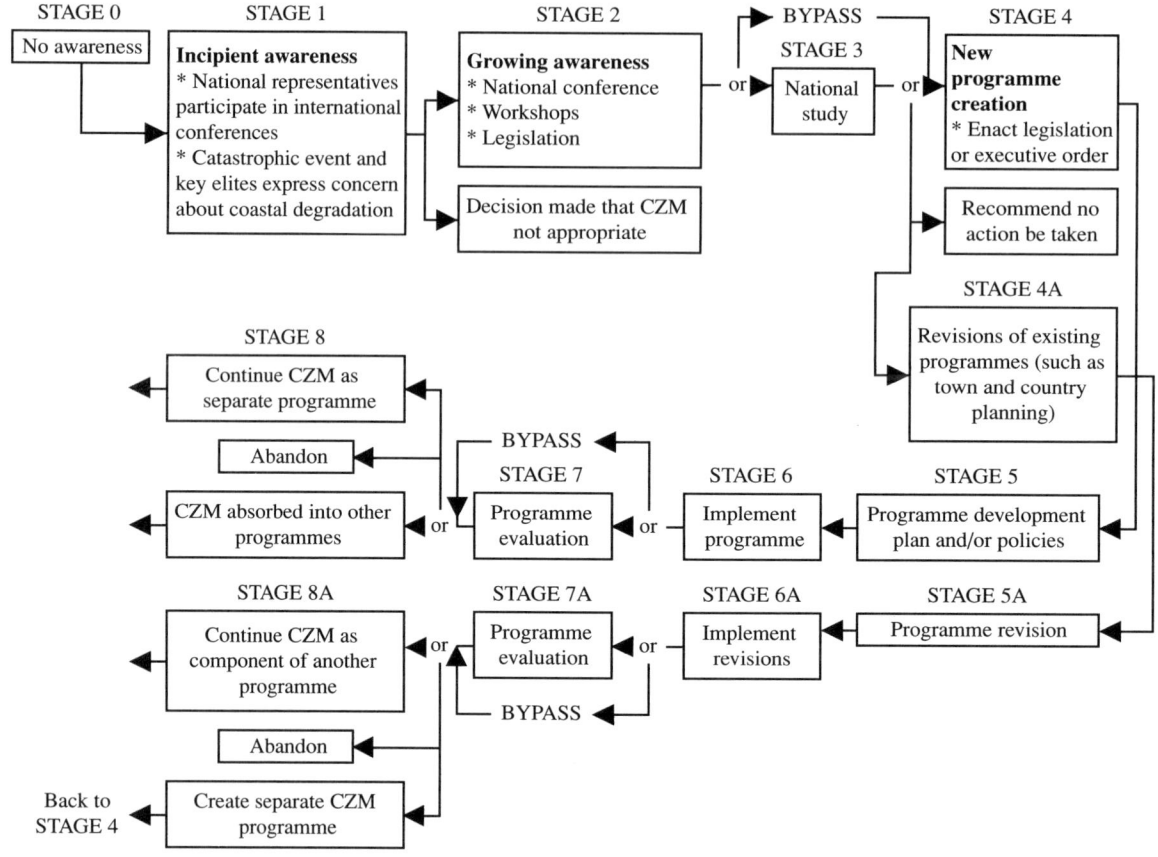

Figure 2.1 Evolution of coastal management: from concept to practice
Source: Sorensen, 1993

Table 2.3 Reasons for the initiation of integrated coastal management in selected countries (in percentages)

Catalyst	All n = 49	Developed n = 14	Middle developing n = 15	Developing n = 20
Depletion of resources	18	0	27	25
Pollution	20	21	13	25
Ecosystem damage	18	21	13	25
Economic benefits from coast and ocean	22	36	13	20
New economic opportunities in coast and ocean	6	7	13	0
Damage from coastal hazards	10	14	7	10
Other	4	0	13	0
Uncertain	0	0	0	0

Source: Cicin-Sain and Knecht (forthcoming)

underway it is possible to identify some common issues that require discussion and resolution for the successful development and implementation of ICZM.

Defining the coastal zone

An inevitable early focus of ICZM programmes is to define the coastal zone. The outcome usually depends on the main reasons for the programme and therefore no universal definition exists except to say that it includes some coastal land, the foreshore and an area of inshore waters (e.g. Sorensen and McCreary, 1990; Gubbay, 1991; HMSO, 1992).

Debate often centres on whether the boundaries should be decided on the basis of environmental considerations or administrative arrangements (Figure 2.2). The outcome is usually some combination of the two with a useful approach being to use environmental factors as the starting point and refine this in light of administrative and

practical considerations. In the Barbados ICZM plan for the west coast, for example, the inland boundary is the first coastal road or the limit of the predicted 100-year storm surge flooding, whichever is further inland. The seaward limit is the 100 m isobath or 200 m seaward of the outer edge of the bank reef, whichever is further seaward (Delcan, 1995).

Trying to set precise boundaries for the coastal zone can dominate early discussions about ICZM and the experience of Canada's programme provides a salutary lesson. There was a great deal of early debate on what comprised the coastal zone but it was a difficult question to resolve because it was being considered without any clear idea of the objectives of the ICZM programme. This was one of a number of factors identified as hindering CZM initiatives in Canada in the 1980s (Hildebrand, 1989). A flexible approach, depending on the issue to be addressed, can therefore be an advantage but a precise definition will be needed if the programme is to be enshrined in law.

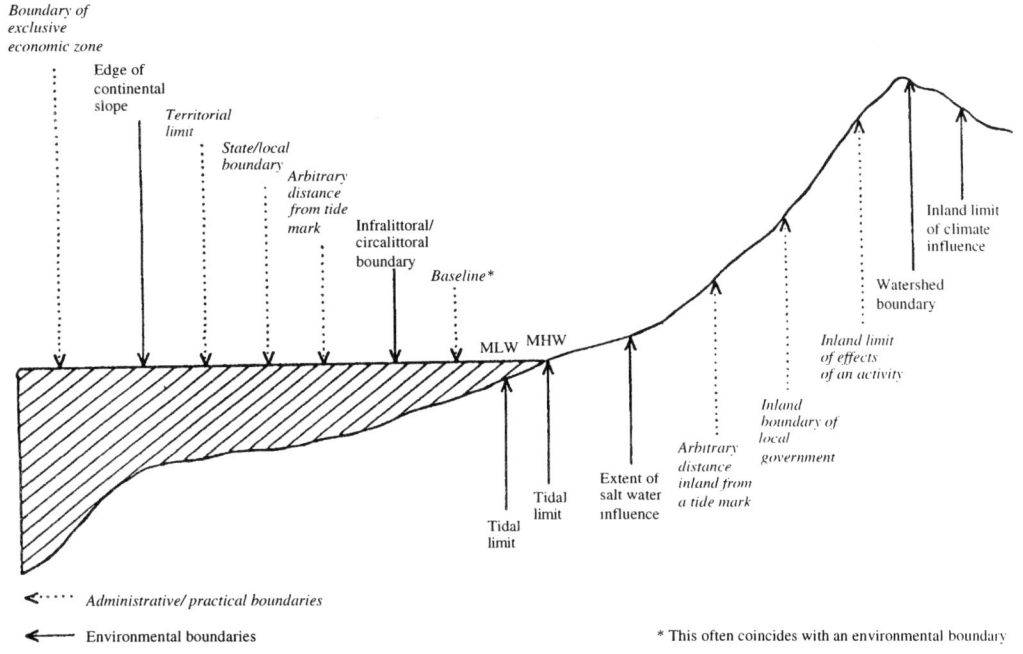

Figure 2.2 Examples of boundaries that may be used to mark the limits of 'the coastal zone' in CZM programmes
Source: Gubbay, 1991

Legislative backing

Coastal management invariably involves some legislation. In general the laws tend to target particular practices or interests. They may be specific to an activity taking place in the coastal zone or apply widely. Existing legislative controls will need to be incorporated into the ICZM programme but a frequently asked question is whether these controls should be brought together under a single, all-embracing piece of legislation.

The situation in the USA is very clear-cut. The 1972 Coastal Zone Management Act provides a legal framework for ICZM throughout the country and individual states develop their own programmes within this framework. These programmes are subject to approval by the federal government and financial support to implement them is linked to that approval. The advantages of such an approach included consistency, a clear requirement to act, financial support and government commitment to ICZM. This is therefore the approach that is favoured by many nations but it is not exclusively the case. The UK government, for example, has chosen to rely on a mix of controls directed at the different sectoral interests. There are no plans for co-ordinating legislation to integrate existing laws and to provide a national framework for ICZM and no statutory requirement for local authorities to prepare ICZM plans (DoE, 1995). A major disadvantage of the latter approach is that there is no clear framework for ICZM planning and, although much can be achieved through the voluntary actions of organisations and coastal managers, priority will always be given to statutory requirements, especially when funding is limited. A voluntary system also means that there is no safeguard to give continuity to an ICZM programme.

Scale up or scale down

ICZM programmes operate at a number of levels with international, national, regional and local elements. The question of where the emphasis should be placed is another frequently considered issue. There does not appear to be any clear correlation with the length of coastline in a particular country. Australia, Brazil, the USA, the Netherlands and Denmark all provide a national context for their programmes and regional and local initiatives work within this framework, albeit with different degrees of independence. Very few countries build their ICZM programmes from a local base. Two examples are the UK and France. In the UK, local planning authorities are taking a key role in the development of ICZM plans. There is planning policy guidance relating to the coast but no overarching legislation. In France, local planning of inshore waters is being promoted through Schema de Mise en Valeur de la Mer. In Australia a review of coastal management in Western Australia supported a shift from the production of local coastal management plans to more emphasis on regional level strategic planning (Kay *et al.*, 1995). Local ICZM programmes have a clear role in the process as they facilitate implementation. They can also create the momentum for a national ICZM programme as was the case with the work of the San Francisco Bay Conservation and Development Commission which led to the introduction of the California Coastal Conservation Act and ultimately a much more extensive ICZM programme in the USA.

Experience from around the world on this matter suggests that both national and local elements are necessary. The national perspective can provide strategic direction, commitment to the idea, statutory backing and support. To implement ICZM plans, a local element is also essential. The preparation of local plans and strategies provides opportunities for involvement from interest groups and users as well as regulators and managers, a means of identifying and acting on local concerns and issues and incorporating local knowledge about the coastal environment into the planning and management process, all of which make the successful implementation of ICZM in the locality more likely. As programmes mature they inevitably have national, regional and local components.

Dealing with existing arrangements

One of the critical questions to consider in the development of a ICZM programme is how to deal with the procedures, laws, agreements and traditional uses that already exist. A need to rationalise and integrate a plethora of existing arrangements may in fact be one of the reasons that ICZM is being promoted and it is important to consider how to do this. In the case of the Netherlands, the National Policy on Water Management provides a framework and is complemented by other policies such as the National Environmental Policy Plan.

In the UK interest in ICZM has given additional momentum to planning procedures which already existed for particular sectors of coastal activity as well as encouraging new ones to be developed. The need to link the different plans has been recognised and there is much literature on interactions between the major types of coastal plans (Gubbay, 1994; DoE, 1996). An ideal situation would be to rationalise all existing statutes relating to the coast; however, as this is likely to be a considerable task a more feasible approach is to draw up framework legislation on key elements such as the objectives of an ICZM programme, the administrative and financial arrangements, and the links to secondary, existing legislation.

Development control

Development control tends to be a strong theme of ICZM programmes and is often the main reason for ICZM being considered. A common method is to define a 'set back' line which prohibits all or certain types of development within a coastal strip. In Denmark, the 1970 Urban and Land Zone Act limited the building of new summer cottages to specifically zoned areas at the coast. It was followed in 1981 by the establishment of a special protection zone up to 3 km wide where special planning requirements apply which look at whether a proposed development requires a coastal location. These regulations were revised in 1994 to distinguish between open coastal areas, summer cottage areas and coastal cities. In the case of summer cottage areas, these buildings are prohibited within 3 km of the coast. Set back lines can be very effective as they can be easily measured and clear to enforce if applied consistently. On the other hand, such an approach has the drawback of limited flexibility and only relates to one aspect of development control. If they are to be used it is also essential to have other types of coastal planning guidance outlining policies on issues such as zoning, environmental standards and specifying inappropriate developments. Set back lines are insufficient if used in isolation as a means of development control in the coastal zone.

Offshore issues

The need to incorporate marine elements into ICZM programmes raises the issue of how landward and marine planning systems can be integrated. The system in Sweden relies on three zones with local, regional and national government agencies variously taking the lead. The National Marine Resources Commission has been the key agency in developing the marine aspects of the programme but the link with the terrestrial side has been in the close work with the National Board of Physical Planning and Building and the National Environment Protection Board. The planning authorities often have a major role in creating this link. In Israel, the National Board for Planning and Building called for the preparation of national plans for all coast and territorial waters in 1970. There are 65 Local Planning and Building Commissions which are responsible for this work at the local level along with a special national commission which approves plans and permits for development in territorial waters (Brachya, 1993). A linked system such as this has the advantage of ensuring that the planning process is suited to the coastal environment as it reduces the risks associated with fragmented decision-making across the coastal zone, an approach that can have major repercussions along the coast (for example, by creating coastal erosion problems) or offshore (for example, by damaging important areas for fisheries as a consequence of sedimentation caused by construction work on land).

The emphasis on a ICZM plan

The preparation of a plan is the main focus of most programmes but the scope of individual plans varies a great deal. Some are developed at national level and are wide-ranging, incorporating policies and programmes as well as the more traditional aspects of planning. These can be viewed as strategies. In other cases there is an emphasis on a programme of action. The draft Thames Estuary Management Plan is an example of this approach working from general principles through to actions on specific issues (Table 2.4).

Table 2.4 Examples of coastal policy guiding action (from the draft Thames Estuary Management Plan)

Principle	Agreement should be reached between the parties involved in future provision for the long-term disposal of dredging ashore to ensure that commercial shipping berths are maintained to required depths
Recommendations	Jointly finance and prepare a long-term dredging disposal strategy for the Thames Estuary. This should review the work on smooth water reaches undertaken previously by the PLA, extend into the rough water reaches, take account of natural coastal processes, and include criteria that meet the needs of all parties (e.g. capacity, location accessibility from the river, wildlife value, hydrology, contaminated land, etc.)
Time-scale	Short
Partners and players	PLA – Port of London Authority, EN – English Nature, County Councils, LPAC – London Planning Advisory Committee, LAs – local authorities, MAFF – Ministry of Agriculture, Fisheries, and Food, EA – Environment Agency, landowners

Source: Thames Estuary Project, 1996

General statements can provide a useful overview but are difficult to implement. Therefore, if ICZM is to have an real effect on the coast the latter approach, with timetables, actions and parties identified to carry out such actions, is essential.

Administrative arrangements

The issue of whether control of a ICZM programme should be centralised or localised is greatly influenced by the political system. Some countries have set up specific agencies to take an overview of the whole ICZM programme. In the USA, for example, the CZM Act is administered at the national level by the Office of Ocean and Coastal Resource Management of the National Oceanographic and Atmospheric Administration (NOAA). The ICZM plans of the individual states have to meet a number of procedural and other requirements set by the national government but they tailor their programmes to their own situation. This may mean variation between states in the extent of the coastal zone, the focus of their policy objectives and how they hope to meet these objectives but it allows both national and local interests to be accommodated. In Sri Lanka, a Coast Conservation Department was set up in 1983 to ensure consistency of coastal management by co-ordinating the actions of several agencies. They take the lead in the development of Sri Lanka's ICZM plans. In the UK, local authorities take the lead at present. There is no central government agency co-ordinating the ICZM programme although there is

a standing forum which has the potential to develop into a co-ordinating body. A number of different models are listed in Table 2.5. Different approaches are appropriate to the different circumstances in each country; however, there is much to be gained by having some administrative focus for ICZM at national level. This will help draw together the many different interests involved in ICZM, as well as the many legal provisions relating to the coast that probably already exist, and provide the strategic overview necessary to build a strong programme.

Integration

Achieving integration is a critical part of ICZM and needs to be taken forward at several levels. The need for integration among the different sectoral interests (fisheries, recreation, mineral extraction, etc.) is often the most apparent but it is also relevant between marine and terrestrial planning and management arrangements, between levels of government, between countries and between different disciplines. In examining this question Cicin-Sain (1993) notes that integration needs to be viewed as a continuum and describes the following five stages:

1 *Fragmented approach* – a situation characterised by the presence of independent units with little communication among them.
2 *Communication* – a forum for periodic communication/meeting among the independent units.
3 *Co-ordination* – independent units take some actions to synchronise their work.

Table 2.5 Institutional arrangements for ICZM

	Degree of sharing of decision-making responsible among actors			
LOW ———————————————————————→ HIGH				
	Broaden planning process	Project review and comment	Policy-making review and comment	Policy-making or project approval
Permanent institutional arrangements				
Existing agency	+	+		
Existing agency with advisory council	+	+		
Expanded agency	+	+		
Expanded agency with advisory council	+	+		
New agency		+	+	+
New agency with advisory council		+	+	+
Interministerial council of equals		+	+	+
Temporary supplements				
Ad hoc panel		+	+	+
Facilitated policy dialogue		NA	+	+
Mediated negotiation		NA	+	+
Arbitration		NA	+	+

Source: Sorensen and McCreary, 1990

4 *Harmonisation* – independent units take actions to synchronise their work, guided by a set of explicit policy goals and directions, generally set at a higher level.

5 *Integration* – there are more formal mechanisms to synchronise the work of the various units which lose at least part of their independence as they must respond to explicit policy goals and directions (this often involves institutional reorganisation).

Achieving integration may mean breaking down traditional barriers between organisations and involves co-operation and co-ordination. It can be especially difficult where there has been a long tradition of many different agencies and organisations involved in the management of coastal activities but there are ways in which integration can be built up with inter-ministerial groups, advisory councils and memoranda of understanding between the different parties. These sorts of initiatives do not mean that sectoral planning is no longer seen as necessary but rather that the balance should shift to a more integrated approach.

Consultation

The involvement of local communities and users is recognised widely as being one of the most critical factors in the success of ICZM programmes (e.g.

Taussik and Mitchell, 1996) and here too there have been a variety of approaches. In Alaska, a state-wide Coastal Policy Council was set up to oversee the programme with more than half the membership consisting of local representatives. The Council defined the boundary of the coastal zone and general coastal development standards but these were only interim, to be replaced by more specific plans written at the local level. Four Coastal Resource Service Areas were created to involve local residents in coastal planning where no local level of government existed (Caulfield, 1991).

The procedure used to develop management plans for the Great Barrier Reef Marine Park is an example of consultation for a marine situation. There were at least three stages of consultation. In the first, basic information was presented and sought in order to identify the sorts of goals and objectives that should address the expectations of the community. A second phase sought reactions to specific proposals and the third stage assessed the strength and degree of acceptance of the revised plans before finalisation. In common with management plans all over the world, they are to be reviewed on a regular basis.

The benefits of good consultation are numerous and include creating a sense of joint ownership, developing solutions across a broad consensus and providing a strong basis for implementation of any

ICZM proposals. This has been achieved through establishment of communty groups as a channel for consultation, appointment of project officers to facilitate the process in the field, as well as through public meetings, seminars and workshops to present, debate and review proposals. The methods used need to be appropriate to the social and cultural characteristics of the area and should be instituted from the outset of the programme.

International initiatives

As interest in ICZM has grown at a national level so there has been a greater recognition of the significance of this issue at an international level. The 1987 report from the World Commission on Environment and Development (Brundtland Commission) was one of the early international documents to give the matter some attention. The prospect of climate change and sea level rise associated with global warming has also given the subject a high profile with the Intergovernmental Panel on Climate Change considering ICZM essential for the management of the coast. ICZM is a central part of the proposals for the management of coasts and oceans in Agenda 21 (Chapter 17); and the ideas of ICZM will be essential for implementation of parts of the Law of the Sea Convention.

All these calls have recognised the value of collaboration between countries to improve ICZM

BOX 2.1   ICZM IN THE UK

The recent interest in ICZM in the UK can be traced to events in the early 1990s when there was considerable lobbying of government by local authorities, conservation groups, planners and coastal managers on the need to provide a framework of ICZM for the UK. One outcome was the decision by the House of Commons Environment Select Committee to conduct an inquiry into coastal zone planning and protection. The Committee reported in 1992 and, in light of the evidence received, concluded: 'We recognise the benefits of the approach known as Coastal Zone Management, and we recommend that such an approach be adopted as the framework for all coastal zone planning and management practices in the UK' (HMSO, 1992). These and other recommendations were welcomed widely. The government responded that it was 'firmly committed to the effective protection and planning of our coast' but was also of the view that no major actions needed to be taken. Since the Committee reported there has been progress with ICZM in the UK by means that could be considered to be 'evolutionary rather than revolutionary' and based on a 'voluntary approach' rather than statute.

There is no national programme on ICZM in the UK but separate countrywide initiatives for England, Wales, Scotland and Northern Ireland. Some liaison is achieved between them through an Inter Departmental Group which brings together the government departments involved in coastal management, although its role remains to be clarified.

Coastal Planning Policy Guidance has been issued for England and Wales and discussion papers on the way forward published by the Scottish Office and the Department of the Environment (Northern Ireland). A coastal forum for England was established in 1994 which has more than sixty member organisations with wide-ranging interests in the coastal zone. A public meeting twice a year provides an opportunity to hear about the work of the different organisations and debate issues of concern. More detailed work may be carried out through topic groups. Coastal fora are due to be set up in Wales, Scotland and Northern Ireland.

The limited progress with the development of ICZM policy at national level is in contrast with interest in ICZM at a local level and indeed this is where most has been done to promote ICZM in the UK. The current emphasis is on the preparation of ICZM plans and 90 examples of coastal planning and management initiatives in England are described in a directory published in 1994 (King and Bridge, 1994). County and district planning authorities have been particularly active in the preparation of coastal strategies, estuary management plans and coastal management plans. A major feature has been the interest and involvement of all major coastal user groups as well as those involved in the management of activities at the coast. Project officers have helped facilitate the process at a number of sites and funding has come either through joint commitments or through seed funds from organisations such as English Nature in its 'Estuaries Initiative' and Scottish Natural Heritage through its 'Focus on Firths' programme.

The UK is also participating in a three-year European Union demonstration programme on ICZM whose objectives are 'to show the practical conditions that must be met if sustainable development is to be achieved in the European coastal zones in all their diversity'.

BOX 2.2   ICZM IN CHINA

The combined coastline of mainland China and its offshore islands exceeds 32,000 km. There are more than 6,000 islands and about 37,000 km of tidelands and wetlands. The population in the coastal zone is more than 460 million, approximately 40 per cent of the national total. Major problems that have been identified are conflicts of use between different industries, poor planning regarding the locations of some uses so that the coast is not used to its full potential, coastal erosion and sea water intrusion, and discharges of large amounts of industrial and domestic sewage.

Planning of the coastal zone and its considerable natural resources operates through the State Planning Commission which is responsible for drawing up national plans and monitoring their implementation. Management of the marine sector is through about twenty ministries, commissions and ministry-level companies and seven independent agencies and, until the early 1980s, there was little co-ordination between them in relation to policies, projects and programmes related to marine development. Lu (1990) notes that 'the growth of the marine sector, the influence of the Third UN Conference on the Law of the Sea, and the experiences of the countries in ocean management prompted China to initiate studies on integrated marine management'.

A survey of the resources and their use in the coastal zone was carried out between 1980 and 1986 by the State Oceanic Administration with 15 ministries, commissions and bureaux as well as the provinces and municipalities. The area covered extends 10 km inland and seawards to the 15 m isobath with information collected on climate, hydrology, environment, land use and social economics amongst others. This information was used for pilot projects. A Coastal Zone Management Act and Coastal Zone Management Regulations are being drawn up and there is also legislation at a regional and local level. The Jiangsu Province, for example, issued CZM regulations in 1991 (Guochen, 1993).

The nearshore environment has been classified into 'marine functional zones' based on the natural resources, environmental conditions and geographic position, ocean development and use, and socio-economic needs. Five major types have been identified with a view to unifying the economic, social, environmental-ecological and resource benefits. These are: development and utilisation zones, regulation and protected zones, nature reserves, special functional zones and reserved zones. The Guangdong Province, which includes the Pearl River Delta is at the final stages of introducing a Regulation on the Use of Sea Areas. Functional zoning of the region has resulted in the identification of 1,254 functional zones and in 1994 the provincial government initiated the Pearl River Delta Economic Region Modernisation and Development Planning. This views the delta as a whole in economic, social and environmental matters to enable an overview to be taken to solve problems that are difficult to resolve by municipalities acting on their own (Cao, 1996).

The reforms were introduced to establish an administrative and legal framework conducive to economic development and resource and environmental management. In China, ICZM is a branch of ocean management with the State Ocean Bureau responsible for policy formulation, planning and supervision of ocean management, and the provincial ocean department responsible for the coastal zone and shallow sea. Detailed management is with local ocean administrative bodies, for example, the Lands Department and the Environmental Protection Agency. An important element is a permit system and user-pays system for marine space and resource users.

---

programmes as well as for action by individual governments. International support has had a significant influence in initiating ICZM programmes and supporting their development especially through aid programmes from the United Nations Environment Programme, the World Bank, the European Commission and the Inter-American Development Bank.

CONCLUSIONS

The combination of a superb natural asset, valuable commodities and pressure of use on the coast has, quite rightly, focused attention on the coastal environment.

The relatively recent interest in ICZM does not mean that there was a vacuum in coastal planning in the past. What it has done, however, is changed the emphasis of such work. Two of the most important developments are, first, the shift to more integrated planning and management, and second, for the resulting policies and programmes to focus on a 'coastal zone'. Previous policies have therefore been recognised but adapted and superseded by new measures. As described in this chapter, the outcomes are extremely varied. In some cases there is a great deal of emphasis on legislation, in others on voluntary collaboration. The principal direction for ICZM comes from the national level in some countries and the local level in others. This variety suggests that although there are some common themes to ICZM programmes there is no single model that can be advocated as 'the best way'. Instead it is necessary to examine the different approaches, then adopt what is appropriate, and adapt other ideas to suit the particular circumstances in the country in question.

Coastal environments are important economically and have great natural wealth with internationally important areas for marine and coastal wildlife, wilderness areas, endemic species and spectacular coastal scenery. In the last decade, reports such as those produced by the Brundtland Commission and the OECD have highlighted the links between economic and environmental well-being. If managed wisely, it is possible to safeguard the natural environment of the coast, and to make a long-term contribution to the wealth of coastal nations. If managed poorly, short-term gain will be at the cost of economic potential, and the loss of some of the world's most impressive natural heritage. Integrated coastal zone management is a realistic and achievable way to help avoid such a loss.

REFERENCES

AMBIO (1995) 'Research and Capacity Building for Sustainable Coastal Management', Special Issue, *AMBIO* XXIV, 7–8, Royal Swedish Academy of Sciences.

Brachya, V. (1993) 'Environmental Assessment in Land Use Planning in Israel', *Landscape and Urban Planning* 23: 167–1881.

CAMPNET (1989) *The Status of Integrated Coastal Zone Management: A Global Assessment*, Report of a workshop convened at Charleston, South Carolina, USA, 4–9 July 1989.

Cao, J. (1996) 'Coastal Zone Management in the Pearl River Delta', in J. Taussik and J. Mitchell (eds) *Partnership in Coastal Zone Management*, Portsmouth, UK: Samara Publishing Ltd, pp. 213–220.

Caulfield, J. (1991) 'The Alaska Coastal Management Program. Involving People in Coastal Resources Management Decisions', in B. Needham (ed.) *Case Studies of Coastal Management: Experience from the United States*, Rhode Island, USA: Coastal Resources Center, University of Rhode Island.

Cicin-Sain, B. (1993) 'Sustainable Development and Integrated Coastal Management', *Ocean & Coastal Management* 21: 11–43.

Cicin-Sain, B. and Knecht, R.W. (forthcoming) *Integrated Coastal and Ocean Management: Concepts and Cross-National Experiences*, Paris, France: UNESCO.

Cicin-Sain, B., Knecht, R.W. and Fisk, G.W. (1996) 'Growth in Capacity for Integrated Coastal Management since UNCED: An International Perspective', *Ocean & Coastal Management*.

Commission of the European Communities (CEC) (1994) 'Economic Development and Environmental Protection in Coastal Areas. A Guide to Good Practice', Brussels, Belgium: ECOTEC Research & Consulting Ltd.

Delcan (1995) *The Integrated Management Plan for the South and West Coast of Barbados*, Barbados: Report to the Government of Barbados, Ministry of Tourism, International Transport and the Environment, Coastal Project Unit.

Department of the Environment (1995) *Policy Guidelines for the Coast*, DoE 95 CCG 218.

Department of the Environment (1996) *Coastal Zone Management. Towards Best Practice*, Report prepared by Nicholas Pearson Associates, London, UK: HMSO.

Grenon, M. and Batisse, M. (eds) (1989) *The Blue Plan – Futures for the Mediterranean Basin*, Oxford, UK: Oxford University Press, UNEP.

Gubbay, S. (1990) *A Future for the Coast. Proposals for a UK Coastal Zone Management Plan*, Ross-on-Wye, UK: Marine Conservation Society.

Gubbay, S. (1991) *A Definition of the 'Coastal Zone' for UK Coastal Zone Management Programmes*, Marine Conservation Society Discussion Paper CZM/1, Ross-on-Wye, UK: Marine Conservation Society.

Gubbay, S. (1994) *Local Authorities and Integrated Coastal Zone Management Plans*, WWF Marine Update 18, Godalming: World Wide Fund for Nature.

Guochen, Z. (1993) 'Coastal Zone Management in China', in *Proceedings of the World Coast Conference*, Vol. 2, The Hague, The Netherlands: Coastal Zone Management Centre, pp. 733–740.

Hildebrand, L.P. (1989) *Canada's Experience with Coastal Zone Management*, Halifax, Nova Scotia: Oceans Institute of Canada.

HMSO (1992) *Coastal Zone Protection and Planning*, House of Commons Environment Committee report, 17–1, London, UK: HMSO.

Holmes, B. (1996) 'Blue Revolutionaries', *New Scientist*, 7 December 1996: 32–36.

Houghton, J.T. (ed.) (1996) *Climate Change 1995 – The Science of Climate Change*, Intergovernmental Panel on Climate Change, Cambridge, UK: Cambridge University Press.

Kay, R., Alder, J. and Anutha, K. (1995) 'Recent Initiatives in Coastal Planning and Management in Australia', in M.G. Healy and J.P. Doody (eds) *Directions in European Coastal Management*, Samara Publishing Limited.

King, G. and Bridge, L. (1994) *Directory of Coastal Planning and Management Initiatives in England*, National Coasts and Estuaries Advisory Group, Cheltenham, UK: Countryside Commission.

Lu, K. (1990) 'Marine and Coastal management in China: A Planning Approach', *Coastal Management* 18: 365–384.

Nelson, J.S. (1984) *Fishes of the World*, New York, USA: John Wiley & Sons.

Nijkamp, H. and Peet, G. (eds) (1994) *Marine Protected Areas in Europe*, Amsterdam, The Netherlands: AID-Environment.

OECD (1993) *Coastal Zone Management. Integrated Policies*, 2 vols, Paris, France: OECD.

Pernetta, J.C. and Elder, D.L. (1993) *Cross-Sectoral, Integrated Coastal Area Planning; Guidelines and Principles for Coastal Area Development*, Gland, Switzerland: IUCN.

Sorensen, J. (1993) 'The International Proliferation of Integrated Coastal Zone Management Efforts', *Ocean & Coastal Management* 21: 45–80.

Sorensen, J.C. and McCreary, S.T. (1990) *Institutional Arrangements for Managing Coastal Resources and Environments*, Coastal Publication No.1, Renewable Resources Information Series, Narragansett, RI, USA: Coastal Resources Center, University of Rhode Island.

Taussik, J. and Mitchell, J. (eds) (1996) *Partnership in Coastal Zone Management*, Centre for Coastal Zone Management, University of Portsmouth, Portsmouth, UK: Samara Publishing Ltd.

Thames Estuary Project (1996) 'Thames Estuary Management Plan', Consultation Draft, July 1996, London, UK: English Nature.

UNEP (1995) *Guidelines for Integrated Management of Coastal and Marine Areas with Special Reference to the Mediterranean Basin*, UNEP Regional Seas Reports and Studies No. 161, Nairobi, Kenya: UNEP.

World Bank (1993) *Noordwijk Guidelines for Integrated Coastal Zone Management*, Washington DC, USA: World Bank.

World Coast Conference (1994) *Preparing to Meet the Coastal Challenges of the 21st Century*, Conference Report, World Coast Conference 1993, Intergovernmental Panel on Climate Change.

SUGGESTED READING

Clark, J.R. (1995) *Coastal Zone Management Handbook*, USA: Lewis Publishers.

Journal of Ocean and Shoreline Management.

SELF-ASSESSMENT QUESTIONS

1 Environmental principles should be used in the ICZM process
   (a) on an occasional basis;
   (b) solely to support nature conservation initiatives;
   (c) to underpin decision-making in all sectors.
2 The coastal zone should ideally be defined on the basis of
   (a) administrative considerations;
   (b) environmental considerations;
   (c) a combination of administrative and environmental considerations.
3 Integration is critical to successful ICZM and can be achieved through
   (a) institutional arrangements;
   (b) co-ordination of management programmes;
   (c) memoranda of understanding between organisations regarding their work at the coast;
   (d) legislation;
   (e) all of the above.

3

RIVER AND INLAND WATER ENVIRONMENTS

David N. Wilcock

SUMMARY

River and lake ecosystems are responses to the flow through them of materials (water, sediment and solutes) and energy. In any lake or river, these flows are essentially determined by the combination of climatic, geological and topographical characteristics in the tributary drainage basin. Under natural conditions and over long periods of time, the productivity, plant communities and food chains of river and lake ecosystems become adjusted to streamflow, sediment transport rates and nutrient supplies. The rate and character of many material fluxes, however, has recently changed, and river and lake ecosystems today are affected by such diverse environmental impacts as acidification, eutrophication, channelisation and water abstraction, all of them directly or indirectly attributable to changes in modern methods of industrial and agricultural production. Understanding these environmental impacts requires careful measurement of streamflow, sediment and nutrient fluxes before and after the impact occurred. The most appropriate type of environmental monitoring needs to be carefully considered for each type of environmental impact, but quantification of environmental impact is important if subsequent management of the ecosystem is to be effective. Particular types of environmental impact may be mitigated by use of self-regulating and recycling functions inherent in all ecosystems, if these can be identified. New environmental management strategies for rivers and lakes are developing based on new eco-engineering technologies and on changing attitudes towards the value of rivers and lakes.

The focus of this chapter is on how river and lake ecosystems are adjusted to water, sediment and nutrient budgets within drainage basins, how these budgets are modified by modern agriculture and industry and how river and lake ecosystems respond. In addition, the main principles on which the environmental management of rivers and lakes should be based are identified.

ACADEMIC OBJECTIVES

On completion of this chapter, the reader should be able to understand the fluxes of water, sediment and nutrients in drainage basins; appreciate the links between (a) water, sediment and nutrient budgets, (b) instream channel, floodplain and lake processes and (c) river and lake food chains; define the main processes and structural elements of freshwater food chains; identify and explain the major contemporary environmental impacts on rivers and lakes; and evaluate environmental management policies on river and lake systems with which he or she is familiar.

RIVERS

Catchment areas

The ecological character of rivers and lakes is determined by flows and transformations of energy, water and sediments. Because water and sediment flows are in turn very much determined by catchment topography and geology, each river and lake system has to be understood within the context of its own unique drainage basin characteristics (Figure 3.1).

Some of the precipitation (mainly rainfall or snowfall) falling on a drainage basin is evaporated and transpired back into the atmosphere and effectively lost to the river system. A fraction may be transmitted rapidly across steep slopes or through surface soil, entering the river as quickflow, while another

**THE DRAINAGE BASIN
AND ITS CATCHMENT AREA**

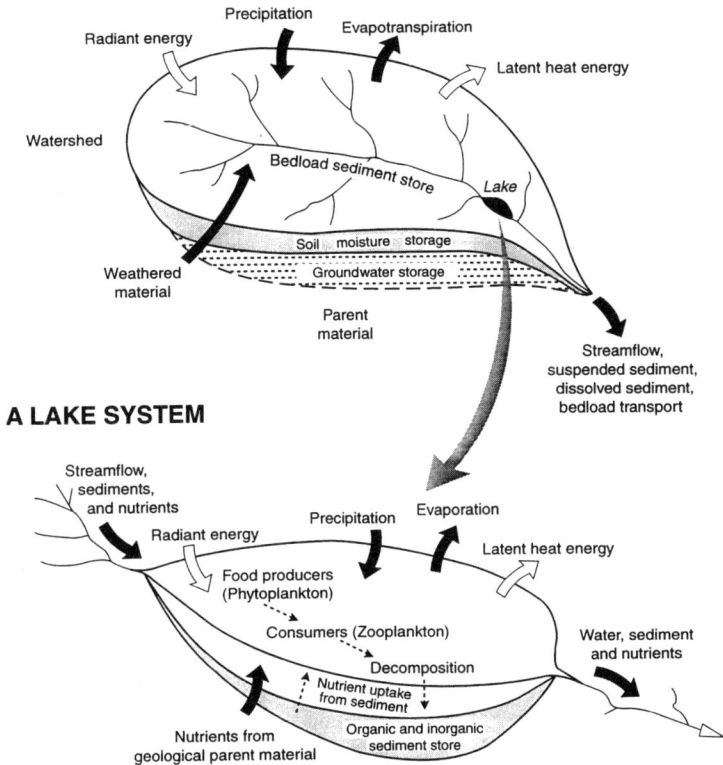

Figure 3.1 Water, sediment and nutrient flows in river and lake systems
The terms 'drainage basin' and 'catchment' are considered synonymous.

fraction may infiltrate into soil moisture and ground-water storage, to be released as river flow hours, days or weeks later. Climate and geographical location, of course, essentially determine the amount and intensity of solar radiation and precipitation, but drainage basin geology, topography and stream network characteristics combine to translate the climate inputs into a particular set of streamflow characteristics in each drainage basin.

Streamflow characteristics are very important and much underemphasised ecological factors in river ecosystems. A river basin characterised by permeable soils and geology permits water to infiltrate easily into the soil moisture and groundwater stores. Passage of water through these stores delays runoff, buffers the river and its plant and animal life against the physical effects of high flood flows, and sustains a high dry-weather flow, often rich in dissolved solids. On the negative side, high infiltration rates may make groundwater in such catchments vulnerable to pollution from agricultural fertilisers. Impermeable soils and/or steep-sided catchments, by contrast, induce rapid runoff, high flood peaks *and* low dry weather flows. Such catchments may be vulnerable to bed and bank erosion, to rapid movements of bed material and to pollution in summer when low flows may be unable to dilute effluent loads from industry and/or agriculture. Of course, it is unwise to be too simplistic or to overgeneralise. The important point to note is that the typical sequence of seasonal flow in any drainage basin, that is the annual hydrograph, is an important factor in river ecology, and is essentially determined by the geological, topographical and vegetational character of the drainage basin.

The *water balance* summarises the inputs, outputs and storage of water for any given catchment and is a basic water management tool for assessing the impact of land-use change and/or river engineering works on river flows. In a simplified form, it may be summarised as follows:

$$\text{precipitation} = \text{streamflow} + \text{evapotranspiration}$$
$$+/- \text{groundwater storage} +/- \text{soil moisture storage}$$
$$+/- \text{measurement errors}$$

Water, sediment and nutrient budgets as environmental management tools

In the same way that a river's streamflow characteristics are largely determined by drainage basin characteristics, so also is its sediment load. A river's total sediment load is conventionally divided into three categories: bedload, suspended load and dissolved load. The exact proportion of each fraction in the total load of any river system has an important bearing on stream or lake ecology, and reflects drainage basin geology and the nature of weathering products.

Any stretch of river receives from upstream tributaries and from adjacent slopes a mixture of sediment and water. These move downstream in the channel under the influence of gravity. Channel form – width, depth, velocity and slope – may be seen as the responses made by the channel in order to transport the water and sediment provided from the catchment. In this view of a river system, discharge and sediment load are essentially independent or external variables of the channel system, determined by climate, catchment characteristics and land-use practices within the drainage basin. Any changes to these variables, brought about by climate change or changes in land-use, may force changes in channel form. Discharge and sediment, therefore, are two of several *forcing functions* on rivers and lakes. These ideas (Leopold *et al.*, 1964) are at the heart of any understanding of a river system in relation to its catchment.

Effective understanding and management of any environmental system ideally requires quantification of its throughputs of matter and energy. Not all drainage basins can be understood in detail, and *representative* and/or *experimental* catchment studies have, therefore, been established in different parts of the world in which components of the water, sediment and nutrient budgets are monitored independently of each other (Dooge, 1973). Major studies are those at Hubbard Brook in the USA (Bormann and Likens, 1979), at the Coweeta Hydrologic Laboratory, North Carolina, USA (Webster *et al.*, 1992), at Plynlimon in the United Kingdom (Clarke and McCulloch, 1979) and in the Alptal valley, Switzerland (Keller, 1990). Representative catchments are established to determine typical regional values for individual components of water, sediment and nutrient budgets. Experimental catchments monitor specifically the effect of an environmental change, for example, afforestation, deforestation, urbanisation, channelisation, on water, and/or sediment and/or nutrient flows in a catchment under strictly controlled experimental conditions. Such studies often involve a pair of adjacent catchments only one of which undergoes the change being monitored. In this way, the effect of the particular land-use change can be isolated from changes due to climate which affect both catchments.

Another important concept in the understanding of river systems is the principle of *indeterminacy* (Leopold *et al.*, 1964). Individual reaches on any river system attempt to adjust their form in such a way as to provide just the energy distribution within the channel required to transport the load provided from upstream. Such a stream is in equilibrium with its catchment and is graded (Mackin, 1948; Newson, 1984). A stream unable to transport the load with the discharge provided from upstream deposits sediment on its bed – it aggrades – and one with surplus energy erodes its bed and channel walls – it degrades. In trying to sustain a graded condition, however, a river system is free to adjust a large number of variables including slope, width, depth, channel perimeter roughness, sinuosity, bedload size and bed form. As to the nature of any specific set of adjustments that may be made, these are impossible to determine. This indeterminacy leads to a wide diversity of morphology, sediment and flow characteristics in natural river channels. These in turn provide a wide variety of ecological habitats within the natural river environment. Understanding these connections between physical and ecological processes lies at the heart of successful environmental management of rivers.

Productivity, habitat, community and niche

Solar energy sustains aquatic life. Green plants, bankside vegetation, for example, or phytoplankton such as blue-green algae and diatoms, use sunlight in the process of photosynthesis to produce organic carbon and oxygen from carbon dioxide and water. They thus produce their own food: they are *autotrophs*. The materials required to fix organic carbon in this way come from nutrients dissolved in water. Organisms that cannot produce their own food are *heterotrophs*. Phytoplankton in rivers and lakes are grazed by heterotrophs such as zooplankton and other herbivores such as insects, snails and fish. These in turn are consumed by carnivores. Such a succession of *producers* and *consumers* constitutes a *grazing food chain*. A second type of food chain, equally important in the life of rivers and lakes is the *decay food chain* in which dead vegetation and animals in a river or lake are consumed by micro-organisms. The end products of the decay food chain are carbon dioxide and mineral matter. The decay food chain consumes large quantities of oxygen in the process of respiration by the decomposers. Such *biological oxygen demands* (BODs) often constitute serious environmental management problems. Food chains vary in their *productivity* depending on the availability of sunlight and nutrients. A shaded stretch of river, for example, is usually less productive than one in full light. Rivers in limestone catchments tend to be more nutrient rich than ones on acidic rocks.

The *habitat* of any particular section of river depends on its precise combination of physical, chemical and biotic factors. Physical factors include channel shape, bed material characteristics and the range of flow velocities. Chemical factors include nutrient content, pH and alkalinity. Biotic factors include the competition for food, territory and shelter. These interactions determine the range of organisms (i.e. the plant and animal *communities*) that live there. A river channel and its adjacent floodplain, for example, might contain areas of fast- and slow-moving water, deeper pools and shallow riffles

of unconsolidated sediments, pockets of marsh, vertical banks, overhanging shrubs and trees. The larger the number of interfaces between mineral matter, vegetation, water and air the greater the number of *niches* (living spaces) and the greater the number and diversity of invertebrates, the heterotrophic building blocks of the grazing food chain. Some invertebrates inhabit the unconsolidated bed material, others the water itself or the water surface, others the splash zone where this exists, and yet others the drier banks. Invertebrates can reach very high densities (e.g. 150,000 per m^2) on the bed of productive rivers (Palmer, 1984) and represent an important food supply for fish. Fish, in turn, require habitats that provide adequate conditions for reproduction, food supplies and protection from predators (Milner, 1984). All these conditions are most likely to be found in river systems with a diversity of sediments, channel morphology and flow characteristics.

Rivers as spatially organised hydraulic and ecological systems

The idea that streamflow and sediment supply determine the downstream geomorphology of river systems was first explored by Leopold and Maddock in 1953. They discovered, and subsequent researchers confirmed, that the width, depth and mean velocity of rivers all increase in the downstream direction in relation to streamflow. These hydraulic geometry relationships, similar on different rivers (Park, 1977), were also found to describe 'regime' canals, that is canals that did not scour or fill over long periods of time. The equations, it was argued, define the quasi-equilibrium condition towards which rivers naturally tend to adjust.

The balance between instream photosynthesis and respiration may also vary systematically in the downstream direction according to the 'river continuum' concept (Vanotte *et al.*, 1980). This concept argues that in upland rivers, where bedload is coarse and streamflow is 'flashy', total community respiration exceeds instream photosynthesis and energy, therefore, must be imported into the river from vegetation adjacent to the channel. This organic material, coarse in composition, is broken down by 'shredder' invertebrates to become a food supply for 'collector' invertebrates downstream. Vanotte and his colleagues argued that 'shredder' and 'collector' invertebrate communities may be interrelated to such an extent that 'shredders' breed slightly before the 'collectors'. In the middle parts of a river, that is in the zone of sediment transport (Newson, 1986), sediment inputs may equal outputs, channel stability is more likely, the river is wider and shading per unit area of river bed is usually less than upstream. In such situations, instream photosynthesis exceeds respiration. In the lowest reaches, where high suspended sediment concentrations hinder light penetration, the energy fixed by photosynthesis may once again diminish relative to that consumed in community respiration. The river continuum concept is an elegant idea but is not accepted by all ecologists (see Moss, 1988).

Less contentious is the general rule that man-modified rivers have less diverse morphology, sedimentology, hydrology and ecology than natural rivers. This is because engineered rivers are usually designed to be hydraulically efficient systems for transporting water. Often, though not always, vegetation is removed, flow ranges diminished and sinuosities straightened. We shall examine some of these effects in more detail later.

UPLAND RIVERS

Streamflow, sediment and ecology

In upland catchments river gradients are steep, streamflow is turbulent and well oxygenated, and sediment load is coarse. Some of the bedload is mobile but much is unable to move, even during bankfull flows. Mosses and liverworts are typical autotrophs of upland rivers, usually the first colonisers of rock surfaces. Invertebrate larvae find food and shelter in the mosses and liverworts, in the gravel interstices where gravel pockets exist, and on the underside of the more stable particles of the bedload. Invertebrates have evolved a range of adaptations to survive the frequent flood spates of an upland stream's high energy environment. Some have developed flattened, streamlined bodies. Others have suckers with which to cling to slippery surfaces.

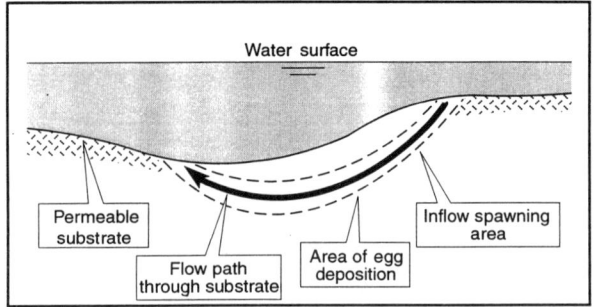

Figure 3.2 Preferred locations for salmon spawning in an upland stream in relation to bed morphology and bed sediments
Source: after Milhous, 1982

Above invertebrates in the instream food chain are fish, and upland rivers are important breeding grounds for many species, notably salmon. Salmon (e.g. *Salmo salar*) require cool, well-oxygenated water, low in suspended sediment. They select deposits of gravel with quite specific size characteristics in which to lay their eggs. Milhaus (1982), for example, observed that in deposits of gravel ranging in size from 0.5 mm to 100 mm salmon showed a distinct preference for spawning only in material ranging between 7 mm and 20 mm in diameter. Eggs are laid in redds 30 cm deep, and after spawning the female covers the eggs with gravel. Water can pass through this gravel to supply dissolved oxygen to the eggs (Figure 3.2). It is important that such gravels are not covered with finer silts which block gravel interstices and cut off oxygen supplies. Silt deposits are associated with a shift in the insect population away from mayfly and caddis larvae towards more mud-loving species such as midgefly larvae (O'Brien, 1987). Not enough is known about the intricate interrelationships that exist between streamflow, sediment, bed-form, water chemistry and fisheries requirements in upland rivers. Much of what we do know comes from environmental impact studies into the effects of three processes currently transforming the aquatic ecology of upland streams throughout the world: reservoir construction, afforestation and deforestation, and acidification.

Environmental impacts on upland streams

Reservoirs

Upland reservoirs usually have one or more of three functions: hydroelectric power generation, water supply and flood control. Their environmental impact on upland rivers results from their interruption of the continuous downstream transfer of water and sediment that normally takes place. This interruption often brings about changes to channel flow regime, sediment transport rates, channel morphology, water quality, water temperatures and dissolved oxygen levels below the reservoirs. The 'knock-on' effects are to the ecology of the river itself and its floodplain. Petts (1984) distinguishes three levels of impact. First order impacts occur as a direct result of impoundment and affect water, sediment and energy transfers downstream. Second order impacts may take 1 to 100 years to manifest themselves and involve changes to the river's morphology and primary productivity by autotrophs. Third order impacts are to invertebrates, fish and bird communities and may take the longest period of time to become obvious. These ideas essentially summarise a time-dependent transfer of effects up the food chain. Petts' summary of the possible effects of river impoundment is shown in Figure 3.3 which illustrates the important interconnections between different processes and the way in which environmental effects may be transmitted downstream through the river system, even to the coast itself.

Forestry

In the European Community between 1972 and 1974 afforestation took place at the rate of about 1,875 km² per year, most of it in France, Italy and the United Kingdom (Briggs, 1987). Much of this afforestation, though by no means all, used quick-growing conifers the environmental effects of which have been extensively monitored in the United Kingdom. Before 1970, water managers in Britain and elsewhere regarded upland coniferous afforestation as good management practice. Afforestation encourages infiltration into the soil, reduces flood

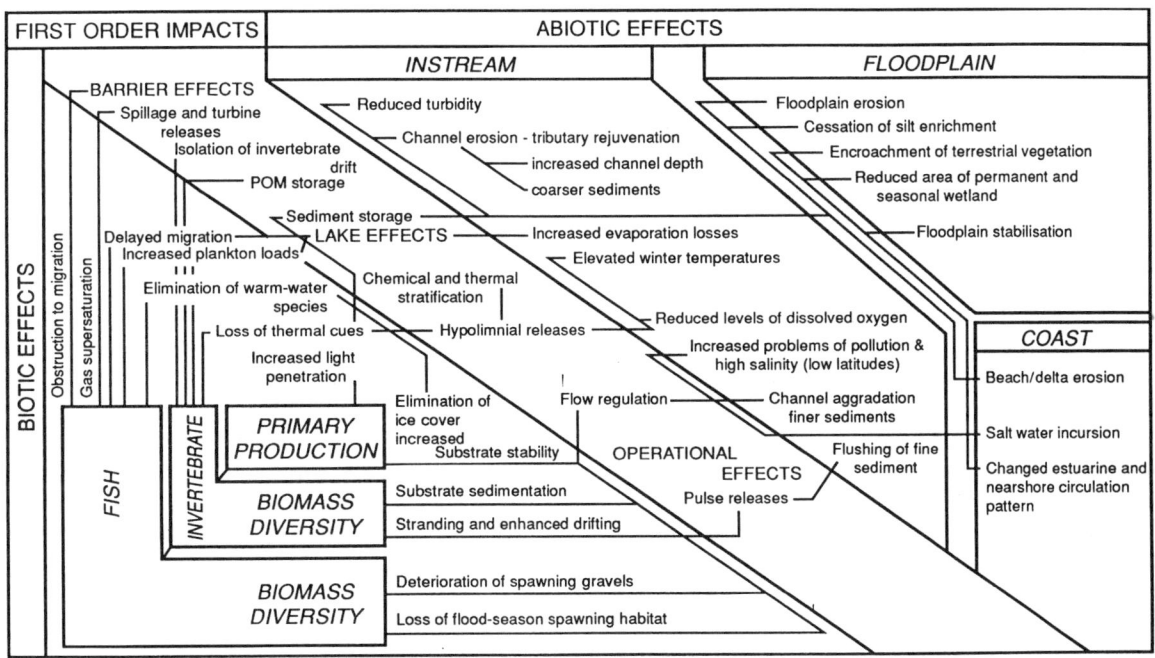

Figure 3.3 The physical, chemical and ecological impacts of river impoundment
Source: after Petts, 1984

flows into reservoirs, augments groundwater supplies and improves water quality. In the 1970s, however, research in the United Kingdom confirmed that coniferous tree canopies intercept and evaporate back into the atmosphere more water than the vegetation they typically replace. Upland reservoirs in the United Kingdom may lose 275 mm of rainfall in this way. The effect is particularly noticeable in dry years when flows on the forested catchments may be 26 per cent less than on the non-forested catchments. This compares with a difference of only 15 per cent in normal years (Clarke and McCulloch, 1979).

Afforestation, in fact, has different hydrological, geomorphological and ecological effects at different stages of tree growth. Initial ploughing increases surface runoff and sediment production. This period produces increased flood flows and sediment supply into streams and reservoirs downstream of the afforested area. Aquatic ecological effects may be complex. Mayfly nymphs in upland streams, for example, typically migrate vertically into the substrate of streams to escape the high velocities of flood peaks. If flood flows are significantly increased, bedload may be removed and with it the refuges for much of the mayfly population (Ormerod et al., 1986). Salmon redds are similarly vulnerable. If these are removed by increased flood flows in the initial period of afforestation, several generations of salmon recruitment to the river may be lost.

At canopy closure, soils are once again protected from heavy rainfall, total runoff diminishes and soil loss is reduced. When plantations are clear-felled, perhaps fifty years after planting, runoff and peak flows again increase, together with sediment supplies. Similar effects occur, often with serious socio-economic consequences, when mature, natural forests are clear-felled. In such complex ways is the management of an afforested area linked to the streamflow, sediment flux and aquatic ecology of upland streams. Planted (and natural) forests do not have one effect on streamflow. They have several effects, each related to different stages of forest growth and management.

Acidification

All rainfall contains dissolved carbon dioxide from the atmosphere and is to some extent acidic.

Increasingly, however, rainfall contains sulphur dioxide, nitrogen dioxides and particulate matter derived from the burning of fossil fuels in modern society. These additional impurities increase rainfall acidity and are, undoubtedly, one of the principal causes of increased acidity in upland waters as far apart as Sweden, Wales and New England. But other factors appear to be associated with increased acidity in upland streams. Coniferous canopies, for example, are efficient collectors of atmospheric dust which is subsequently washed down the trees into the soil and watercourses. For these reasons, conifer plantations are often regarded as a cause of acidification in upland waters, particularly when planted on acidic soils with low buffering capacity.

In Wales, stream acidification increased quite markedly in the upper Wye valley following afforestation (Ormerod et al., 1986). Although the process is poorly understood, acid runoff is most marked at high flows and in winter when soil storage is minimal. Improved drainage in conifer plantations appears to accelerate the uptake of base cations like calcium and magnesium by trees and in some, as yet, unspecified way accelerates the release of aluminium from the soil. High aluminium concentrations at pH values of about 5.0, and the toxicity of this element to many aquatic plants and gill-breathing animals like fish and invertebrates, lies at the heart of the acidification problem. At low pH bacterial decomposition slows down, organic debris builds up and nutrient cycling is reduced. The number and composition of phytoplankton species is altered, invertebrates may fail to take up calcium and salmon fail to hatch. All trophic levels appear to be adversely affected (Gjessing et al., 1989).

But acidification is a complex problem and afforestation is evidently not a precondition for it. Work in Scotland, for example, has shown that it may occur on both forested and non-forested catchments (Flower and Battarbee, 1983; Flower et al., 1987). Whatever its causes, liming of streams and of upland catchments is often seen as a partial solution to acidification. Indeed, the problem may itself be partly related throughout the United Kingdom to the decline in agricultural grants for liming in the 1960s (e.g. Wilcock, 1979). Given the serious ecological effects of acidification, it is sensible policy not to allow further afforestation in areas with low weathering rates and/or buffering capacity.

LOWLAND RIVERS

Streamflow, sediment and ecology

There are four essential differences between rivers in their middle and lower stages. Lowland rivers have (a) greater depth, more sediment and fewer submerged plants, (b) more emergent plants and deposit-feeding animal communities, (c) more meanders and (d) more phyto- and zooplankton (Moss, 1988). For the most part, lowland rivers have gentle gradients, naturally sinuous channels and are separated from the drainage basin's steeper valley sides by extensive areas of gently sloping land. Sometimes wide floodplains border the river channel. From an environmental management point of view, floodplains have to be seen as integral elements of the river channel, interacting closely with its hydrological, geomorphological and ecological processes. Lowland rivers receive their discharge and nearly all their sediment from upstream reaches and, in natural conditions and over long periods of time, their geomorphology is adjusted to maintain the orderly downstream transfer of water and sediment without systematic erosion or deposition. Scour and fill might occur after individual floods. This is normal and needs to be distinguished, though it often is not, from long-term, systematic channel change (erosion and/or deposition) which indicates a channel not presently in equilibrium with the independent catchment variables of streamflow and sediment discharge. This view of a river, as a series of interconnected reaches, the flow of sediment and discharge in any one of which affects the reach downstream, is epitomised in the hydraulic geometry concept described earlier. It emphasises that adjustments towards equilibrium in any individual reach have to be set in a spatial (i.e. catchment) and a temporal context, the latter extending over several generations (perhaps 100–200 years). Newson (1984, 1986) contrasts this holistic view of a river system with the traditional view of a civil engineer, trained to see adjustments only within individual reaches, 'average flows of both water and sediment (being) considered unvarying in the long term' (Newson, 1984:4). Modern river engineering, certainly in Britain, is happily moving towards a reconciliation of these two contrasting points of view (Gardiner, 1991).

Environmental impacts on lowland rivers

Lowland rivers provide a range of environmental problems associated primarily with the management of extreme events. Floods, for example, are a natural phenomenon. Most natural lowland rivers reach their bankfull capacity once every one to two years, and larger floods spill over the channel walls depositing sediments and their adsorbed nutrients on to the floodplain, thus maintaining its natural fertility. Viewed in this way, the floodplain is a natural flood discharging element of the drainage basin.

Over generations, however, fertile soils and easy communications have attracted large settlements to floodplains. It was not always like this. As Wellcome (1992) points out, western civilisation tended to avoid the lakes and swamps of floodplain areas right up to Renaissance times when water meadows, together with residual lowland flood forests, were integrated into local economies. With the industrial revolution, new engineering technologies were developed to control floods and shape channels, and these led to a progressive invasion of the floodplain by towns and cities. The earlier perception that floodplains were simply the channel system 'in flood' was lost and cities extended to the very edge of river channels. Flood protection is now an important modern element of river management, and channels have to be enlarged and/or constrained. Alternatively, flood flows downstream may be reduced by upstream flood regulation reservoirs.

At the other extreme are low flows and associated problems of pollution control. The addition of large volumes of organic waste from cities and farms into lowland rivers creates locally high biological oxygen demands (BODs). A typical oxygen sag curve downstream of an organic waste outfall is shown in Figure 3.4. An important aspect of waste management, therefore, is to ensure that effluent dilution is adequate at all times, particularly during dry weather flows.

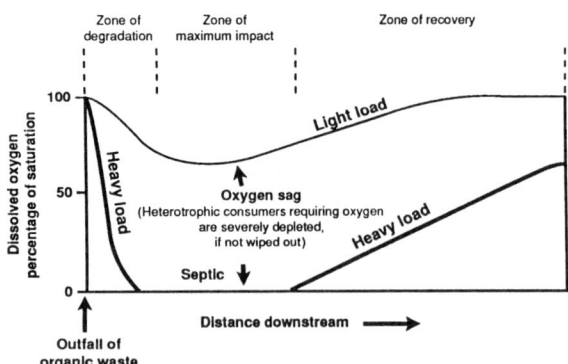

Figure 3.4 The effects of organic waste on dissolved oxygen levels in a river downstream of an outfall

Channelisation

Channelisation has affected many lowland rivers in the world for a variety of reasons – land drainage improvement, the maintenance of navigation, flood control, highway protection – and illustrates many of the conflicts that have to be reconciled in effective management of the total river environment. The process has been one of the most sustained and extensive of all environmental impacts on rivers, affecting over 320,000 km of river in the USA and nearly 100 per cent of the drainage network in Denmark (Brookes, 1988).

In the United Kingdom and the Republic of Ireland the process has been an integral element of an agricultural strategy since 1945 designed to increase agricultural output from areas where either flooding is frequent and/or field drainage is poor. Underlying all drainage strategies is the engineering imperative to evacuate water as quickly as possible from any given area. Consideration of Manning's formula, one of several equations relating stream velocity to channel form, shows how this might be done. Manning's formula states that

$$v = \frac{R^{2/3}S^{1/2}}{n}$$

where v is velocity, R can be taken as mean depth in a wide trapezoidal channel, s is channel slope and n is a roughness coefficient incorporating the effect of all characteristics in the channel (channel vegetation, channel curvature, bed forms) that retard

streamflow. Velocity is clearly increased by increasing depth and/or slope, and by decreasing roughness. Channelised rivers, therefore, are widened, deepened and straightened. Bankside vegetation is removed or thinned. All these changes make the channel a more efficient system for moving water downstream, though the precise effects of field drainage as opposed to channelisation on the hydrology of the catchment are sometimes difficult to disentangle (Robinson, 1990).

The general environmental impact of channelisation is to reduce hydraulic, morphological, sedimentological and ecological diversity in streams. Figure 3.5 summarises the environmental impacts of channelisation on instream processes and aquatic ecology. To these effects can be added others. Tenfold increases in the concentration of suspended sediment, from 100 mg/l during floods to more than 1,000 mg/l during average flows, have been measured. Little detailed work on the precise effects of enhanced suspended sediment yields have been undertaken, but 80 mg/l has been identified as a threshold beyond which it is difficult to maintain good freshwater fisheries (Alabaster and Lloyd, 1982). Nor are the ecological effects of channelisation confined to instream habitats. Floodplain wetlands are *ecotones* (see p. 56) and owe their special characteristics to regular inundation. Channelisation curtails flooding (Figure 3.6) and deprives floodplain fens of essential nutrients (Wilcock and Essery, 1991). It may take time for the ecological effects of such changes to appear, for flood events are irregular occurrences in the natural order of things, and floodplain ecosystems are adjusted to long periods of drought. But the systematic removal of floodplain flooding causes a gradual transition to drier ecosystems and systematically diminishes areas of wetland habitat.

Instream habitats, fortunately, now receive more protection from the environmentally damaging effects of channelisation. Single-bank working to maintain bank form, plant diversity and shade, the careful timing of mowing operations so as to minimise (or encourage) changes in riparian plant communities, the reinstatement of pools and riffles, and the establishment and maintenance of scrub and woodland as corridors alongside river banks to facilitate wildlife development are just some of the practices now being integrated into modern river

BEFORE CHANNELISATION

Overhanging vegetation provides shade,
optimal water temperature and cover for fish life.
Litterfall provides food supply.

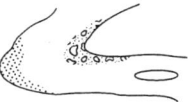

Sorted gravels provide diverse habitats through their
effects on river morphology and stream velocities.
Riffles assist oxygenation.

AFTER CHANNELISATION

Loss of overhanging vegetation. Increased range of
water temperatures, loss of cover,
reduced leaf material inputs.

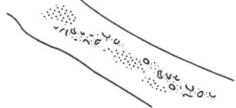

Unsorted gravels. Reduced habitat diversity.
Less diverse populations of invertebrates.

THE POOL ENVIRONMENT

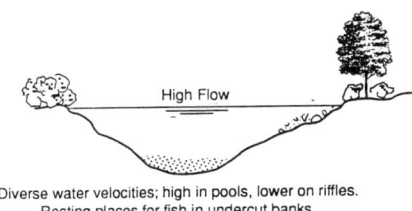

Diverse water velocities; high in pools, lower on riffles.
Resting places for fish in undercut banks.

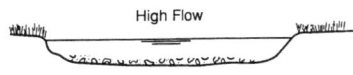

Increased stream velocities;
higher than some aquatic life can tolerate.

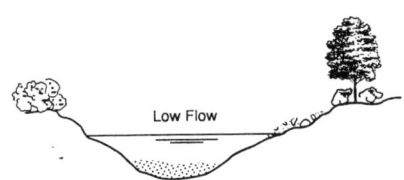

Sufficient depth to support fish life in dry seasons.

Much reduced depth of flow in dry seasons.
Fish territories severely constrained.

Figure 3.5 The ecological impacts of channelisation on a river channel
Source: after Brookes, 1988

management techniques (Lewis and Williams, 1984; Gardiner, 1991; Boon *et al.*, 1992).

Water quality

Lowland rivers are used for water supply, for the decomposition of domestic sewage and industrial wastes, for recreational bathing and as feeding and breeding grounds for invertebrates, birds and fish.

Reconciling these functions requires that water quality be monitored carefully, and most European Union countries now attempt systematic classification of water quality in their major catchments. Different classifications have been developed, an early example being that designed for the Saar–Moselle system on the borders of France and Germany in 1981. Five water quality classes were recognised. Unpolluted water of the highest quality (Class 1) was defined by

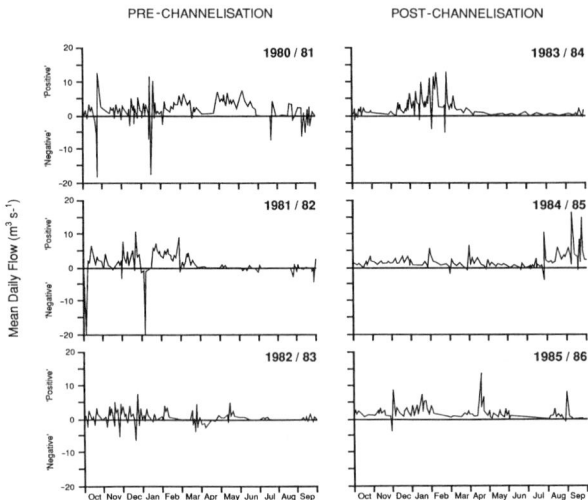

Figure 3.6 Mean daily streamflow from a floodplain area of fens and raised bogs in Northern Ireland, before and after channelisation

Flows below the zero datum are 'negative' flow days on which streamflow into the floodplain river from upstream exceeds downstream outflow. Such a pattern of flows adds nutrients to the instream and floodplain system. 'Negative' flow days decrease markedly after channelisation

Source: after Wilcock and Essery, 1991

the following physico-chemical properties: temperature below 20°C, dissolved oxygen above 7 mg/l, BOD below 3 mg/l, chemical oxygen demand (COD) below 20 mg/l, ammonium below 0.1 mg/l, phenols below 5 μg/l, and total cyanides below 10 μg/l. Excessively polluted water (Class 5) was defined as follows: temperature above 30°C, dissolved oxygen below 1 mg/l, BOD above 25 mg/l, COD above 80 mg/l, ammonium above 8 mg/l, phenols above 500 μg/l, and total cyanides above 50 μg/l. The entire length of the Saar between St Avoid and Saarburg was in Class 5 (Briggs, 1987).

Most physico-chemical classifications of rivers are based on periodic samples at a limited number of sites. Samples are often collected only at two-weekly intervals. Such surveys often fail to detect short-lived pollution incidents between successive samples, even though such incidents might eliminate invertebrate and fish populations in a long stretch of river. To be effective, physico-chemical water quality classifications

need to be derived from continuous sampling. If this is too expensive, and it often is, biological monitoring of instream vegetation and animal communities is an alternative approach. Since changes in plant and animal communities, with time, can easily be used as bioindications of changes in water quality, regular but discontinuous monitoring can yield important evidence about trends in water quality. Many biologists regard biological monitoring as the sounder of the two methods.

THE POLITICAL DIMENSION OF RIVER MANAGEMENT

Water is an important economic resource, especially in arid or semi-arid areas, and few river systems have been more modified for irrigation and hydroelectric power generation than those of the south-west USA (Figure 3.7). The effects on river flows are dramatic, streamflow on the Colorado River at the international border between the USA and Mexico being only 33 per cent of that at the border between Utah and Arizona. The ecological effects of impoundment on this scale are serious and include reduced freshwater flows into the Gulf of California and the frequent supersaturation of water below the high dams with dissolved oxygen and nitrogen. Cold water released from the dams often makes the river unfit for native fish species, and non-native species have been introduced with implications for competition between species and the integrity of food chains. The Colorado below Glen Canyon Dam apparently remains too cold for most native species for a distance of over 400 km (Petts, 1984).

In river systems of any significant size there are always conflicts between *upstream* and *downstream* users. Few such conflicts, however, have been more dramatic than that between the states of California and Arizona on the one hand, and Washington, Oregon and Idaho on the other, over proposals in 1968 to build two dams on the Colorado in the Grand Canyon to supply water to the Central Arizona irrigation project. Arizona developed industrially later than California. Claiming the upstream user's right to use water resources in its own territory, Arizona put forward plans to build two additional dams on the Colorado in the mid-1960s. Concerned at the

BOX 3.1   CHANGING RIVER MANAGEMENT PRIORITIES. ENGINEERING ECOLOGICAL DIVERSITY IN A RECENTLY CHANNELISED RIVER – THE RIVER BLACKWATER IN NORTHERN IRELAND

The catchment area of the river Blackwater covers 1,500 km^2 on the border between the Republic of Ireland and Northern Ireland. Now complete, the scheme was designed to improve agricultural production in 59 km^2 of the catchment and to accommodate within the engineered river all flows up to a three-year recurrence interval. It cost £19 million at 1984 prices, involved widening and deepening 607 km of river channel and was largely funded by the European Union (EU) at a time when the EU's Common Agricultural Policy (CAP) still focused on agricultural expansion and when flooding and poor soil drainage in many parts of Northern Ireland was seen as the greatest single impediment to agricultural development.

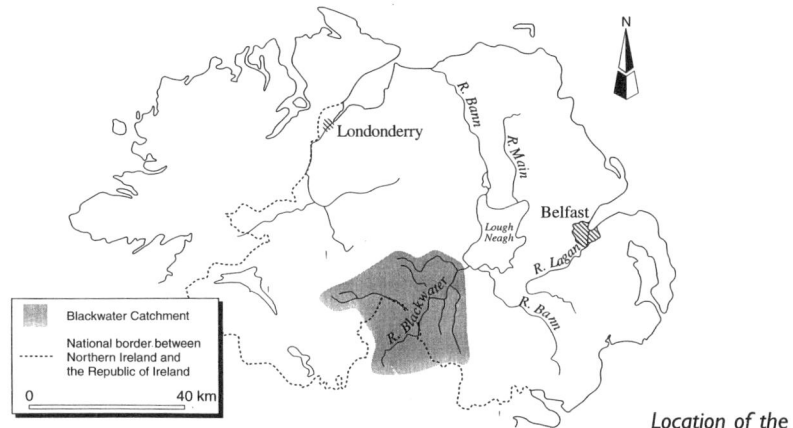

Location of the River Blackwater in Ireland

In the early and middle 1980s EU support faded for river engineering works on the Blackwater scale on grounds of economy and relevance at a time of growing agricultural surpluses (Bowers, 1983). But the greatest criticisms were on environmental grounds. Such schemes reduced geomorphological, hydrological and ecological diversity and removed instream and flood plain habitat. The Blackwater had been a significant environmental resource:

Gravel-bedded, sinuous and lined with earth cliffs, alder, oak and ash, the Blackwater was an attractive river, important for game fishing. Kingfishers, sand martins, dippers and grey wagtails were common. Otters were often seen. The floodplain was dotted by small, boggy wetlands important for breeding waders, especially snipe and lapwings. A few of these wetlands provided refuges for the declining corncrake. In winter, flood waters attracted whooper swans and Greenland white-fronted geese.

(Williams and Browne, 1987: 12)

It was against this background that the need was acknowledged for 'rehabilitation work, including restoration of the rivers fishery aspect' and the reinstatement of tree cover and bank vegetation 'where critical to fishery habitat' (Department of Agriculture for Northern Ireland, in Ward et al., 1994: 306). To this end, holding pools up to 2 m deep have been excavated in the river bed for game fish (salmon and trout), spawning and nursery gravel beds have been added, and engineered rock structures installed to oxygenate low flows and to stabilise sediment movement. Willows and alders have been planted on many banksides. Different pool and rock structure designs have been used at 300 sites and are now being tested. Introducing morphological and sedimentological diversity in this way represents a welcome recognition by river engineers of the geomorphological, sedimentological and ecological diversity loss inherent in channelisation but still fails to acknowledge that physical processes in river channels are inextricably interconnected with catchment processes (especially the delivery of water and sediment to the channel) and need to be managed with these links always in mind. The basic conflict in drainage engineering is between the drainage requirement for smoothness and hydraulic efficiency and the environmental requirements for roughness and high morphological, sedimentological and ecological diversity. The natural evolution of diversity in a river takes thousands of years. It remains to be seen if engineered diversity is an adequate substitute.

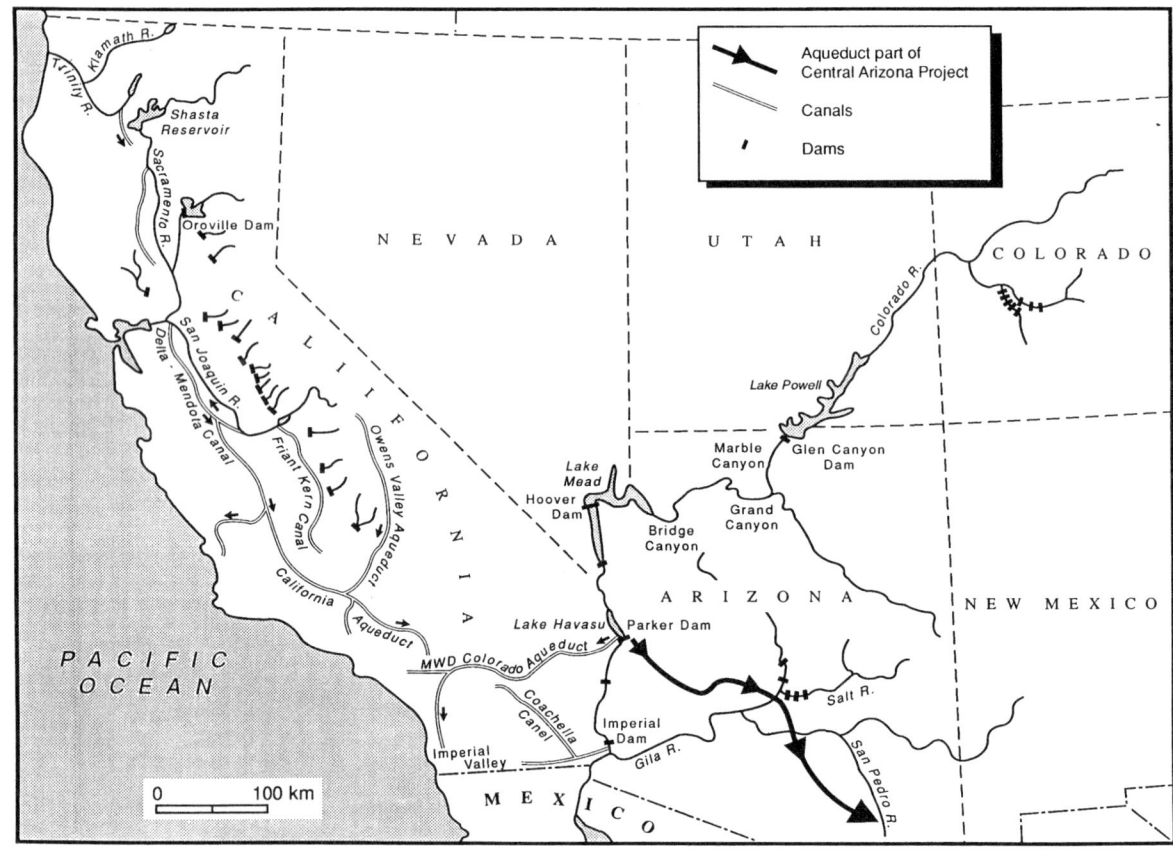

Figure 3.7 The pattern of water impoundments and transfers in the south-west USA
Source: after Wilcock *et al.*, 1976

implications for her own energy and water supplies, California claimed the right of 'prior use' and opposed Arizona's proposals. The Central Arizona Project, as eventually proposed, was supported by California, and included plans to divert water from the Columbia–Snake system in the Pacific north-west states to the Colorado river to compensate for streamflow removed by the Central Arizona Project. The states of Washington, Oregon and Idaho opposed these plans, fearful for their own future economic development. Environmentalists also opposed proposals to flood the Grand Canyon, one of the country's primary scenic resources. In the event, a modified version of the proposal was accepted by the federal government and all reconnaissance studies proposing water transfers into the Colorado basin from outside were prohibited for ten

years. The conflict is seen retrospectively by many environmental scientists as one of the important turning points in the debate about the need for environmental impact statements. The United States' National Environmental Policy Act, including provision for environmental impact statements, was implemented in 1970 (Wilcock *et al.*, 1976).

In the European Union equally important political, economic and environmental conflicts are presented by management of the Rhine. Parts of the drainage basin lie in Switzerland, Germany, the Netherlands, France and Luxembourg. Pollution levels increase downstream with serious water quality implications for the Netherlands and the North Sea. In the 1970s and 1980s, sulphates rose by 75 per cent between Switzerland and the Netherlands. Over the same length of river nitrate levels doubled, phosphates and

BOX 3.2 INTERNATIONAL RIVERS – THE NEED FOR NEW LEGAL FRAMEWORKS

The Tigris and Euphrates rivers rise in the relatively well-watered mountains of eastern Turkey and flow south-eastwards into Syria and Iraq, two very much drier countries both of which have been heavily dependent on natural river flow from Turkey for centuries (Roberts, 1991). Indeed, the harnessing of spring floodwaters on the Tigris–Euphrates system, resulting from snowmelt in the headwaters of both catchments, was the basis of early civilisations in Mesopotamia, today in modern Iraq. Turkey, however, lacks fossil fuel energy resources and since the mid-1960s has developed a series of hydroelectric power stations on the Euphrates. The first such dam became operational at Keban in 1974. The second, at Karakaya, came on stream in 1988. These dams do not affect total flow volumes downstream but may adversely affect downstream flow regimes, reducing the dilution afforded by natural snowmelt in spring and early summer. More recently and seriously, dams built for irrigation purposes at Atatturk in southern Turkey and at Tabqa in Syria abstract 45 per cent of Euphrates flow, seriously threatening the viability of irrigation-based agriculture in Iraq, principally through its negative effects on base flows and water quality. Iraq's short-term management solution, to divert Tigris water into the Euphrates, only transfers a Euphrates catchment management problem into the Tigris. A recently proposed alternative solution, to formally consolidate and expand existing exchanges of oil for water between Turkey and Iraq, also avoids specifically addressing the Euphrates water management problems of changed flow regimes and deteriorating water quality.

This case study, like that described for the Blackwater, illustrates a more enlightened engineering approach to water management problems than prevailed 20 years ago. This approach is to be welcomed. But the concepts of sustainable development and the conservation of biodiversity recognised by governments at the United Nations Conference on Environment and Development require that environmental systems are managed with regard to the natural energy flows and material cycles within them. In the case of water management this requires management on a drainage basin basis. Few water management schemes anywhere in the world comply with these principles. Water development policies in one nation which cut off 'downstream' nations from their historical natural gravity-fed water supplies threaten future political instability as well as sustainable development and biodiversity: 'the problem of developing world rivers which cross international boundaries appears to need a negotiated revision of inter-national law' (Newson, 1992: 185).

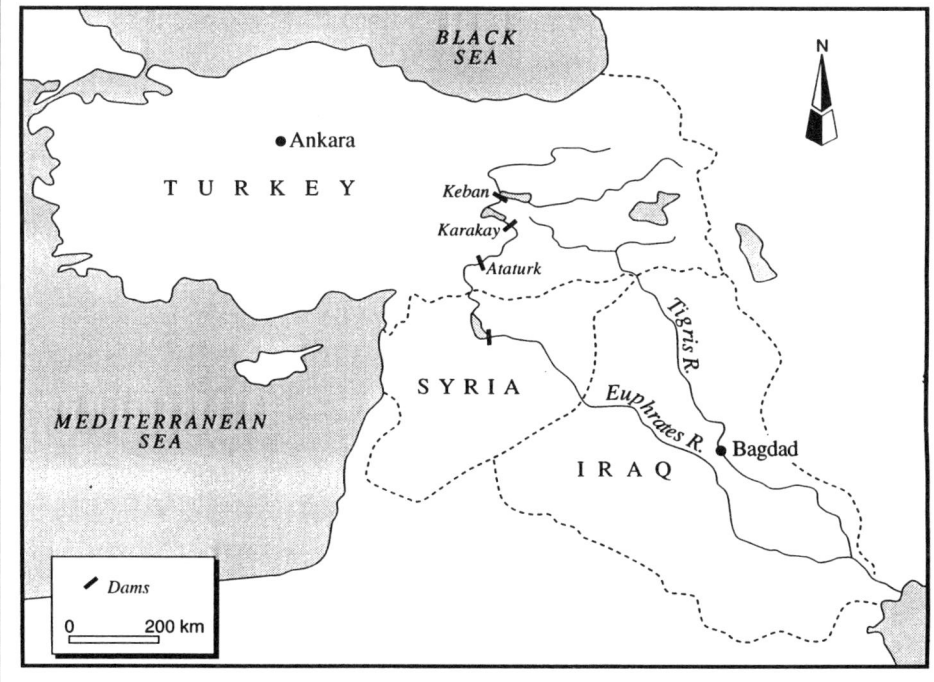

The Tigris and Euphrates drainage basins in relation to political boundaries

mercury levels trebled, and cadmium rose nearly nine-fold! When streamflow was brought into the calculations, the downstream increases appeared even more dramatic, the orthophosphate load, for example, increasing from 630 tonnes per year in Switzerland to 22,000 tonnes per year in the Netherlands (Briggs, 1987). Such problems could only be solved internationally and the Rhine Action Programme, established in 1986, now looks at the way the river is managed from source to mouth. It has set the objective of returning salmon to the river by the year 2000 (Byrne, 1989). Already, some improvements are evident, the story for cadmium being fairly typical: 'cadmium levels are presently declining; however, the levels are still high, and additional measures may be needed to reduce cadmium concentrations further' (Stanners and Bourdeau, 1995: 100).

LAKES

Ecology

Lakes, both natural and of human origin, present a range of environmental management problems. As with rivers, lakes are often used for a range of functions, some of which, like recreation, water supply and waste disposal, may be mutually incompatible.

As physical, chemical and ecological systems, lakes are related to their catchments in much the same way as rivers. Water, sediment and solutes cascade from the catchment hillslopes through river channels and floodplains to the deltas at the head of the lakes and finally into the open water. Sediments, solutes, dead organisms and pollen accumulate in the bed material, the decomposition products of which may be an important secondary source of nutrients in the lake and for the development of vegetation. These relationships are represented in a flow diagram in Figure 3.8.

Lakes function as ecosystems essentially as follows: during summer, solar radiation warms the surface waters where phytoplankton, such as diatoms and algae, photosynthesise, releasing dissolved oxygen. The warmer surface water (the epilimnion) is separated from cooler water at depth (the hypolimnion) because solar radiation is nearly all absorbed in the top 1 or 2 metres of the water column (Figure 3.9). Photosynthesis does not occur at depth because of insufficient light, and the hypolimnion, therefore, is usually oxygen-deficient. Stratification is a feature of lakes in temperate climates during summer only, when temperature differences between epilimnion and hypolimnion may be 16°C. In winter, surface waters cool and sink, and the distribution of temperature and dissolved oxygen is more uniform throughout the water column. In warmer climates, temperature differences between epilimnion and hypolimnion may only be 2–3°C but stratification may none the less be very stable, because water density changes more rapidly with temperature in warm than in cold water (Moss, 1988). The decomposition of organic matter, sinking to deeper layers, creates an enhanced demand for oxygen in the hypolimnion which further depletes dissolved oxygen in this layer.

Lakes with steep shorelines have less marginal vegetation than shallow lakes. Nutrient release from decomposition is limited, the lakes tend to be *oligotrophic* (low in nutrition) and oxygen levels are high. Shallow lakes, in contrast, often have extensive fringing vegetation which releases large amounts of nutrients on decomposition. Bed currents in shallow lakes may also assist in the recycling of nutrients from lake sediments during summer and this adds to the nutrient status of such lakes which are frequently *eutrophic* (rich in nutrients). Lakes of intermediate nutrition are *mesotrophic*. As with rivers, therefore, the ecology of lakes results from an intricate combination of physical, chemical and biotic factors, and each lake has to be understood as the unique response to the prevailing interrelationships between it, its tributary rivers and its catchment area.

Lakes can be classified on several bases. Nutrient status is one criterion. Oligotrophic lakes are usually characterised by very clear water and small amounts of phytoplankton which can exist at considerable depth. Marginal vegetation tends to be limited, the clean, stony margins of oligotrophic lakes being their most immediately distinguishing characteristic. Submerged vegetation also exists, some stoneworts (e.g. *Nitella opaca*) being found at depths of 8 metres in oligotrophic lakes in Scotland. In general, however, plants and invertebrates in oligotrophic lakes are limited in number and diversity.

In contrast, eutrophic lakes have abundant and diverse vegetation on their margins. Close to the

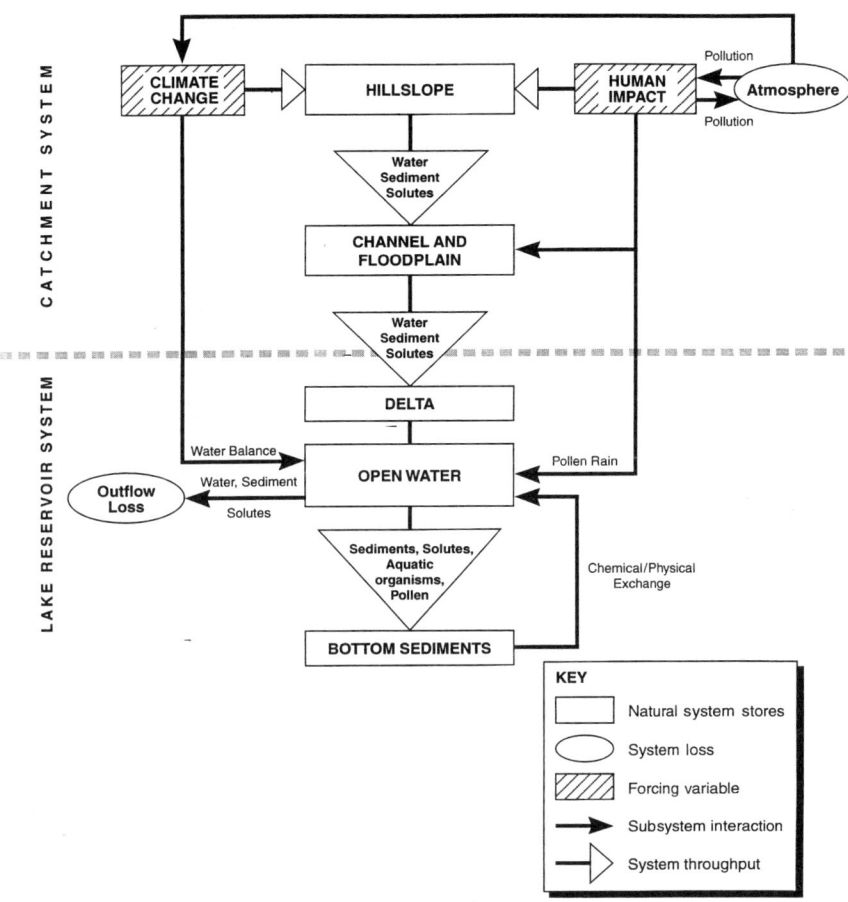

Figure 3.8 Sediment flows in a lake catchment framework
Source: after Foster *et al*, 1988

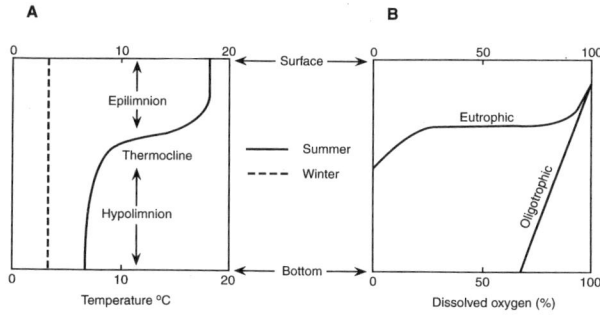

Figure 3.9 **(A)** Summer and winter temperature profiles in a mid-latitude lake. The summer thermocline is a zone of rapid temperature change.
(B) Oxygen profiles in oligotrophic and eutrophic lakes

shore edge is a zone of emergent plants, able to root in sediments and flower above water, such as the British reedmace (*Typha latifolia*). Further out, but still rooted in sediments, are completely submerged species such as Canadian pondweed (*Elodea canadensis*). Such plants rarely reach the water surface. In yet deeper water, rootless plants like Britain's common duckweed (*Lemna minor*) absorb dissolved nutrients in the lake water through fine rootlets. Marginal vegetation provides food and protection for a wide variety of invertebrates, for example, the mayfly (*Ephemera danica*), the stonefly (*Chloroperla torrentium*) and the dragonfly (*Aeshna cyanea*) (Angel, 1981).

Lake sediments as a source of environmental history

Lake sediments are a rich source of data from which to reconstruct the events of recent environmental history. The analysis of pollen grains within the sedimentary record has proved particularly useful in the reconstruction of vegetational history, often back into the late Quaternary period. Different plants release diagnostic pollen grains and spores into the atmosphere. Carried aloft by wind, pollen grains eventually fall to the ground and become trapped in sediments. Sediments low in oxygen, like those in a well-stratified lake, are particularly good environments in which to preserve pollen as they inhibit biodegradation by micro-organisms. Sediment cores are extracted from the lake, distinctive horizons matched to similar horizons in other columns, and each type of pollen grain is identified and counted. Pollen grains in the sediment cores are matched against pollen produced today by vegetation in different parts of the world, and thus the sequence of vegetational and climatic change in the catchment can be reconstructed.

Changes in acidity within a lake can be unravelled from the diatom record (ter Braak and van Dam, 1989) while the record of industrial pollution can be reconstructed from analysis of heavy metals like lead, zinc and copper. Dating the horizons of the sedimentary column is possible using a variety of techniques, most notably Carbon-14 dating. It is important that accurate dates are established in order to determine the fluxes at which sediments accumulated at different periods.

Changes in phosphorus concentrations in sediment cores from Lough Neagh in Northern Ireland are presented in Figure 3.10. Increasing concentrations in the top layers are probably related to the increasing use of phosphates in detergents during the 1960s. The profiles in Figure 3.10 are shown in their approximate geographical locations within Lough Neagh. Also shown are the major tributaries of the Lough, their mean annual streamflow, and their annual loadings of total phosphorus (TP) and particulate phosphorus (PP) in tonnes per annum.

It is evident that recent accumulation of phosphorus is most marked in the south-west of the Lough

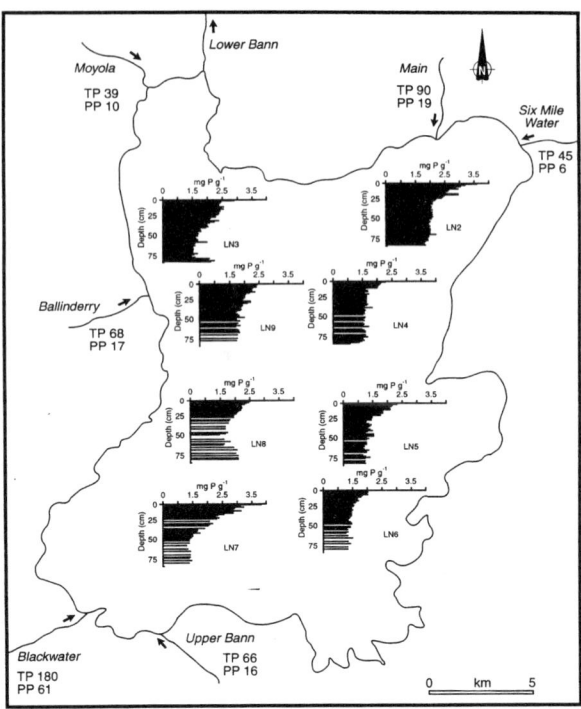

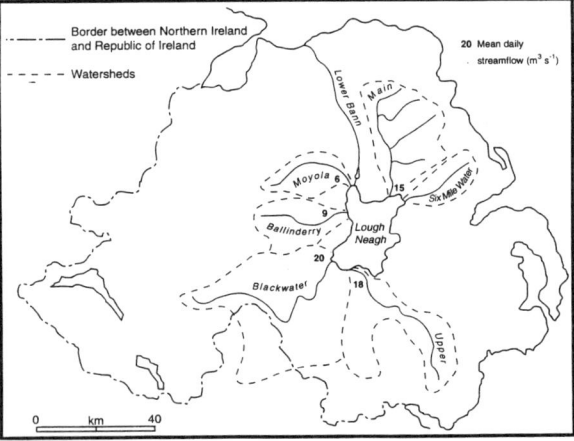

Figure 3.10 Changing phosphorus concentrations in sediment cores from Lough Neagh, Northern Ireland
Source: after Fletcher, 1990

adjacent to the two largest tributaries: the River Blackwater and the upper River Bann. Lower rates of accumulation occur on the eastern side of the Lough where large tributaries are absent. Fletcher's work (1990), from which most of this section on lake sedimentation has been derived, stresses the difficulties of relating contemporary catchment budgets of plant nutrients and/or heavy metals to the rates of accumulation in the sedimentary record. The emphasis on accurate calibration in such studies is important if contemporary social and economic impacts on environmental systems are to be understood in detail and management plans derived to counter their effects.

Contemporary environmental management problems

Limnologists commonly recognise four contemporary management problems for lakes: acidification, cultural eutrophication, oxygen depletion and toxic pollution. Problems of lake acidification are similar to those of upland rivers which have already been examined. Perhaps the main management problem of lakes is *cultural eutrophication*, that is the nutrient enrichment of lakes from agricultural, domestic and industrial sources. Two nutrients are principally involved, nitrogen and phosphorus. Of the two, phosphorus is probably the more important in determining productivity (Moss, 1988). Where nutrient enrichment is high, phytoplankton production may be stimulated. This sometimes takes the form of floating mats of algae blooms on the surface of the lake. On other occasions phytoplankton is distributed throughout the water column, giving the surface waters a green, cloudy appearance. Though phytoplankton also produce oxygen through photosynthesis, excessive phytoplankton production reduces light penetration and limits photosynthesis below the surface layers. The decomposition of large phytoplankton populations also withdraws more oxygen in bacterial respiration than is produced by photosynthesis, and dissolved oxygen levels are depleted. Anaerobic conditions may set in, especially in the hypolimnion during summer, and fish kills may result from this, as well as from elevated pH values and high ammonia levels.

Today, phosphorus compounds in lakes and lowland rivers come mainly from detergents and domestic wastes. Nitrogen compounds mainly come from artificial fertilisers, the use of which has rapidly increased in Western Europe and North America since 1945.

The most notable example of lakes made eutrophic by industry and agriculture are the Great Lakes of North America. Some, but not all, ecologists distinguish cultural eutrophication from natural eutrophication, that is the natural transition of an oligotrophic lake to a eutrophic condition through the addition of nutrients adsorbed on to sediments and the associated enrichment thus brought about. According to Giddings (1973), cultural eutrophication rates in Lake Erie are 300 times faster than natural eutrophication rates in the lake.

Oxygen depletion is a third environmental management problem in many of today's lakes. Often the result of eutrophication, well-stratified lakes being particularly vulnerable, it may also be the direct result of agricultural pollution, particularly from silage and slurry discharges. Modern livestock-rearing systems are based on nitrogen-fertilised, high yield grass swards and require the collection and storage of animal slurries before these are spread on the land. Slurries, however, have very high BOD values, often exceeding 4,000 mg/l, and silage liquor BOD may be even higher, 12,000–16,000 mg/l. Leakage of these effluents from storage tanks or their release on to pasture when soils are wet and have little infiltration capacity can result in massively elevated BOD values in lakes and rivers. Since the BOD of a lake or river has to be maintained at no more than about 6 mg/l for salmon to survive, the effect of such effluents on fish life in lakes and lowland rivers is only too obvious. The BOD values quoted above for silage liquor illustrate the importance of sustaining dry weather streamflow in order to provide an acceptable level of dilution in lakes and rivers.

Toxic pollution of lake waters from pesticides and heavy metals is also a contemporary management problem in many lakes. As with eutrophication and deoxygenation, the results vary with the severity of the incident. Small-scale pollution often affects only small numbers of autotrophs whereas large-scale incidents might eliminate higher components of the food chain.

Various methods have been devised to deal with some of these problems. Any practice that raises oxygen levels in the bottom waters of a eutrophic lake is to be encouraged. Iron in anaerobic sediments, for example, is more soluble and releases phosphorus more readily than iron in aerobic sediments. Progressive deoxygenation of the hypolimnion and bottom sediments, therefore, leads to the progressive release of phosphorus and further eutrophication. The breaking of this self-perpetuating cycle may require artificial aeration on the lake bed or, in extreme cases, the physical removal of phosphorus-enriched lake sediments as in Lake Trumenn, Sweden (Jorgenson and Jorgenson, 1989).

MANAGING RIVERS AND LAKES

It is appropriate to conclude by identifying some important principles that should be applied to the management of rivers and lakes. The first, and most important, is that managing rivers and lakes is largely an exercise in *managing drainage basins*, since the drainage basin is the spatial framework within which streamflow, sediment and nutrient loads are determined. It is more difficult to apply this maxim to large than to small catchments, since in large catchments the multiple effects of different management practices are superimposed on one another and their individual effects are difficult to identify. We also know very little, in a quantified way, about the downstream attenuation (or amplification?) of upstream changes to any of the forcing functions.

The above points are illustrated by the evolving problem of managing soluble reactive phosphorus (SRP). SRP is the type of phosphorus available for algae growth in rivers and lakes, and most appears to come from urban sources. In line with this thinking, phosphorus reduction programmes in urban sewage treatment works were installed in many areas during the 1980s. Despite attempts to reduce SRP loadings in this way recent studies in Northern Ireland suggest that SRP in some rivers and lakes may still be increasing, mainly because of small but persistent leakages of SRP from steadily accumulating levels of phosphorus in the soil derived from fertiliser applications over many years (Foy *et al.*, 1995). Soil tests have recently shown that more than half of all soils tested for P in Northern Ireland have more than enough P for plant growth. Surplus P is removed from the soil in runoff and soil water and finds its way into rivers and lakes, promoting algae growth in rivers and lakes rather than grass growth on farms.

Observations like these illustrate the need for ongoing reviews of strategies to reduce phosphorus loadings on rivers and lakes from all catchment sources.

A second management principle is to manage catchment land-use so as to preserve those ecosystems that contribute naturally to the control of pollution. Wetland soils and plants, for example, often have a high phosphorus take-up potential. The engineering of marginal wetlands round a lake and the harvesting of biomass produced in them might significantly reduce phosphorus inputs into the lake itself. Such a strategy has been explored for western Lake Erie in the Great Lakes system of North America where existing wetlands are thought to retain about 85 tons of phosphorus out of an estimated annual non-point phosphorus loading of approximately 2,000 tons. It is argued that the development of 1,000 km^2 of wetland on the margin of the lake might reduce phosphorus inputs to the lake by 24–33 per cent (Mitsch *et al.*, 1989). Riparian forests can have a similar function as nitrogen filters in agricultural catchments if regularly harvested to maintain young stands of timber (Lowrence *et al.*, 1984). Similar effects may be achieved by river marginal wetlands (Baker and Maltby, 1995). Such management strategies attempt to harness the 'self-designing', natural recycling capacities of non-riverine, non-lacustrine ecosystems within a drainage basin as a means of regulating nutrient loads on the rivers and lakes themselves.

Several recent river and lake management texts stress the importance of *ecotones* as indicators of ecological stability in drainage basins. Ecotones have been defined as 'zones of transition between adjacent ecological systems, having a set of characteristics uniquely defined by space and time scales and by the strength of interactions between adjacent ecological systems' (Risser, 1990: 8). Examples of ecotones are floodplain wetlands (products of alternating dryness and wetness), littoral lake zones (products of dynamic sedimentation), and areas of significant groundwater–surface water exchange. Such areas have high conservation value because, as transitional zones, they are particularly sensitive to change in drainage basin processes and act as bioindicators of change which might later affect adjacent ecosystems.

Petts (1990) identifies two important aspects of aquatic ecotone management. The first is the need to ensure that the duration, frequency and magnitude of flows necessary to sustain nutrient cycling in aquatic ecotones is maintained. The second is the need to identify the minimum size necessary for maintenance of the ecotone as a viable ecological unit. This latter point is important, as the robustness of delicate ecosystems tends to increase with size. Petts' point about flow duration, frequency and magnitude introduces another essential element of sound river and lake management: the need for *continuous measurements of river flows* and the integration of such measurements into the development of environmental policy. Accurate flow measurements might sensibly be used both to define drainage basin management policy and to check its implementation. In channelisation schemes, for example, the engineered river channel capacity is designed to accommodate a flood discharge of a certain recurrence interval. In agricultural areas of the United Kingdom this is often the five-year flood, that is the flood expected to recur once every five years. Definition of the design flood is important in cost–benefit analysis, and a channelised scheme designed to accommodate only a three-year flood at the bankfull stage would provide less protection to agriculture, industry and housing on the floodplain than one designed for a five-year flood. It would also be less expensive. Although the design flood is always specified before channelisation, less reference is made to it if flooding occurs after channelisation. It is often the case, for example, that the first post-channelisation inundation of the floodplain provokes renewed calls for yet more channel works. In such cases, sound environmental management would assess the frequency of the post-channelisation event which caused the overbank flooding. If this is a smaller and more frequent event than the design flood, maintenance is in order. If not, the call for maintenance should be resisted since the designed channel was not meant to withstand the effect of such floods.

The growing recognition of continuous measurement as an integral element of river basin management has led to the development of high quality hydrological and ecological databases. The large volumes of data held in these databases have encouraged the development of environmental models and these, in turn, have sometimes facilitated the prediction of environmental impact. Early hydrological models based on large national datasets were developed to predict flood flows and low flows from catchment characteristics (United Kingdom Natural Environment Research Council, 1975; United Kingdom Institute of Hydrology, 1980). More recently, the UK Institute of Hydrology's lumped catchment model, HYRROM, has been successfully used to extend the length of runoff records (Blackie and Eeles, 1985), and to quantify the effects of channelisation on daily runoff (Wilcock and Wilcock, 1995). In the field of aquatic ecology, the River Invertebrate Prediction and Classification System (RIVPACS), developed by the UK Institute of Terrestrial Ecology, uses survey data from 438 unpolluted sites to establish how particular invertebrate groups can be related to particular channel characteristics of discharge, width, depth, substrate, alkalinity, altitude, slope and distance downstream. Measurement of these site characteristics at individual reaches in a river allows the user of RIVPACS to predict the suite of invertebrates that would exist at any particular site, if it were unpolluted. These can then be compared with what is actually present to derive an Ecological Quality Index for the site (Armitage and Petts, 1992). An even more elaborate modelling package, the Instream Flow Incremental Methodology (IFIM) developed by the Aquatic Systems Branch of the US Fish and Wildlife Service (Bullock *et al.*, 1991), integrates concepts of hydrology and aquatic ecology to predict how different flow durations affect available physical habitat. This model can be used to predict ecological impacts of channel change and the availability of physical habitat in a particular river for the different life stages of fish species. Modelling packages like HYRROM, RIVPACS and IFIM require high quality data and encourage more uniformity in data collection. They also help identify gaps in existing knowledge and facilitate more straightforward methods of transferring specialised knowledge to planners and other decision-makers.

Other important ecological engineering principles, many of them appropriate to river basin management, are summarised by Mitsch and Jorgenson (1989). Perhaps the most important management principle, however, is the *realistic pricing of environmental resources*. Decision-making in free market economies relies on relative cost as the best indication of efficient investment. But markets find it very difficult accurately to reflect the long-term costs and benefits of environmental resources. One approach to this problem is to assess the total economic value of a contemporary environmental resource as the sum of present value plus benefits forgone if the environmental resource is destroyed. In Northern Ireland such an approach has been applied to an assessment of the true economic value of upland salmonid streams. Replacing young salmon after an agricultural pollution incident in such streams might cost only £1 per m². But young salmonids live in very high densities and an eventual adult crop has an estimated value of £8.50 per m². This includes its value to commercial fisheries, angling and as a supply of eggs for future generations. This latter estimate grosses to a value of £85,000/hectare of river bed in upland streams, approximately 4 to 8 times the value of the best adjoining agricultural land (Kennedy, 1986). Is there a better argument in favour of managing catchment activities so as to protect the ecology of the river bed?

REFERENCES

Alabaster, J.S. and Lloyd, R. (1982) *Water Quality Criteria for Freshwater Fish*, London, UK: Butterworths.

Angel, H. (1981) *The Natural History of Britain and Ireland*, London, UK: Book Club Associates.

Armitage, P.D. and Petts, G.E. (1992) 'Biotic score and prediction to assess the effects of water abstractions on river macroinvertebrates for conservation purposes', *Aquatic Conservation: Marine and Freshwater Ecosystems* 2: 1–17.

Baker, C.J. and Maltby, E. (1995) 'Nitrate removal by river marginal wetlands: factors affecting the provision of a suitable denitrification environment', in J. Hughes and L.H. Heathwaite (eds) *Hydrology and Hydrochemistry of British Wetlands*, Chichester, UK: Wiley, pp. 291–313.

Blackie, J.R. and Eeles, C.W.O. (1985) 'Lumped catchment models', in M.G. Anderson and T.P. Burt (eds) *Hydrological Forecasting*, Chichester, UK: Wiley, pp. 311–346.

Boon, P.J., Calow, P. and Petts, G.E. (eds) (1992) *River Conservation and Management*, Chichester, UK: Wiley.

Bormann, F.H. and Likens, G.E. (1979) *Pattern and Process in a Forested Ecosystem*, Berlin, Germany: Springer-Verlag.

Bowers, J.K. (1983) 'Cost–benefit analysis of wetland drainage', *Environment and Planning* A, 15: 227–235.

Briggs, D. (1987) *The State of the Environment in the European Community, 1986*, Luxembourg: Office for Official Publications of the European Community.

Brookes, A. (1988) *Channelised Rivers: Perspectives for Environmental Management*, Chichester, UK Wiley.

Bullock, A., Gustard, A. and Grainger, E.S. (1991) *Instream Flow Requirements of Aquatic Ecology in Two British Rivers*, Report No. 115, Wallingford, UK: Institute of Hydrology.

Byrne, C.D. (1989) 'Community water quality policy for the nineties', in D. Wheeler, M.L. Richardson and J. Bridges (eds) *Watershed 89 – The Future for Water Quality in Europe*, Oxford, UK: Pergamon, pp. 9–14.

Clarke, R.T. and McCulloch, J.S.G. (1979) 'The effect of land-use on the hydrology of small upland catchments', in G.E. Hollis (ed.) *Man's Impact on the Hydrological Cycle in the United Kingdom*, Norwich, UK: Geo Abstracts, pp. 71–78.

Dooge, J. (1973) 'The water balance of bogs', *Proceedings of the International Association of Hydrological Sciences*, Publication No. 105, Paris, France, pp. 273–280.

Fletcher, C.L. (1990) 'The recent sedimentary history and contemporary budgets of zinc, copper and lead in Lough Neagh, Northern Ireland', unpublished Ph.D. thesis, Coleraine, Northern Ireland: University of Ulster.

Flower, R.J. and Battarbee, R.W. (1983) 'Diatom evidence for recent acidification of two Scottish lochs', *Nature* 305: 130–132.

Flower, R.J., Battarbee, R.W. and Appleby, P.G. (1987) 'The recent palaeolimnology of acid lakes in Galloway, Southwest Scotland: diatom analysis, pH trends, and the role of afforestation', *Journal of Ecology* 75: 797–824.

Foster, I.D.L., Dearing, J.A. and Grew, R. (1988) 'Lake catchments: an evaluation of their contribution to the study of sediment yield and delivery processes', *Proceedings of the International Association of Hydrological Sciences*, Publication No. 174, Wallingford, Oxon, UK, pp. 413–429.

Foy, R.H., Smith, R.V., Jordan, C. and Lennox, S.D. (1995) 'Upward trend in soluble phosphorus loadings to Lough Neagh despite phosphorus reduction at sewage treatment works', *Water Research* 29, 4: 1051–1063.

Gardiner, J.L. (1991) *River Projects and Conservation: A Manual for Holistic Approach*, Chichester, UK: Wiley.

Giddings, J.C. (1973) *Chemistry, Man and Environmental Change*, San Francisco, Calif.: Canfield Press.

Gjessing, E.T., Alexander, J. and Rosseland, B.O. (1989) 'Acidification and aluminium – contamination of drinking water', in D. Wheeler, M.L. Richardson and J. Bridges (eds) *Watershed 89 – The Future for Water Quality in Europe*, Oxford, UK: Pergamon, pp. 15–21.

Jorgenson, S.E. and Jorgenson, L.A. (1989) 'Eco-technological approaches to the restoration of lakes', in W.J. Mitsch and S.E. Jorgenson (eds) *Ecological Engineering*, New York, USA: Wiley, pp. 357–374.

Keller, H.M. (1990) 'Monitoring water and nutrient budgets in small mountain basins: collecting data and/or understanding processes', in J.C. Hooghart, C.W.S. Posthumus and P.M.M. Warmerdam (eds) *Hydrological Research Basins and the Environment*, The Hague, The Netherlands: The Netherlands Organisation for Applied Scientific Research, pp. 225–243.

Kennedy, G.J.A. (1986) 'Silage effluent pollution', *Agriculture in Northern Ireland* 60: 402–406.

Leopold, L.B. and Maddock, T. (1953) *The Hydraulic Geometry of Stream Channels and some Physiographic Implications*, Professional Paper, 252, Washington DC, USA: United States Geological Survey.

Leopold, L.B., Wolman, M.G. and Miller, J.P. (1964) *Fluvial Processes in Geomorphology*, Oxford, UK: Freeman.

Lewis, G. and Williams, G. (eds) (1984) *Rivers and Wildlife Handbook*, Sandy, Bedfordshire, UK: Royal Society for the Protection of Birds.

Lowrence, R., Todd, R., Fair, J. Jr., Hendrickson, O. Jr., Leonard, R. and Asmussen, L. (1984) 'Riparian forests as nutrient filters in agricultural watersheds', *Bioscience* 34: 374–377.

Mackin, J.H. (1948) 'Concept of the graded river', *Bulletin of the Geological Society of America* 59: 463–512.

Milhous, R.T. (1982) 'The effect of sediment transport and flow regulation on the ecology of gravel-bed rivers', in R.D. Hey, J.C. Bathurst and C.R. Thorne (eds) *Gravel-bed Rivers – Fluvial Processes, Engineering and Management*, Chichester, UK: Wiley, pp. 819–842.

Milner, N. (1984) 'Section 4: Fish', in G. Lewis and G. Williams, *Rivers and Wildlife Handbook*, Sandy, Bedfordshire, UK: Royal Society for the Protection of Birds, pp. 51–55.

Mitsch, W.J. and Jorgenson, S.E. (eds) (1989) *Ecological Engineering*, New York, USA: Wiley.

Mitsch, W.J., Reeder, B.C. and Klarer, D.M. (1989) 'The role of wetlands in the control of nutrients with a case study of Western Lake Erie', in W.J. Mitsch and S.E. Jorgenson (eds) *Ecological Engineering*, New York, USA: Wiley, pp. 129–158.

Moss, B. (1988) *Ecology of Fresh Waters*, Oxford, UK: Blackwell.

Newson, M. (1984) 'River processes and form', in G. Lewis and G. Williams (eds), *Rivers and Wildlife Handbook*, Sandy, Bedfordshire, UK: Royal Society for the Protection of Birds, pp. 3–9.

Newson, M. (1986) 'River basin engineering – fluvial geomorphology', *Journal of the Institution of Water Engineers and Scientists* 40: 307–324.

Newson, M. (1992) *Land, Water and Development*, London, UK: Routledge.

O'Brien, J.S. (1987) 'A case study of minimum streamflow for fishery habitat in the Yampa river', in C.R. Thorne, J.C. Bathurst and R.D. Hey (eds) *Sediment Transport in Gravel-bed Rivers*, Chichester, UK: Wiley, pp. 921–946.

Ormerod, S.J., Mawle, G.W. and Edwards, R.W. (1986) 'The influence of forest on aquatic fauna', in J.E. Good (ed.) *Environmental Aspects of Plantation Forestry in Wales*, Bangor, Wales, UK: Institute of Terrestrial Ecology, pp. 37–49.

Palmer, M. (1984) 'Section 2: Invertebrates', in G. Lewis and G. Williams (eds), *Rivers and Wildlife Handbook*, Sandy, Bedfordshire, UK: Royal Society for the Protection of Birds, pp. 37–46.

Park, C.C. (1977) 'World-wide variations in hydraulic geometry exponents of stream channels: an analysis and some observations', *Journal of Hydrology* 33: 133–146.

Petts, G.E. (1984) *Impounded Rivers: Perspectives for Ecological Management*, Chichester, UK: Wiley.

Petts, G.E. (1990) 'The role of ecotones in aquatic landscape management', in R.J. Naiman and H. Decamps (eds) *The Ecology and Management of Aquatic-terrestrial Ecotones*, Paris, France: Parthenon, pp. 227–261.

Risser, P.G. (1990) 'The ecological importance of land-water ecotones', in R.J. Naiman and H. Decamps (eds) *The Ecology and Management of Aquatic-terrestrial Ecotones*, Paris, France: Parthenon, pp. 7–21.

Roberts, N. (1991) 'Geopolitics and the Euphrates' water resources', *Geography* 76, Part 2: 157–159.

Robinson, M. (1990) *Impact of Improved Land Drainage on River Flow*, Wallingford, Oxfordshire, UK: Institute of Hydrology.

Stanners, D. and Bourdeau, P. (eds) (1995) *Europe's Environment – The Dobris Assessment*, Luxembourg: European Environment Agency.

ter Braak, C.J.F. and van Dam, H. (1989) 'Inferring pH from diatoms: a comparison of old and new calibration methods', *Hydrobiologia* 178: 209–223.

United Kingdom Institute of Hydrology (1980) *Low Flow Studies Report*, 3 vols, UK: United Kingdom Natural Environment Research Council.

United Kingdom Natural Environmental Research Council (1975) *Flood Studies Report*, 5 vols, London, UK.

Vanotte, R.L., Minshall, G.W., Cummings, K.W., Sedell, J.R. and Cushing, C.E. (1980) 'The river continuum concept', *Canadian Journal of Fisheries and Aquatic Sciences* 37: 120–137.

Ward, D., Holmes, N. and Jose, P. (eds) (1994) *The New Rivers and Wildlife Handbook*, Sandy, Bedfordshire, UK: Royal Society for the Protection of Birds, National Rivers Authority and the Royal Society for Nature Conservation.

Webster, J.R., Golloday, S.W., Benfield, E.F., Meyer, J.L., Swank, W.T. and Wallace, J.B. (1992) 'Catchment disturbance and stream response: an overview of stream research at Coweeta Hydrologic laboratory', in P.J. Boon, P. Calow and G.E. Petts (eds) *River Conservation and Management*, Chichester, UK: Wiley, pp. 231–254.

Wellcome, R.L. (1992) 'River conservation – future prospects', in P.J. Boon, P. Calow and G.E. Petts (eds) *River Conservation and Management*, Chichester, UK: Wiley, pp. 453–462.

Wilcock, D.N. (1979) 'Post-war land drainage, fertiliser use and environmental impact in Northern Ireland', *Journal of Environmental Management* 8: 137–149.

Wilcock, D.N. and Essery, C.I. (1991) 'Environmental impacts of channelisation on the River Main, County Antrim, Northern Ireland', *Journal of Environmental Management* 32: 127–143.

Wilcock, D.N. and Wilcock, F.A. (1995) 'Modelling the hydrological impacts of channelisation on streamflow characteristics in a Northern Ireland catchment', *Proceedings of the International Association of Hydrological Sciences*, Publication No. 231, Wallingford, Oxfordshire, UK: pp. 41–48.

Wilcock, D.N., Birch, B.P. and Cantor, L.M. (1976) 'Changing attitudes to water resource development in California', *Geography* 61: 127–136.

Williams, G. and Browne, D. (1987) 'The drainage of Northern Ireland', *Ecos* 8, 3: 8–15.

Brookes, A. (1988) *Channelised Rivers: Perspectives for Environmental Management*, Chichester, UK: Wiley.

Gardiner, J.L. (1991) *River Projects and Conservation: A Manual for Holistic Approach*, Chichester, UK: Wiley.

Mitsch, W.J. and Jorgenson, S.E. (1989) *Ecological Engineering*, New York, USA: Wiley.

Lewis, G. and Williams, G. (eds) (1984) *Rivers and Wildlife Handbook*, Sandy, Bedfordshire, UK: Royal Society for the Protection of Birds.

Moss, B. (1988) *Ecology of Fresh Waters*, Oxford, UK: Blackwell.

Naiman, R.J. and Decamps, H. (eds) (1990) *The Ecology and Management of Aquatic-terrestrial Ecotones*, Parthenon.

Petts, G.E. (1984) *Impounded Rivers: Perspectives for Ecological Management*, Chichester, UK: Wiley.

SUGGESTED READING

Boon, P.J., Calow, P. and Petts, G.E. (eds) (1992) *River Conservation and Management*, Chichester, UK: Wiley.

ACKNOWLEDGEMENTS

My thanks to Kilian McDaid and Nigel McDowell for preparation of the text figures.

SELF-ASSESSMENT QUESTIONS

1 Which of the following statements about the water balance concept for a drainage basin is correct?
   (a) Rainfall on a drainage basin equals streamflow (i.e. stream discharge) +/– evapotranspiration.
   (b) Precipitation on a drainage basin equals the sum of streamflow (i.e. stream discharge) and evapotranspiration.
   (c) Precipitation on a drainage basin equals the sum of streamflow (i.e. stream discharge) and evapotranspiration +/– changes in subsurface storage.
   (d) Precipitation on a drainage basin equals the sum of streamflow (i.e. stream discharge) and evapotranspiration +/– changes in subsurface storage +/– errors of measurement in any of the individual water balance components.
2 Which of the following variables in a channel system is unable to adjust in order to transport bedload or suspended load?
   (a) Slope.
   (b) Width.
   (c) Streamflow.
   (d) Velocity.
3 Which of the following are not necessary to produce organic carbon?
   (a) Sunlight.
   (b) Oxygen.
   (c) Nutrients.
   (d) Carbon dioxide.
   (e) Water.
4 The effects of afforestation in an upland drainage basin usually include:
   (a) a progressive increase of sediment supply from the catchment as the forest matures;
   (b) an initial increase in peak streamflow values followed by a decrease;
   (c) the enhancement of oxygen flows through gravel bed deposits;
   (d) a progressive increase in streamflow volumes as a percentage of rainfall;
   (e) an increase in the pH of stream water in the catchment.
5 A stream can increase its velocity only by increasing its slope.
   (a) True.
   (b) False.

6 Which of the following statements is correct?
   (a) Channelisation reduces the size of flood peaks.
   (b) Channelisation increases channel depth, bed slope and roughness and therefore reduces flood peaks.
   (c) Channelisation decreases the incidence of overbank flooding by accommodating increased flood peaks in a larger and, usually, steeper channel.
7 Oxygen is often deficient at depth in lakes because of:
   (a) insufficient light;
   (b) insufficient respiration by decomposers;
   (c) decomposition of sinking organic matter;
   (d) too much mixing with surface waters.
8 How many times higher are the BOD values of farm slurries and silage liquor than those typically found in rivers supporting salmonids?
   (a) 5.
   (b) 5–10.
   (c) 10–100.
   (d) 100–1,000.
   (e) 1,000–2,000.
   (f) 2,000–10,000.
9 Sound ecological management requires information on which aspects of streamflow in a drainage basin?
   (a) Magnitude (i.e. the actual volume of water in a given stream discharge).
   (b) Frequency (i.e. the number of times a given stream discharge recurs over a period of years).
   (c) Duration (i.e. the length of time a given stream discharge lasts).
   (d) All of the above.

4

WETLANDS

Paul P. Schot

SUMMARY

Wetlands are found where dryland meets water. They include such areas as beaches, tidal flats, lagoons, mangrove swamps, estuaries, floodplains, marshes, fens and bogs. Until recently wetlands were seen as wastelands associated with disasters such as floods, and diseases such as malaria and schistosomiasis. Therefore wetlands were drained for agricultural and health purposes, and dams and dikes constructed for flood protection, irrigation and hydropower production. During recent decades the importance of wetlands has been recognised. They support extensive agriculture, fisheries and wildlife, and perform a large number of functions for society such as flood attenuation, coastal protection, groundwater recharge, water purification, fuel production and recreation. Extensive alteration has resulted in considerable loss of these functions and has made wetlands one of the most threatened ecosystems in the world. Their conservation and sustainable use are now high on the international agenda. Wetland management must be based on a landscape-ecological analysis of the relations between abiotic and biotic systems within the wetland and their dependency on processes upstream and downstream within the catchment. As socio-economic reasons underlie wetland loss, a participative approach involving consultation and participation of local communities is essential for sustainable wetland development.

ACADEMIC OBJECTIVES

The aim of this chapter is to provide insight into issues related to environmental management of the wetlands of the world.
  After examination of the text the reader should have a basic understanding of:

- the definition and classification of wetlands;
- wetland functions, values and benefits;
- factors leading to deterioration of wetlands;
- principles and approaches for the management of wetlands;
- ways to integrate the protection of wetland benefits in policy.

INTRODUCTION

Wetlands are places where dryland meets, or is inundated by, water. They are among the most productive areas on Earth; their soils are extremely fertile, they support extensive fisheries and wildlife and they perform vital hydrological functions such as flood attenuation and water purification. Many civilisations developed near wetlands and were dependent upon their fertility.

  However, until recently wetlands were seen as wastelands, and their functions and benefits were not appreciated, especially by planners and developers. Drainage of wetland allowed access throughout the year, and more intensive cultivation. Wetlands are also associated with disasters like floods, and diseases such as malaria and schistosomiasis. Therefore dams and dikes were constructed to regulate waterflow and many areas near to human habitation have been drained to improve public health. These human interventions have led to an enormous loss of wetland area as well as of their functions, values and benefits.

During recent decades the importance of wetlands has been recognised. In 1971, wetlands became the first ecosystem type with its own global international conservation treaty: the Ramsar Convention, named after the town of Ramsar, Iran, where the Convention was ratified on 2 February 1971. Under this Convention signatories agree to include wetland conservation in their national planning and to promote their sound utilisation (Maltby, 1986). The principal concern that led to the Ramsar Convention was the conservation of habitat for waterbirds. However, it stimulated interest in wetlands in general and contributed to a growing awareness of the importance of wetlands in the landscape. Wetlands are now recognised as one of the most threatened ecosystems on the planet. Their conservation and sustainable use are high on the international agenda which is evidenced among others by the declaration of 1985 as the Year of the Wetland by the International Union for Conservation of Nature and Natural Resources (IUCN) and by Agenda 21, from the UNCED conference in 1992, which identifies the conservation of wetlands as a priority 'owing to their ecological and habitat importance for many species and taking into account social and economic factors' (OECD, 1996).

This chapter will provide some basic insight into wetland issues. It addresses the definition and classification of wetlands, their functions, values and benefits, as well as deterioration, management and policy aspects of wetlands.

DEFINITION AND CLASSIFICATION

Definition

Because wetlands occur in many different climatic zones and geographical regions, they have been given various names dependent on characteristics of soil, vegetation, water regime, etc. 'Wetland' is a relatively new term for landscape that many people knew before under different names (Williams, 1991), such as swamp, fen, bog, marsh, mire, moor, mangrove, salt marsh, pan, dambo, sebkha, etc. Figure 4.1 gives examples of the different wetlands of the world.

In the Ramsar Convention, wetlands are defined as:

> Areas of marsh, fen, peatland or water whether natural or artificial, permanent or temporary, with water that is static or flowing, fresh, brackish or salt, including areas of marine waters, the depth of which at low tide does not exceed six meters.

This broad definition, however, does not incorporate the adjoining dryland ecotone and deeper water, which have obvious significance for watershed habitat conservation. The Convention therefore also provides that wetlands 'may incorporate riparian and coastal zones adjacent to the wetlands and islands or bodies of marine water deeper than six metres at low tide lying within' (Breen, 1991).

The Ramsar definition is an international definition that caters for the interests of conservationists in general, but particularly of those concerned with waterbirds. Lefor and Kennard (1977) showed that different definitions can be formulated by the geologist, soil scientist, hydrologist, biologist, systems ecologist, sociologist, economist, political scientist, public health scientist and lawyer. This variance arises from differences of emphasis in the definer's training on the one hand, and the way individual disciplines deal with wetlands on the other.

Wetland definitions often include three main components (Mitsch and Gosselink, 1993):

1 The presence of water, either at the surface or within the root zone.
2 The presence of hydric soil conditions.
3 The presence of vegetation adapted to wet conditions (hydrophytes, phreatophytes) and absence of flooding-intolerant vegetation.

However, combining these hydrological, soil and vegetation characteristics to obtain a precise definition is difficult because wetlands form part of a continuous gradient between dry upland and open water. It is therefore not surprising that the problem of definition usually arises at the edges of the wetland.

The frequency of flooding is the variable that has made the definition of wetlands particularly controversial. Although water may be present in an area for at least part of the time, the depth and duration of flooding vary considerably from wetland to wetland and from year to year, especially in arid zones. Breen (1991) discusses the example of Etosha Pan, a mudflat in northern Namibia which is one of the

a. Floodplain of the Ob river, West Siberia
(photo W. Bleuten)

d. Lake Titicaca, Peru
(photo M. Wassen)

b. Mangroves on the Venezuelan coast
(photo H. Olde Venterink)

e. Rice paddies on Java, Indonesia
(photo P. Schot)

c. Okavango inland river delta, Botswana
(photo P. Schot)

f. Tidal wetlands in the Wadden sea, The Netherlands
(photo M. Wassen)

Figure 4.1 Examples of different wetlands of the world

driest countries in the world. Etosha is famous for its abundant wildlife which is attracted by the presence of numerous shallow waterholes. Intervals between periods when excess water is present can be very long (of the order of several years) making the Pan look like dryland rather than wetland to the inexperienced viewer. However, Etosha Pan can be accorded wetland status both on the basis of a temporary presence of shallow water (using the international Ramsar definition) as well as on the basis of soil development and biota (using the widely accepted scientific definition of the US Fish and Wildlife Service; see below).

Therefore the boundaries of wetlands cannot be always determined by the presence of water at any one time. As a result, in any definition the upper and lower limits of wetland excursion are arbitrary boundaries.

Precise wetland definitions are needed for two distinct interest groups: (1) wetland scientists and (2) wetland managers and regulators (Mitsch and Gosselink, 1993). The scientist is interested in a flexible yet rigorous definition that facilitates classification, inventory and research. The wetland manager is concerned with laws or regulations designed to prevent or control wetland modification and thus needs clear, legally binding definitions.

One of the most widely accepted scientific definitions is that of Cowardin *et al.* (1979) which was adopted by the US Fish and Wildlife Service (USFWS). It states that a wetland is:

> Land where an excess of water is the dominant factor determining the nature of soil development and the types of animals and plant communities living at the soil surface. It spans a continuum of environments where terrestrial and aquatic systems intergrade.

The USFWS definition has its main utility in scientific studies and inventories and has been more difficult to apply to the management and regulation of wetlands. Nevertheless, it has been accepted as the official definition of wetlands in India and has been used in proposed wetland legislation by some states in the USA. However, the US government regulatory definition of wetlands is based on regulations by the US Army Corps of Engineers. Since 1989 the term 'jurisdictional wetland' has been used for legally defined wetlands in the USA to facilitate distinction

from wetland delineation by means of other definitions (Mitsch and Gosselink, 1993).

Thus, there are many definitions of wetlands. No single definition will accommodate the wishes of all those concerned with wetlands and therefore different definitions will be found throughout the literature dependent on author, interest group, region, etc. In this chapter the USFWS definition will be adopted.

Classification

To deal with wetlands on a regional scale, wetland scientists and managers have found it necessary (1) to define the different types of wetlands that exist and (2) to determine their extent and distribution, activities that are called *wetland classification* and *wetland inventory* (Mitsch and Gosselink, 1993). A primary goal of wetland classification is to 'impose boundaries on natural ecosystems for the purposes of inventory, evaluation, and management' (Cowardin *et al.*, 1979). These authors identified four major objectives of a classification system:

1 To describe ecological units that have certain homogeneous natural attributes.
2 To arrange these units in a system that will facilitate decisions about resource management.
3 To identify classification units for inventory and mapping.
4 To provide uniformity in concepts and terminology.

As with the definition of wetlands, no universally agreed classification of wetlands exists. This is related to the different backgrounds, perceptions and goals of the users of wetland classification systems.

Cowardin *et al.* (1979) elaborated a categorisation for the USFWS structured on attributes of wetlands and not on the particular requirements of users (see Figures 4.2 and 4.3). Because wetlands were found to be continuous with deepwater ecosystems (i.e. deeper than 6 metres at low water) both categories were addressed in this classification. It is thus a comprehensive classification of all continental aquatic and semi-aquatic ecosystems (Mitsch and Gosselink, 1993).

On the highest level of distinction, five major ecological *systems* are recognised based on geomorphological, hydrological and biological characteristics:

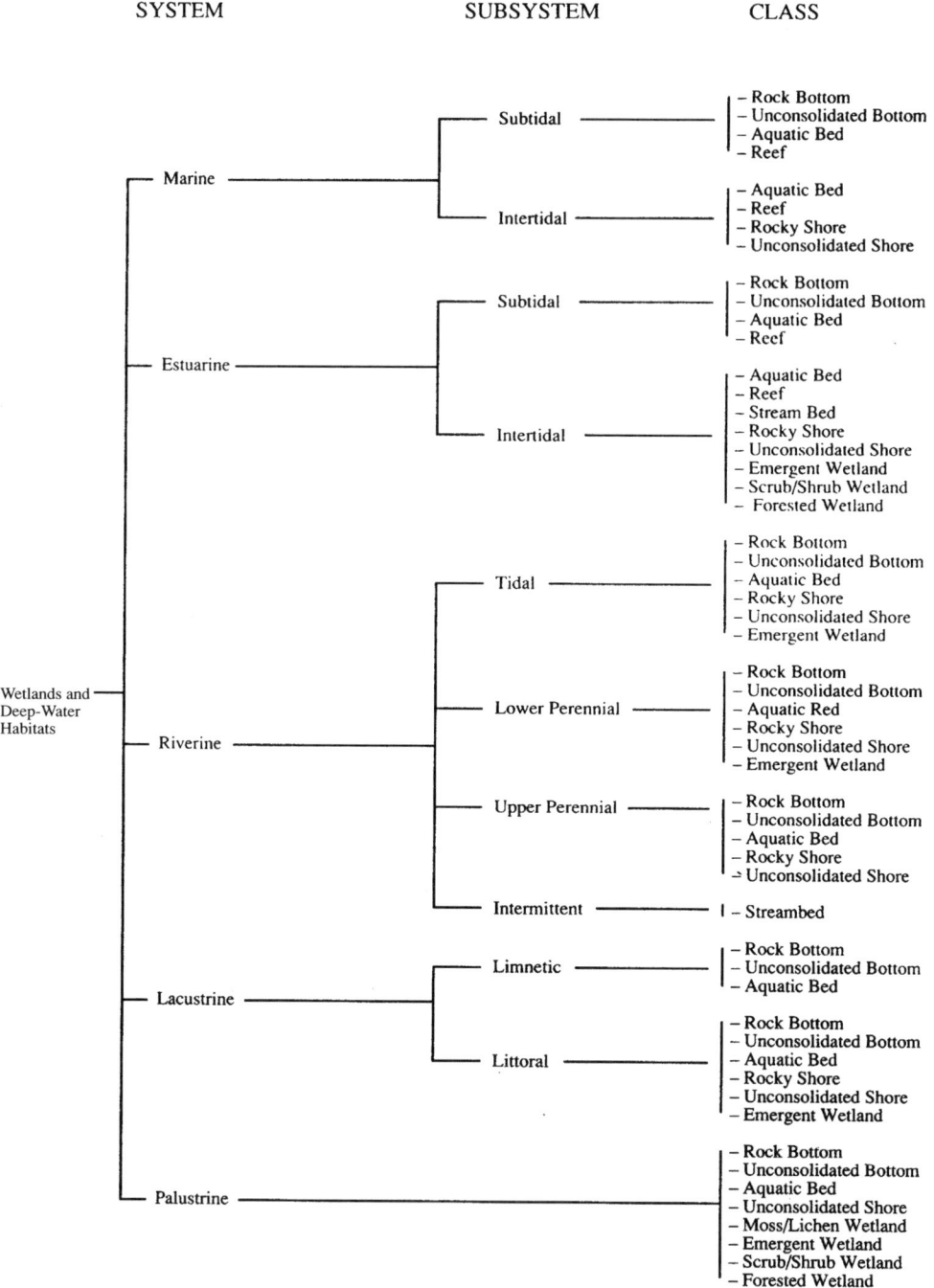

Figure 4.2 The hierarchical wetland classification scheme of Cowardin *et al.* (1979)

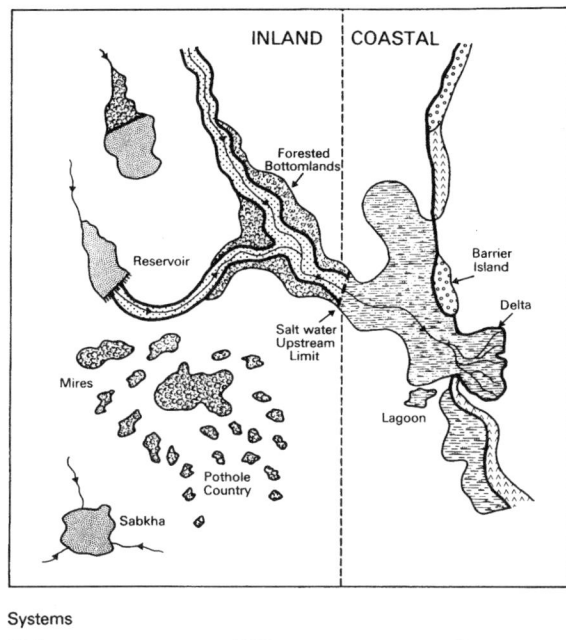

INLAND | COASTAL

Forested
Bottomlands

Reservoir

Barrier
Island

Delta

Salt water
Upstream
Limit

Mires

Lagoon

Pothole
Country

Sabkha

Systems

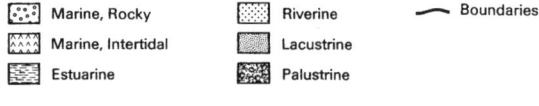

Marine, Rocky   Riverine   —— Boundaries

Marine, Intertidal   Lacustrine

Estuarine   Palustrine

Figure 4.3 Diagrammatic sketch of wetland types
Source: Williams, 1991, after Tiner, 1984

marine, estuarine, riverine, lacustrine and palustrine wetlands. The first four systems include wetlands and deepwater habitats, the palustrine system includes only wetland habitats.

1 *Marine* wetlands pertain to open ocean overlying the continental shelf and its associated high-energy coastal line. They include intertidal areas such as beaches, open shores, rocky shores, lagoons and some coral reefs. Open shores are not subject to the influence of river water or lagoon ecosystems. Coastal energy from waves and currents is low and they may include mangroves and mudflats.

2 The *estuarine* system pertains to deepwater tidal habitats and adjacent tidal wetlands that are usually semi-enclosed by land but have open, partially obstructed, or sporadic access to the ocean and in which ocean water is at least occasionally diluted by freshwater runoff from the land. The estuarine system includes coastal wetlands like salt and brackish tidal marshes, intertidal flats, mangrove swamps, bays,

sounds and coastal rivers. There is a daily tidal cycle whilst salinity varies between the fresh water of the outflowing river and the saline waters of the sea. In addition, there may be a distinct seasonal cycle of increased freshwater river discharge due to heavy rainfall followed by saline intrusion during the dry season.

3 The *riverine* system is restricted to wetland and deepwater habitats contained within a channel. However, the following channel habitats are excluded: (1) wetlands dominated by trees, shrubs, persistent emergents, emergent mosses or lichens, and (2) habitats with water containing ocean-derived salts in excess of 0.5 parts per thousand.

4 *Lacustrine* systems contain wetlands and deepwater habitats with all of the following characteristics: (1) situated in a topographic depression or a dammed river channel; (2) lacking trees, shrubs, persistent emergents, emergent mosses or lichens with greater than 30 per cent areal coverage; and (3) total area exceeds 8 ha (20 acres). Similar wetland and deep water habitats totalling less than 8 ha are also included when an active wave-formed or bedrock shoreline feature makes up all or part of the boundary, or when the depth in the deepest part of the basin exceeds 2 m (6.6 feet) at low water.

5 The *palustrine* system includes all non-tidal wetlands dominated by trees, shrubs, persistent emergents, emergent mosses or lichens, and all such wetlands that occur in tidal areas where salinity stemming from ocean-derived slats is below 0.5 parts per thousand. It also includes wetlands lacking such vegetation but with all of the following characteristics: (1) area less than 8 ha; (2) lack of active wave-formed or bedrock shoreline features; (3) water depth in the deepest part of the basin of less than 2 m at low water; and (4) salinity stemming from ocean-derived salts of less than 0.5 parts per thousand. The palustrine system includes the majority of inland marshes, bogs, swamps, floodplains, etc. Palustrine wetlands occur in continental interiors and consist largely of freshwater systems, but also include salt and brackish marshes of arid and semi-arid regions (Gopal *et al.*, 1990).

On the next level the systems are divided into several *subsystems*. The marine and estuarine systems are subdivided on the basis of the substrate being continuously submerged (subtidal) or exposed and flooded by tides (intertidal). The riverine system is subdived on the basis of tidal influence (tidal), low or high gradient (low and high perennial) and the absence of flow for part of the year (intermittent). The lacustrine system is divided into limnetic (deepwater) and littoral (shore habitats to a depth of 2 m below low water) subsystems. The palustrine system has no subsystems. The subsystems are further divided into *classes* on the basis of substrate or dominant vegetation. Further descriptions of the wetlands are possible through the use of *subclasses* (persistent or nonpersistent), *dominance types* (plant or animal species) and *modifiers* (water regime, salinity, pH, soil).

Many more classification systems are possible based on vegetative life forms, hydrologic regime, environmental forcing functions, wetland function and value, etc. The adoption of a particular classification will depend on one's personal goal and interest, taking into account the classification system's scope and limitations.

World distribution

Estimates of the percentage of the Earth's land surface occupied by wetlands lie in the order of 3 to 8 per cent (Söderlund and Svensson, 1976; Maltby, 1988; Williams, 1991; Lugo *et al.*, 1990). Exact determination of the total wetland area is hampered by the lack of a universally applied definition of wetlands and a lack of data, especially in the developing world. Wetlands are quite evenly distributed over temperate and tropic/subtropic regions, respectively 44 and 56 per cent. The humid areas of both tropics and subtropics contain by far the largest proportion of this, some 70 per cent, while the remaining 30 per cent occur in arid and semi-arid regions (OECD, 1996). Williams (1991) estimates that inland wetlands occupy about three-quarters, and coastal wetlands one-quarter, of the world's wetlands, the palustrine and estuarine wetlands accounting for the majority of them. Figures 4.4 and 4.5 show the global distribution of organic wetlands (mires) and coastal wetlands.

FUNCTIONS, VALUES AND BENEFITS

Wetlands perform a large number of functions that are valued by society. Functions, values and benefits of wetlands are thus interrelated and generally no clear distinction can be made between them. An overview is presented below, following the approach of Williams (1991), who employed four broad categories of functions of wetlands to summarise values and benefits to society: physical/hydrological, chemical, biological and socio-economic functions.

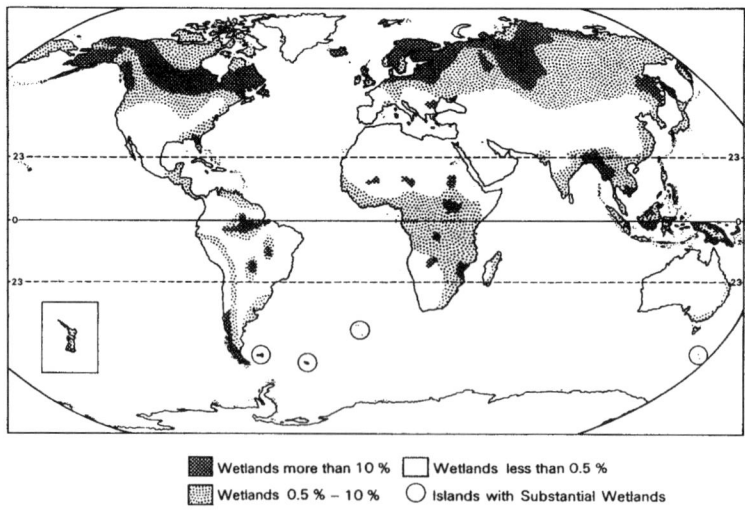

Figure 4.4 Global distribution of mires
Source: after Gore, 1983, from Williams, 1991

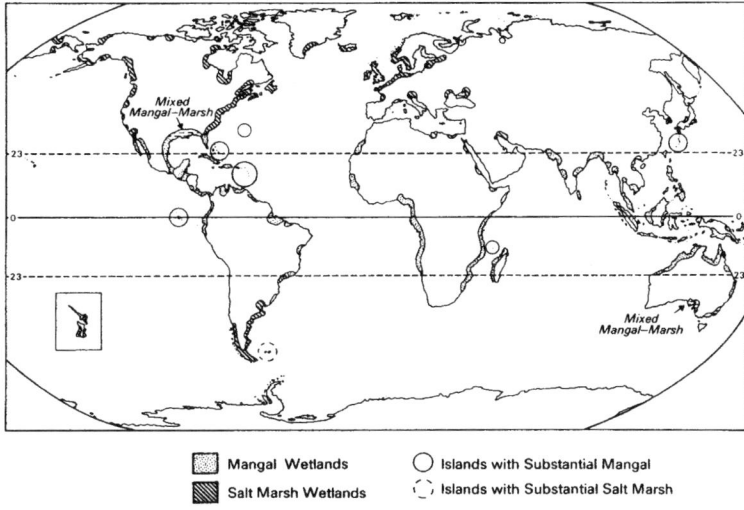

Figure 4.5 Global distribution of salt marsh and mangrove
Source: after Chapman, 1977, from Williams, 1991

Physical/hydrological functions

1 Flood attenuation

Wetlands can quickly store runoff water which is subsequently released at a much slower pace than that at which it was originally stored. In this way wetlands reduce peak river flows following extreme precipitation events, thus preventing flooding in lowland areas and protecting human occupation. This capacity of wetlands has long been recognised in land drainage engineering and river regulation works, where temporary spillover areas in water meadows and pastures are provided to reduce water levels during peak discharges in order to protect urban areas further downstream from flooding (Williams, 1991). In the case of the Charles river near Boston, Massachusetts, the value of this natural mitigation characteristic has been quantified. The US Corps of Engineers considered four alternative flood alleviation schemes: storage reservoir construction, extensive embankments, major flood protection works and the purchase of 3,408 ha of wetlands. An analysis of a major downpour after 5 inches (13 cm) of rain revealed a peak discharge from the 477 km² of the Charles river (with extensive wetlands) of only 91 m³/s and took a month to dissipate, whereas discharge from the adjacent Blackstone catchment (which had no wetland) peaked at 480 m³/s and took a week to run off (US Corps of Engineers, 1972; Carter *et al.*, 1979; Sather and Smith, 1984). Wetlands were the key to the different flood performances.

Another advantage of the slow pace at which water is released from the wetland after rainfall has ceased is that it ensures that river water levels drop only gradually, keeping them navigable for longer periods and so reducing the need for infrastructural works such as weirs.

2 Coastal protection

Erosion by waves during storms may cause shorelines to retreat, resulting in loss of land and property. Coastal marshes absorb wave energy and protect coasts from erosion. More than 50 per cent of wave energy is dissipated within the first 2.5 m, 80 per cent at 10 m, while it is virtually eliminated at 30 m (Knutson, 1988). Mangrove swamps, with their extensive large root system, provide excellent protection by dissipating erosive water forces. The process is self-reinforcing as mangroves trap sediment on which new mangroves can form, thus enlarging the protection zone. In Bangladesh the government is encouraging the replanting of mangroves to protect embankments in over 65,000 ha (Wijkman and Timberlake, 1984; Maltby, 1985).

3 Recharging and discharging aquifers

Groundwater and surface water in wetland areas are usually strongly interdependent, with the groundwater table under the land surface being continuous with the water level in pools, ponds and lakes. Depending on the differences in groundwater and surface water levels, wetlands may perform discharge or recharge functions for groundwater aquifers. Throughflow may also occur, with discharge from the aquifer on one side of the wetland and recharge on another side (Winter, 1989; Schot *et al.*, 1988; Schot, 1991). Groundwater discharge into wetlands is usually more stable in time than the inflow by surface water, which may be of considerable importance in arid and semi-arid regions providing drinking water for humans and animals and soil moisture for the growth of plants used for food supply. When surface water levels are high, for example, after flooding events, wetlands may recharge aquifers through infiltration. People at great distances downstream of the wetland area may benefit from this process by abstraction of groundwater for drinking water or agricultural purposes from the aquifers recharged in the wetland.

4 Sediment trapping

When fast flowing water reaches wetlands, its velocity drops and sediments settle out and form deposits on the water bottom. In estuaries, where fresh and salt water meet, flocculation of suspended clay particles can occur (Boto and Patrick, 1979). These sedimentation and flocculation processes may lead to improvements in water quality as the precipitating clay particles and dissolved organic matter absorb nutrients, heavy metals and organic micro-pollutants. Thus wetlands perform a function in water quality improvement. Furthermore, the clearer water leaving the wetland enhances the performance of sluices and channels and the lifespan of dams as sedimentation downstream is reduced.

Chemical functions

1 Water quality and wastewater treatment

Water loaded with nutrients from sewerage works or from agricultural practices may cause rapid growth of undesirable aquatic plants and algae (eutrophication). The decay of these plants and algae may cause oxygen shortages in slow-moving water as a result of oxygen used for the decomposition of organic carbon. This in turn may lead to the death of fish and other aquatic life and affect drinking water quality and recreational activities. Wetlands have the capacity to reduce nutrient concentrations in water passing through them, amongst others through uptake by plants, chemical processes like absorption, precipitation and reduction, and settling in anaerobic sediments. In a similar manner, other forms of pollution, like pesticides, heavy metals and organic mico-pollutants may be removed from the water as it moves through a wetland area. Boto and Patrick (1979) reported reductions of coliform bacteria from faeces (by 86 per cent), biological oxygen demand (80 per cent), chemical oxygen demand (44 per cent), turbidity (43 per cent), suspended solids (29 per cent) and total phosphorus (13 per cent) from a Wisconsin marsh charged with domestic waste. Suitability for pollutant removal, however, may vary between wetlands. Wetlands with a low nutrient status and/or a short flood period may be less suitable for wastewater treatment as they become susceptible to invasion by other species that can utilise those resources, but more research is needed before general conclusions can be drawn.

2 Atmospheric and climatic functions

Wetlands play an important role in the biogeochemical cycles of carbon, nitrogen, sulphur and oxygen (Lugo *et al.*, 1990). The waterlogged conditions favour bacteria which reduce the atmospheric oxidised forms CO_2, SO_4 and NO_3 to CH_4, S^{2-} and N_2 or NH_4. There are indications that temperate organic soil wetlands act as a net carbon sink through fixation of carbon dioxide (CO_2) by wetland plants through photosynthesis. Because decomposition processes are slowed by waterlogging, undisturbed wetland areas may accumulate organic plant remains (peat formation), thus acting as sink for the major cause of the greenhouse effect (Armentano, 1980). Armentano (1980) suggested that temperate organic soil wetlands absorbed 57–83 million tons/yr of carbon before widespread draining reduced their ability to do this. Since the end of the eighteenth century peat burning and draining might have added 590–700 million and 4,140–5,600 million tons of carbon to the atmosphere respectively. These are significant amounts on a world scale, but the effects of these changes on global warming and increases in sea level are difficult to calibrate.

Biological functions

1 Productivity
Wetlands are among the most productive natural ecosystems in the world, rivalled only by some tropical rain forests and the most intensively cultivated areas of land. In wetlands some 24 per cent of the world's net primary production is produced on only 6 per cent of the Earth's land surface (Leith, 1975). This productive capacity is related to the ample availability of moisture and of nutrients provided by repeated flooding. Many wetland plants are perennials and are nearly all leaf with little or no wood or thickened tissues. Therefore they are constant and efficient converters of solar energy to fix carbon (photosynthesis) and create biomass.

2 Habitats
The efficiency of wetlands in producing plant biomass forms the basis for several food webs. The greatest food value comes from the death of the plants on which organisms such as larvae, fungi, bacteria and protozoa thrive. These organisms constitute the aquatic food web of high yielding animals and fish such as salmon, crabs, shrimps and worms (Crow and MacDonald, 1979; de la Cruz, 1979; Murkin and Wrubleski, 1988). Larger fish, waterfowl and larger animals subsequently feed on these high yielding fish and animals making wetlands the habitat for a large variety of birds and animals. Many of these organisms use the wetlands for only part of the year, as spawning, nursing or breeding grounds or as a repository on their migration routes. The sight of migrating waterfowl landing, resting, feeding and taking off from wetland habitats was the first and probably most powerful cause of positive action of wetland preservation (Williams, 1991).

Socio-economic functions

1 Agricultural food production
Consumption of plants from undisturbed wetlands is relatively limited (e.g. fruits, cranberries, mushrooms, algae). The value of wetlands for agriculture lies primarily in their drained form, making them extremely productive agricultural lands as a result of the presence of organic soils rich in nutrients and the availability of water. They are used as pastures for livestock and for the growth of fodder and human food crops. More than half of the world's population is supported by rice, a domesticated tropical reedswamp grass. A large portion of its production is based on paddy culture in which the young seedlings are planted into shallow water, which as such represent a form of agricultural wetlands. Other products grown on drained wetland soil include maize, soybeans, grain feeds, cotton, peppers, onions and carrots (Etherington, 1983).

2 Fish, fowl and fauna
Wetlands are rich in fish and other animal life. Some 60–65 per cent of the world's fish and shellfish are probably attributable to wetlands, mainly coastal and estuarine wetlands (Williams, 1991). The mangroves in South and South-East Asia and in other tropical regions provide excellent opportunities for shrimp fishing and farming. Inland wetlands provide fish mainly for local markets. In arid zones, thousands of relatively small artificial wetlands have been formed over the past decades as a result of the construction of small dams in wadis and river beds; Schot and Blankwaardt (1994) reported a number of 'retenues' (artificial reservoirs) in Burkina Faso (West Africa) alone of about 1,000. These dams store the rapid runoff from rain falling during the (short) rainy seasons or during rare storm events. Temporary artificial wetlands are formed which provide fish, with much needed protein for local people. Water from the wetland can be used for small-scale irrigation and to provide drinking places for livestock. During recent decades natural wetlands are increasingly used for aquaculture, for example, shrimp and salmon farming, oyster culture, etc. Combined culture systems have evolved in South-East Asia, such as fish cultivation combined with rice fields (Ruddle, 1980). Apart from fisheries, a number of valuable mammals are harvested from wetlands, for example, muskrat, alligators, mink. Furthermore, wetland fowl are hunted for food supply and recreational purposes.

3 Energy

Peat has been excavated and dried for centuries to provide fuel for heating, especially in Northern Europe and Russia. Nowadays large machines are used for large-scale industrial peat harvesting for fuel, although in many regions the application of peat in horticulture has surpassed that for fuel. Peat production is especially large in the northern hemisphere, notably in Russia, Ireland and Finland (Bord Na Móna, 1985), although during recent years production has also started in tropical regions like Kalimantan. Energy from wetlands is also obtained from wood gathering and logging, which is used for direct burning or charcoal production. Furthermore, a large number of wetland areas have been converted to artificial reservoirs by the construction of dams for hydroelectric power generation. Finally, wetlands are used for small-scale power generation using methane from peat deposits, solar ponds and heat pumps in shallow lakes (Löffler, 1990).

4 Building materials

Forested wetlands provide timber for the commercial building industry, as well as for small-scale local construction purposes. In the southern USA, hardwood bottomland forests (cypress swamps) were felled at a rate of 175,000 ha/yr between 1960 and 1975, after which the land was converted to soybean fields (Sternitzke, 1976; Turner et al., 1981). The mangrove forests in South-East Asia provide construction wood for poles and scaffolding. Wetlands also provide building materials such as gravel, sand and clay, which in some areas are constantly being replenished by river floods or tidal streams.

5 Non-consumptive benefits

Apart from the wealth of functions that can be measured to some degree in terms of monetary units, there are a number of benefits of wetlands that are difficult to quantify. These include recreational, educational, aesthetic, archaeological and gene pool benefits. Natural wetlands generally have a high diversity in flora and fauna, and thus rich gene pools. This value is hard to quantify although its potential may be enormous. The medical potential of wetland plants is an almost untouched field. Similarly, the possible use of invertebrates for protein production is only at its beginning and each year economically valuable species are detected that are useful for agri-cultural application (Löffler, 1990). Wetlands also provide a large potential from an archaeological point of view. Waterlogged conditions favour the preservation of fragile organic materials which are usually lost from dryland sites. Furthermore, the archaeological evidence is often of very good quality as sites are preserved when flooding and subsequent peat growth preserves them from disturbance by more recent human occupation (Coles, 1991).

ADVERSE EFFECTS OF WETLAND ALTERATION

Wetlands are dynamic ecosystems changing continually in their natural setting. The movement of water is the primary process bringing about these changes. Tidal currents constantly change water levels and sediment patterns in coastal zones and estuaries. Flooding by rivers changes the area submerged by water and deposits gravel, sand and clay in distinct zones on the floodplain. Even at relatively constant discharge rates, river flow gradually changes the position of the stream channel and meanders through continuous erosion and sedimentation. Plant growth and the presence of fish, fowl and fauna follow the changes in morphology, hydrodynamics and sedimentation, each in their own specific manner, so creating a varied and ever-changing landscape.

Thus wetland alterations are a natural phenomenon. In addition to these natural alterations, human activities have brought about changes in wetland ecosystems which are significant on a world scale. Human alterations pertain mostly to drainage for agricultural purposes and to construction works for flood protection, industrialisation and urbanisation. These activities have brought a wealth of benefits to humans, such as increased food production to sustain an ever-growing number of people, as well as increased incomes following economic growth.

However, during recent decades there has been a growing awareness that wetland alterations brought about by human activities have also had a number of negative impacts for society. An overview of a number of important adverse effects of wetland alteration will be given below. Insight into the nature and causes of these adverse effects will provide the basic knowledge necessary for future wetland management directed towards sustainable use of wetland functions and resources.

Agricultural impacts

Drainage

Drainage for increased agricultural production is the principal cause of wetland loss world-wide; by 1985 it was estimated that 56–65 per cent of the available wetland had been drained for intensive agriculture in North America and Europe. The figures for tropical and subtropical regions were 27 per cent for Asia, 6 per cent for South America and 2 per cent for Africa (OECD, 1996).

Wetland drainage for intensive agriculture has an enormous impact on the geography of the area. When waterlogged soils become aerated, oxygen from the atmosphere will cause oxidation of the organic compounds present. This process of mineralisation will result in nutrients present in the organic matter becoming mobile. They can dissolve in rainwater percolating through the wetland soil which transports the nutrients and other compounds from mineralisation towards draining channels. Thus, runoff water in draining channels and in brooks and rivers further downstream becomes loaded with nutrients which may lead to eutrophication and the associated problem of oxygen depletion, resulting in large-scale death of water organisms.

Through dehydration and mineralisation, peatsoils become compacted and lose their water-absorbing properties. This process is largely irreversible and impairs the wetland flood control function. Moreover, compaction leads to erosion and land subsidence which may be in the order of several metres (Heathwaite et al., 1993). This land subsidence may have been active for several centuries in some areas, especially in Western Europe. To keep the wetland suitable for agricultural practices a continuous cycle has to be maintained, involving lowering of the water table followed by mineralisation and land subsidence, which in turn has to be compensated by additional lowering of the water table etc.

Land subsidence may also result from peat fires to which wetlands may become susceptible after drainage (Maltby et al., 1990).

Drainage may also result in changes in groundwater flow patterns. In the Netherlands, drainage of agricultural lands has led to lowering of water tables in adjacent wetlands managed as nature reserves. To prevent water levels in the nature reserves from becoming too low, external polluted river water is supplied, which has induced changes in vegetation and loss of biodiversity. Furthermore, seepage areas have been turned into infiltration areas following changes in groundwater flow directions. A shift in root zone water quality may result when the water source changes from exfiltrating groundwater (usually non-acidic and relatively high in dissolved solids) to infiltration of precipitation water which is relatively acid and low in dissolved solids. Again, this may induce marked changes in vegetation development (Schot et al., 1988; Schot, 1991; Wassen et al., 1989; see Box 4.2).

Land-use intensification associated with drainage (such as fertiliser application and the construction of houses and roads) leads to pollution by fertilisers, pesticides, heavy metals and oil derivatives. Habitat is lost when wetlands are drained, especially that of migratory waterfowl.

Irrigation

The abstraction of water from wetlands or rivers feeding wetlands for irrigated agriculture may induce a number of problems. In Central Asia irrigation water taken from the Amudarya and Syr-darya rivers has dramatically reduced inflow to the Aral Sea, which has shrunk by 50 per cent in the last 30 years following a drop in water level of 15 m. The salinity of the inland sea has at least doubled through evaporation and the inflow of saline drainage water from the irrigation systems (Hollis, 1978). Fish catches have collapsed, the frequency of dust storms has increased, and the clouds have been charged with salt crystals and residues of pesticides which are deposited over large areas, seriously affecting human health (WWF, 1989). Irrigation may induce salinisation of productive soils by evaporation from the waterlogged soils when inadequate drainage systems accompany the irrigation works, leaving the soil unusable for agricultural production. Ponding by irrigation water may lead to the spread of waterborne diseases like malaria into areas formerly unaffected.

Fisheries and aquaculture

Turner and Boesch (1988) established positive relations between shrimp harvests and the presence of coastal wetlands in locations which include Australia, Malaysia, the Philippines and the Gulf of Mexico. They concluded that shrimp and many other fisheries were directly limited by the quantity and quality of wetland habitat. The large-scale transformation of coastal wetlands into oil palm plantations, rice paddies and aquaculture ponds in the tropics is therefore likely to have disrupted essential food chains. The cutting of mangroves has significantly reduced natural fish spawning areas and fish populations. In fish ponds exotic fish compete with indigenous species, and the use of fertilisers, chemicals and antibiotics leads to water pollution (OECD, 1996).

Fisheries in inland wetland areas are an important renewable resource, especially in developing countries. They are operated on a subsistence basis and are intricately interwoven into the local economy and social structure of the community. Overexploitation of the fish population by means of a commercial harvesting and marketing programme can have immense sociological implications, disrupting local socio-economic benefits and stability (Schrader, 1991; Drijver and van Wetten, 1992).

Logging

Logging for timber, firewood and agricultural clearing has caused increased soil erosion, coastal abrasion and sediment transport by rivers. Forest industry leads to water pollution by organic substances, especially from paper and pulpmills, and toxic wood preservatives.

Civil construction

Dams, dikes, embankments

Dams in river channels are built to create reservoirs for hydropower generation. They act as barriers for migratory fish, and the reservoirs created may lead to an increase of waterborne diseases in the surrounding area. Roggeri (1985) reported an annual rate of increase in the number of malaria cases around the man-made lakes of the Tana river in Kenya of between 14 and 40 per cent, and an increase for bilharzia in the Lake Volta region in Ghana from 1–5 per cent before construction of the dam to 80–100 per cent afterwards.

The stream velocity of river water entering the reservoir is lowered drastically, allowing suspended materials to settle out on the reservoir floor. In the long run this will silt up the reservoir, making it useless for power generation. Water leaving the reservoir on the downstream side is stripped of its fertile sediments. In addition, the reservoir strongly smoothes variation in downstream river discharge, often virtually removing seasonal fluctuations. Flooding of the Nile river plain in Egypt has thus been eliminated as a result of the construction of the Assouan dam, depleting the agricultural fields alongside the riverbanks of their yearly renewal by the fertile sediments that sustained agriculture along the Nile for many centuries. Artificial fertilisers are now needed to keep up productivity, depriving local farmers of part of their income, while excess fertilisers lead to pollution of the river downstream. Furthermore, the enlargement of the Nile delta in the Mediterranean Sea by the settling out of suspended sediment at the river mouth has been brought to a halt. Nowadays the delta shoreline is retreating as a result of wave action and coastal abrasion.

Dikes and embankments along rivers are constructed for flood protection. With the floodplain narrowed and the area for floodwater storage decreased, peak water levels actually become higher during storm events, while higher flow velocities lead to increased erosion. Belt (1975) showed how confining and shortening the Mississippi (by removing meanders) has made the river unstable leading to increasing peak water levels. For the same discharges, river stages were found to be up to 3 m higher after flood construction works, compared to those measured before the construction works. This results from loss of water storage capacity in wetlands on the river plain following embankment. In addition, as river water is disposed of relatively fast, water levels drop quickly after precipitation has ceased, leading to difficulties for river boats as water levels become too low for navigation. In addition, stream channels may become silted leading to enormous costs for dredging.

In the Netherlands a number of projects are now under consideration which concern the reduction of peak floods in stream channels by reconnecting parts of floodplains which had become isolated as a result of dikes and embankments. The increase in active floodplain area leads to a higher storm water storage capacity and thus to lower peak water levels. In addition, wetland habitat is restored, in terms both of area and of restoration of dynamic ecosystem processes such as sedimentation, erosion, meandering, seasonal flooding, etc. Through these processes, abiotic ecosystem gradients in soil, water level, nutrients, shade and the like are restored, both on a spatial and on a temporal scale. It is expected (and in some cases it has already been observed) that the restoration of these gradients will increase biodiversity in the formerly isolated areas to a certain degree of its former natural level.

Port industrialisation

Port expansion is responsible for considerable wetland losses along coastlines and estuaries, resulting in habitat loss for waterfowl and wetland fauna and impairment of the wetland's coastal protection function. Wetland areas are reclaimed for industrial, transshipment and storage purposes, while estuaries and rivers are deepened by dredging to allow navigation by larger ships. Dredging may mobilise toxic muds in areas already polluted. Newly established refineries, power generation plants, fish processing factories, steel factories, aluminium smelters and chemical industries lead to increased chemical, thermal or nuclear water pollution. Changes to the hydraulic regime may lead to bank erosion and salt water encroachment. Electric power lines, and noise generated by industrial activities, roads and rail disturb migratory and nesting birds.

Urbanisation

Major causes of wetland conversion by urbanisation are population growth, port expansion and tourism development schemes, especially along scenic beach coastlines. During recent decades, cities like New York, Singapore and Hong Kong have expanded to a large extent at the cost of wetland areas. In the case of Hong Kong a new airport-seaport is now being constructed just offshore; Amsterdam is planning a new city area in the bordering Lake IJ.

Urbanisation generally leads to increased pollution (sewerage, roads, waste, etc.) and habitat loss and fragmentation, as well as disturbance through noise and through people entering sensitive ecosystems. Fragmentation of wildlife habitats by roads and building sites may lead to blockage of ecological corridors like migration paths of wetland fauna and fish. When spawning areas cannot be reached any more, reproduction is impeded. Traffic may lead directly to the death of large numbers of animals trying to cross roads that divide wetland habitats.

The increase in hard surface area for roads, pavements and houses results in decreased groundwater recharge and increased runoff. Water abstraction for drinking water, washing and garden sprinkling may lead to the significant lowering of water levels in adjacent wetland areas. In the Netherlands, deterioration of fen ecosystems has been attributed to groundwater abstraction for urban drinking water supply (Grootjans, 1985; Wassen et al., 1990).

ENVIRONMENTAL MANAGEMENT

Definition

People have used wetlands for thousands of years for hunting, wood gathering, grazing by livestock, etc. The benefits were acquired fairly directly, without the need for preparatory activities. These activities may be categorised simply as 'wetland use'. At a certain stage in time man started deliberately to change characteristics of wetlands to make them suited for other purposes, for example, draining for agricultural purposes or construction of dikes for flood protection. In these cases the term 'management' may be applied, since a certain control over elements of wetlands was exerted (the word management implies some sort of control over something). The control measures 'draining' and 'construction of dikes' are directed at only one wetland element (groundwater level and river water level, respectively) and therefore may be referred to as 'soil water management' and 'flood management' and the like. The term 'wetland management' is less apposite to these more sectoral control measures.

Wetland management in the context of this chapter is used to indicate the integrated and simultaneous control over multiple features of wetlands, such as river water level, soil water level, fish stock, wildlife habitat, etc. 'Environmental wetland management' is used to indicate efforts to minimise the cumulative adverse effects of wetland use and alteration on society as a whole. It aims at making optimal choices in wetland use and conservation, such as draining parts of wetlands for agricultural production, but at the same time maintaining adequate water-storing capacities and nesting/spawning areas, providing access to hunters and tourists whilst preserving undisturbed habitats and sufficient number of waterfowl, etc.

Basic principles of wetland management

A number of basic principles may serve as starting-points for wetland management. The Ramsar Convention formulated the *Principle of Wise Use of Wetlands* as 'their sustainable utilisation for the benefit of humankind in a way compatible with the maintenance of the natural properties of the ecosystem'. The *Principle of Interdependence* implies that wetlands cannot be considered independently or as an isolated ecosystem. They should be managed on a wider ecosystem basis which may cross political and social barriers and different sectors. The *Precautionary Principle* means that it is preferable to avoid activities with an assumed negative environmental impact, even when these impacts have not been fully scientifically proved. The *No Net Loss Principle* is derived from North America, and states that if a wetland area is lost by a development, the developer should make the best efforts to create or restore the functions of an equivalent area of wetland (compensation principle). However, the created or restored wetland may have different values and functions than the area destroyed, and therefore the preservation of existing wetlands is usually preferable.

Approaches in wetland management

The practical realisation of environmental wetland management must come from the recognition that wetlands are complex ecosystems that are highly dependent on areas upstream and downstream in the watershed. This makes it necessary to view wetlands on the basis of a *systems approach* to the hydrological cycle within whole catchments (Hollis, 1990). Hydrologists and geohydrologists can provide insight into the functioning of natural and human induced surface water and groundwater systems and the mutual relationships between them. Spatial and temporal patterns of the flow of water and solutes are analysed as a function of topography, geology and soils present in the catchment.

The movement of groundwater and surface water forms the main vehicle for the transport of energy and matter through the landscape, directing the development of soils and nourishing vegetation, fish, fowl and fauna. In order to understand the critical factors of the ecosystem, hydrologists and ecologists compare the spatial patterns in soils, water flow and water quality with those of biotic systems within the wetland. Such a *landscape-ecological approach*, in which abiotic and biotic systems in the landscape are analysed in an integrated manner, is essential to an understanding of wetland ecosystems. It addresses the interdependency of wetland functioning and other processes in the catchment.

Besides the abiotic and biotic systems of water, soil, vegetation, fish, fowl and fauna, the human factor is of prime importance in wetland management. Social and economic issues are the main causes underlying wetland loss. Therefore they need particular attention in wetland policies and management. Special attention should be paid to local communities, which are most affected by wetland loss and will benefit from improved wetland management. They constitute a valuable resource of knowledge, skills and labour to maintain a net flow of wetland benefits. A *participative approach* aimed at active participation of local communities is essential to wetland management directed at sustainable use.

The concept of integrated environmental management

During recent years the insight into the interdependency of abiotic, biotic and social systems has led to the concept of integrated management. Integrated environmental management implies consideration of all elements affecting environmental quality, including physical, chemical, biological, social, cultural, legal, economical, infrastructural and communication aspects. For wetlands, integrated management must encompass the whole area physically linked to the wetland by the flow of water, that is the river basin or the coastal zone.

Integrated river basin management (IRBM) tries to make the best use of resources within a river basin, while satisfying the needs of different groups dependent upon the resources of the river and its floodplain or surrounding catchment area. IRBM normally requires the following (OECD, 1996; adapted from Dugan, 1990):

- An understanding of the *hydrological balance of the river basin*, including quantity and quality of surface, underground and coastal waters during conditions of drought, spate and average conditions.
- An assessment of the *values of the major ecosystems* in the river basin and the biophysical processes on which they depend.
- An inventory of *products and services* of the river basin and the minimum requirements for their sustainable use.
- An assessment of *short- and long-term environmental impacts of planned changes* to the system and of appropriate mitigatory measures.
- An analysis of the *socio-economic conditions* of the various community groups, including cultural and traditional behaviour, and the distribution of natural and economic resources.
- A strategic analysis of existing and future *water development projects*.

While the collection and analysis of the data are relatively straightforward, it is more difficult to allocate and manage the water resources equitably to maintain the critical hydrological balance. This requires political decisions and, sometimes, diplomatic negotiations with neighbouring countries over upstream and downstream water rights.

Integrated coastal zone management (ICZM) relies upon an understanding of ecological and resource use constraints. ICMZ requires similar data gathering and analyses as listed for IRBM (understanding hydrology, ecosystem values, products and services, socio-economic conditions, impact of planned changes). In contrast, the boundaries of the coastal zone may not

be so obvious as those of a river basin, and must therefore be defined at the outset using the following (OECD, 1996; adapted from Dugan, 1990):

- *Biophysical boundaries*: geological formation, substrates, ecosystem processes, chemistry, hydrology and oceanography, species distribution.
- *Socio-economic boundaries*: areas in which resource extraction, pollution, overfishing, landfill, etc. have an impact.
- *Legal and administrative boundaries*, including national borders, jurisdiction over landward and seaward sides of the area.

Environmental wetland management in practice

Although a number of basic principles and general approaches have been established for environmental management of wetlands, it is impossible to draw up a detailed management plan that can be applied to all wetlands. This is because of the variation of natural and socio-economic conditions from one place to another. As a consequence various management options should be carefully studied for each wetland.

In practice *experimental management* will be the only instrument available to planners and project managers for identifying interventions that are both productive and environmentally sustainable over the long term (Roggeri, 1995). Through a learning-process approach, information and past experience gained from other wetlands of the same type may help to identify the essential processes and key factors that support the functioning of the wetland. Experience gained from current activities in the wetland under consideration complement the existing knowledge.

Management decisions should be taken in the light of the current values of the ecosystem, the impact of various proposed or observed interventions on these values and the distribution of the costs and benefits of these interventions. For this purpose, planners and project managers must be able to rely on detailed and complete information about the wetland under consideration. This is often unavailable, especially in developing countries. Therefore the first step in wetland management must be the elaboration of an information base.

In general, the main steps in the elaboration of management plans for wetlands on a national or regional scale may comprise the following (Roggeri, 1995):

1 *Building an information base*, including the compilation of a wetland inventory followed by a preliminary evaluation of benefits provided by each wetland, which enables the drawing up of a list of (potentially) valuable wetlands.

2 *Identification of action priorities*, including the identification of wetlands to be maintained and identification of protection priorities for key processes and factors. Protection of external key processes and factors may be the preservation of the hydrological regime of the river feeding the wetland. Protection of internal key factors may include maintaining sufficient area for floodwaters to spread for the purpose of protection of downstream areas.

3 *Situation and problem analysis* for wetlands and key processes and factors to be maintained. This includes inventory of ongoing and planned projects and assessment of their environmental impact, inventory of wetland resource use by local communities and private businesses and their environmental impact, and identification of the impact of national and regional development policies and strategies.

4 *Development of an action plan* aimed at adaptation of ongoing and planned activities threatening wetlands or key processes and factors to be maintained, as well as monitoring and implementation of a strategy for non-priority wetlands.

For a particular wetland site management planning may include the following steps (Roggeri, 1995):

1 *Drawing up a wetland management plan* including:
- delineation of the area to be managed (the wetland itself as well as key parts of the drainage basin);
- elaboration of a wetland environmental profile (wetland benefits and resources, socio-economic surveys, rapid environmental impact assessment of current and proposed land and water uses, etc.);
- selection of an appropriate management option (are there any legal provisions for protective measures or protective zones, are funds avail-able, is the option compatible with the wise use principle, etc.).

2 *Experimental management* by beginning with the development of wetland resources on a small scale. This is necessary as we still know little about the functioning of wetlands and the impact of interventions. By monitoring the impact of the actions undertaken, information is gathered initiating a learning process which will serve to orient (or reorient) wetland management activities. The learning process will often be slow, and therefore wetland management requires a long-term view. However, the slowness of the process and the reduced scale of initial activities facilitates the involvement of local communities. This has the advantage of making empirical wetland knowledge of these communities available to the wetland manager. Furthermore, participation of local people will facilitate rule enforcement.

3 *Monitoring, evaluation and adaptation of activities.* Information obtained through monitoring and evaluation allow for adaptation of the initial management plan. Such information gathered for a particular wetland may also be useful for the adaptation of management plans on a regional or national scale.

From the above it is clear that wetland management must be a continuous and dynamic process which essentially consists of responding to changes observed. These changes may be physical or biological, but just as well social, for example, changes in river bed, deterioration of grasslands, nomadic people starting to claim grazing rights for their cattle. Wetland management involves continuously covering the following cycle:

1 Identification of critical factors.
2 Developing a set of management practices.
3 Monitoring changes in the wetland.
4 Adjusting the management practices appropriately.

Environmental wetland management is always a response to human protests, be it a fisherman seeing his catches diminishing, a naturalist concerned with changes in biodiversity or people suffering the loss of goods and property following flooding. Therefore, a *participative approach* of all those concerned, especially local communities, is essential to the success of sustainable wetland management. If local people are not consulted in the process of developing management plans, they are unlikely to conform to regulations imposed by external parties. Conversely, when they participate from the outset in the process of establishing management plans, these plans will become more realistic and viable. Management plans accepted by local communities will be executed more easily in practice, thus making them sustainable.

In Boxes 4.1 and 4.2 case studies of wetland management are presented. Box 4.1 presents a project in the Inner Niger delta in Mali, aimed at the restoration of degraded wetland resources resulting from combined natural and human causes. It demonstrates a participative approach to establish management systems aimed at sustainable development, with positive economic and environmental impacts.

Box 4.2 presents an example of integrated wetland management in the Netherlands. A landscape-ecological approach provided insight into the relationship between water flows on a catchment scale and the deterioration of water quality and vegetation composition in wetland areas. Substantial adaptations in catchment water management are being implemented to restore and preserve fen ecosystems for biodiversity and recreational purposes.

POLICY GUIDELINES

Governmental actions

Gosselink and Maltby (1991) point out the disparity between the interests of private owners of wetlands and those of society as a whole. Wetland development is usually to the benefit of the owner, or in the interest of government officials who benefit from wetland development. Conservation is to the benefit of the public at large; communities downstream enjoying good water quality, fishermen making a living from wetland fish, even tourists on other continents enjoying the sight of migratory birds.

The solution of this discrepancy is not easy, although its recognition is an essential first step in conserving wetland benefits for society as a whole. It implies the involvement of governments, whose task it is to manage a nation's natural resources and to safeguard the interests of their populations. In order to conserve wetland benefits according to the Wise Use of Wetlands principle, governments should take the following action (OECD, 1996):

1 Formulate and implement national wetland policies and strategies, which should:
   - address *legislation* and government policies affecting wetlands;
   - strengthen wetland *institutions* and organisational arrangements;
   - increase the *knowledge and awareness* of wetlands and their values;
   - draw up a *national inventory of wetlands* and identify management priorities for each site;
   - identify actions to *address problems* at nationally significant sites.
2 Incorporate wetland conservation and sustainable use into sectoral policies, programmes and projects, by reviewing and adapting:
   - sectoral and economic *policy planning and programme planning* in sectors like agriculture, forestry, fisheries, water resources, flood control, water pollution control, energy, transport (notably ports and shipping), tourism and environment, and economic and fiscal policies related to all of these;
   - *land-use and site-specific planning*, emphasising integrated resource management along catchment or ecosystem boundaries rather than along political boundaries (IRBM/ICZM).

Environmental impact assessment

Environmental impact assessment (EIA) is a suitable instrument for protecting wetland resources and values in the planning process, both on the strategic level of policy and programme planning, and on the area-oriented level of land-use and site-specific planning. The importance of threatened wetlands to local people must be included in EIA, including cultural matters. EIA does not end with a report or statement but is a process including the development of alternatives, mitigation measures, monitoring and follow-up measures if necessary. Local communities and NGOs should be involved in this process as much as possible.

BOX 4.1   BOURGOU REGENERATION IN THE INNER NIGER DELTA, MALI

Prior to 1970 a complex socio-economic system existed in the Inner Niger Delta in Mali, involving extensive pastoralism, agriculture, fishing and commerce. One of the key elements in this system was an aquatic grass locally called 'bourgou' (*Echinochloa stagnina*), which depends on the annual floods of the Niger river for its survival. Bourgou was critical for the dry season grazing of livestock. It was also important for rice farmers, who used it as a protective shield against rice-eating fish. During the flood period bourgou was cut and sold for fodder, while seeds provided food for fishermen and the stalks were used to produce sugary syrup. Bourgoutières, the plains in which bourgou grows, acted as breeding and feeding grounds for fish, birds and other wildlife. The economic and ecological importance of the bourgoutières would be difficult to overestimate.

Seventeen years of drought between 1968 and 1985 changed everything. With the drought came a drop in level and duration of the flooding of the Niger river, affecting the growth of bourgou. Along with overgrazing and land clearance by cultivators, the drought resulted in the almost near-complete destruction of many bourgoutières in the late 1970s and early 1980s. This led to a plunge in fish yields and the decline of the pastoral system. Livestock either died or had to migrate south, but not before seriously degrading the land by overgrazing and trampling.

In 1982 the United Nations Sudano-Sahelian Office (UNSO) launched a project to regenerate these former grasslands and establish management systems. Meetings were held with local government administrators and community leaders to investigate interest in regenerating the bourgoutières and to determine the local land tenure and use situation. Nurseries were created to produce bourgou seedlings which were planted in the appropriate soils. Extensive research took place to select potential sites for regeneration of the floodplain vegetation and to determine methods of rehabilitating and managing the land surrounding the bourgoutières. In order to prevent overexploitation, pastoralists were introduced to the concepts of vegetation production, stocking rates and carrying capacity. Management committees were set up to ensure that stocking rates were respected, and labourers were hired for protection and maintenance duties, including replanting where plants had died.

A cost–benefit analysis showed that profits from dairy or fodder production clearly exceed the regeneration expenses. Moreover, bourgou has proven itself one of the most (if not *the* most) economic productive resources in the region, offering a far better return on investment than irrigated rice or recession culture. The health of the human population has improved through enhanced nutrition as a result of increased milk production from cows feeding on bourgou. A change in land tenure system, from free resource use towards usufruct system (only those who contribute to regeneration, maintenance and management should benefit) was essential to secure sustainable use of the regenerated bourgoutières. This shift in the land tenure system has not happened easily, however, as nomadic groups attempted to exploit the regenerated bourgoutières, just as they freely exploit the natural ones. In recent years the project has also worked with nomadic groups to regenerate bourgoutières in their traditional lands. The UNSO project has worked from the philosophy that sustainable development must include participation of the communities involved. This slowed progress to a certain extent, but real progress has nevertheless been made by creation of institutions that will continue to function well beyond the time when external inputs cease.

Source: UNSO, 1990; adopted from Drijver and van Wetten, 1992 and Roggeri, 1995

Economic environmental evaluation, which thoroughly evaluates all true costs and benefits associated with a planned project, must be part of an integral EIA. Many project plans neglect natural wetland benefits or consider only investment costs while neglecting yearly operating costs. The case of flood works in the lower Charles river near Boston in the USA showed that purchase of 17 natural wetlands in an area of 3,400 ha was more effective and less expensive than the construction of storage reservoirs or embankments (US Corps of Engineers, 1972; Carter *et al.*, 1979; Sather and Smith, 1984). Barbier *et al.* (1991) describe how the benefits of irrigated crops

in the upstream Kano river project Phase 1 were estimated at US$ 115/ha (in 1989) which would exceed those from traditional wetland benefits estimated at US$ 58/ha. When operating costs of the irrigation scheme were deducted, the net benefits were lowered to only US$ 2.5/ha. Even more realistic is an evaluation in terms of benefits per unit of water, as water is the scarce commodity. Such analysis showed a net benefit from irrigated crops in the order of US$ 0.0025 per million litres, compared to flood recession agriculture benefits of US$ 2.9 per million litres.

The total value of other wetland benefits may well exceed those of agriculture discussed above (Barbier

BOX 4.2   INTEGRATED WETLAND MANAGEMENT OF THE VECHT RIVER
           PLAIN (THE NETHERLANDS)

The Vecht river plain in the Netherlands comprises an area of some 150 km². The entire river plain consists of polders, areas in which water levels are artificially controlled by means of ditches and by discharge of excess water through pumping. Roughly half of the river plain is used for dairy farming while the remaining part consists of lakes and wetland nature reserves, one of which has been included in the Ramsar Convention. The wetland nature reserves are of international botanical importance owing to the presence of rich fens (fens characterised by minerophilous species) which harbour species-rich vegetation belonging to *Caricion davallianae*. This vegetation type is restricted to Western Europe and is well developed on the Vecht river plain. During recent decades, however, marked changes in vegetation composition have been observed here.

Hydroecological research was initiated to determine the causes of these changes in vegetation composition. Most wetlands are supplied by groundwater seepage at one end, which originates from a sandy ridge bordering the river plain. The seepage water flows through the wetland and subsequently infiltrates at the other end as a result of low water levels in adjacent agricultural polders.

The spatial patterns in vegetation composition in the wetlands were found to be closely related to the flow and quality of groundwater and surface water. The botanically significant fen vegetation is supplied by calcareous, mesotrophic seepage water from the sandy ridge. The deterioration in this type of vegetation was attributed to a decline of groundwater seepage resulting from increased groundwater abstraction on the ridge for drinking water production. It was also found that infiltration from the wetland had increased as a result of ever lower water levels in the surrounding agricultural areas to keep up with land subsidence following peat mineralisation. Both the decline in seepage and the increase in infiltration necessitated the supply of external polluted river water to prevent the wetland from desiccation. This water is nutrient-rich and contains relatively high amounts of chloride, sodium and sulphate and was shown to have detrimental effects on the aquatic vegetation, amongst others through bloom of algae which cover lake surfaces thus blocking sunlight from reaching submerged vegetation.

To counteract the decline in vegetation composition and biodiversity within the wetland nature reserves, a number of management measures were taken. Water treatment plant were built to remove phosphorus from the polluted river water supplied to the wetlands. This did not have the intended effect as phosphorus stored in lake bottom sediments is remobilised in the water column by wind and wave activity. Additional measures were needed to remove polluted soils on the lake bottoms by dredging. Groundwater abstraction on the sandy ridge is being reduced to 50 per cent of its former volume to regenerate seepage in the wetland and stimulate the growth of botanically important fen vegetation. In the surrounding agricultural areas land has been bought from farmers in order to raise water levels, which will reduce infiltration within the wetland. These measures aimed at the (partial) restoration of the natural water flow within the wetland will reduce the need for the supply of external river water.

This case shows the importance of a catchment-wide landscape-ecological approach for integrated wetland management. As wetlands are ecosystems dependent on processes both upstream and downstream, it is not sufficient to 'fence' the area and declare it a nature reserve. Conservation must be based on recognition and management of the key processes involved (here: groundwater and surface water flow), while aiming at restoring these processes within, as well as outside, the wetland conservation area.

Source: after Witmer, 1989; Wassen *et al.*, 1989, 1990; Schot *et al.*, 1988; Schot, 1991; van Liere and Gulatti, 1992; Barendregt, 1993; Barendregt *et al.*, 1995

et al., 1991). Such benefits include groundwater recharge, grazing resources, wildlife habitat, non-timber forest products and protection of adjacent agricultural lands against bird damage by providing natural feeding habitat.

Hollis (1990) points out that EIA should be part of a wider strategy, as EIA is often seen as a negative approach showing only adverse effects, delaying project implementation and thus impeding 'development'. Moreover, where EIA is the only defence for wetland resources, it can be 'bought' by those seeking to change wetlands. Such a wider strategy may be a National Wetland Strategy or a National Conservation Strategy as described above, which integrates actions in a range of fields at a number of levels, thus making wetland less dependent on the outcome of EIA alone.

Underlying issues

Usually the threats to wetlands are caused by underlying issues affecting society and its governance. If these issues are not addressed, wetland loss will continue despite all actions to conserve them. Therefore wetland management and policy should aim to address these issues, which include (OECD, 1996):

- social issues, such as population growth, lack of awareness, land ownership, employment;
- economic issues, such as divergence of private and social benefits of wetlands, multiple user conflicts, recognition for non-market values, subsidies;
- legal issues – providing adequate protection for wetlands coupled with an effective enforcement;
- policy issues – increasing awareness of wetland needs and management options for government officials in order to prepare specific wetland policies which overcome the disregard of wetlands in sectoral policies;
- institutional issues – providing adequate resources for wetland institutions, co-ordination of sectoral institutions, development of joint management methodologies.

CONCLUSIONS

Wetlands have long been regarded as wastelands. Therefore they have been altered to make them suitable for a number of different uses, such as agriculture, water and hydropower reservoirs, fish ponds and industrial and urban areas. With these alterations many of the functions that wetlands provide for society have been lost, resulting in flood aggravation, coastal abrasion, land subsidence, soil erosion, declines in fish harvests and groundwater recharge, and loss of water purification functions, wildlife habitats and genetic pools.

Environmental management is needed to ensure a Wise Use of Wetlands, which aims to maximise the benefits of wetlands for society as a whole rather than for a limited interest group. In practice, such management must start from sound knowledge of the functioning of abiotic and biotic elements of the wetland ecosystem, and of their relations with processes upstream and downstream in the catchment. In addition, socio-economic systems of wetland users need to be analysed, especially those of local communities. In order to attain sustainable use, a participative approach must be applied which engages wetland users in the process of establishing and carrying out management measures.

REFERENCES

Armentano, T.V. (1980) 'Drainage of organic soils as a factor in the world carbon cycle', *Bio Science* 30, 12: 825–830.

Barbier, E.B., Adams, W.M. and Kimmage, K. (1991) *Economic Valuation of Wetland Benefits: the Hadejia-Jama'are Floodplain, Nigeria*, London, UK: London Environmental Economics Centre (IIED and University College London).

Barendregt, A. (1993) 'Hydro-ecology of the Dutch polder landscape', Ph.D. thesis, The Netherlands: University of Utrecht.

Barendregt, A., Wassen, M.J. and Schot, P.P. (1995) 'Hydrological systems beyond a nature reserve, the major problem in wetland conservation of Naardermeer', *Biological Conservation* 72: 393–405.

Belt, C.B. (1975) 'The 1973 flood and man's constriction of the Mississippi River' *Science*, 189: 681–684.

Bord na Móna (Irish Peat Development Authority) (1985) *Fuel Peat in Developing Countries*, World Bank Technical Paper No. 41, Washington DC, USA: World Bank.

Boto, K.G. and Patrick, W.H. Jr. (1979) 'Role of wetlands in the removal of suspended sediments', in P.E. Greeson, J.R. Clark and J.B. Clark (eds) *Wetland Functions and Values: The State of Our Understanding*, Minneapolis, Minn., USA: Water Resources Association Technical Publication, pp. 479–488.

Breen, C.M. (1988) 'Wetlands classification', in R.D. Walmsley and E.A. Boomker (eds) *Inventory and Classification of Wetlands in South Africa*, Foundations for Research Development, Council for Scientific and Industrial Research Occasional Report No. 34.

Breen, C.M. (1991) 'Are intermittently flooded wetlands of arid environments important conservation sites?' *Madoqua* 17, 2, in R.E. Simmons, C.J. Brown and M. Griffin (eds) (1991) 'The status and conservation of wetlands in Namibia', Special wetlands issue, *Journal of Arid Zone Biology and Nature Conservation Research*, Namibia, pp. 61–65.

Carter, V., Bedinger, M.S., Novitzki, R.P. and Wilen, W.O. (1979) 'Water resources and wetlands', in P.E. Greeson, J.R. Clark and J.E. Clark (eds) *Wetland Functions and Values: The State of Our Understanding*, Minneapolis, Minn., USA: Water Resources Association Technical Publication, pp. 344–376.

Chapman, V.J. (1977) *Wet Coastal Ecosystems, Ecosystems of the World I*, Amsterdam, The Netherlands: Elsevier.

Coles, B. (1991) 'Wetland archaeology: a wealth of evidence', in M. Williams (ed.) *Wetlands: A Threatened Landscape*, UK: The Institute of British Geographers, pp. 145–181.

Cowardin, L.M., Carter, V., Golet, F.C. and La Roe, E.T. (1979) *Classification of Wetlands and Deepwater Habitats in the United States*, Washington DC, USA: US Department of the Interior, Fish and Wildlife Service, FWS/OBS-79/31.

Crow, J.H. and MacDonald, K.B. (1979) 'Wetland values: secondary production', in P.E. Greeson, J.R. Clark and J.E. Clarke (eds) *Wetland Functions and Values: The State of our Understanding*, Minneapolis, Minn., USA: Water Resources Association Technical Publication, 146–161.

de la Cruz, A.A. (1979) 'Production and transport of detritus in wetlands', in P.E. Greeson, J.R. Clark and J.E. Clarke (eds) *Wetland Functions and Values: The State of our Understanding*, Minneapolis, Minn., USA: Water Resources Association Technical Publication, 146–161.

Drijver, C.A. and van Wetten, J.C.J. (1992) *Sahel Wetlands 2020. Changing Development Policies or Losing Sahel's Best Resources*, Leiden, The Netherlands: International Council for Bird Preservation, Centre of Environmental Science, CML.

Dugan, P. (1990) *Wetland Conservation – A Review of Current Issues and Required Action*, Gland, Switzerland: IUCN.

Etherington, J.R. (1983) *Wetland Ecology*, Studies in Biology No. 154, London, UK: The Institute of Biology.

Gopal, B., Kvet, J., Löffler, H., Masing, V. and Patten, B.C. (1990) 'Definition and classification', in B.C. Patten *et al.* (eds) *Wetlands and Shallow Continental Water Bodies, Vol. 1*, The Hague, The Netherlands: SPB Academic Publishing b.v.

Gore, A.J.P. (1983) *Mires: Swamp, Bog, Fen and Moor, General Studies, Ecosystems of the World 4A*, Amsterdam, The Netherlands: Elsevier.

Gosselink, J.G. and Maltby, E. (1991) 'Wetland losses and gains', in M. Williams (ed.) *Wetlands: A Threatened Landscape*, The Institute of British Geographers, Oxford, UK: Basil Blackwell Inc., pp. 296–323.

Grootjans, A.P. (1985) 'Changes of groundwater regime in wet meadows', Ph.D. thesis, Groningen, The Netherlands: University of Groningen.

Heathwaite, A.L., Eggelsman, R. and Göttlich, Kh. (1993) 'Ecohydrology, mire drainage and mire conservation', in A.L. Heathwaite and Kh. Göttlich (eds) *Mires, Process, Exploitation and Conservation*, Sheffield, UK: University of Sheffield.

Hollis, G.E. (1978) 'The falling levels of the Caspian and Aral Seas', *Geogr. J.* 144: 62–80.

Hollis, G.E. (1990) 'Environmental impacts of development on wetlands in arid and semi-arid lands', *Hydrological Sciences Journal* 35, 4, 8: 411–429.

Knutson, P.L. (1988) 'Role of coastal marshes in energy dissipation and shore protection', in D.D. Hook *et al.* (eds) *The Ecology and Management of Wetlands, Vol.1*, London, UK: Croom Helm, pp. 161–175.

Lefor, M.W. and Kennard, W.C. (1977) *Inland Wetland Definitions, Report No. 28*, Connecticut, USA: University of Connecticut, Institute of Water Resources.

Leith, H. (1975) 'Primary productivity of the major vegetation units of the world', in H. Leith and R.H. Whittaker (eds) *Primary Productivity of the Biosphere*, Berlin, Germany: Springer-Verlag, pp. 203–216.

Löffler, H. (1990) 'Human uses', in B.C. Patten *et al.* (eds) *Wetlands and Shallow Continental Water Bodies*, Vol. 1, The Hague, The Netherlands: SPB Academic Publishing b.v., pp. 17–27.

Lugo, A.E., Brown, S. and Brinson, M.M. (1990) 'Concepts in wetland ecology', in A.E. Lugo, M. Brinson and S. Brown (eds) *Ecosystems of the World 15, Forested Wetlands*, Amsterdam, The Netherlands: Elsevier, pp. 53–56.

Maltby, E.D. (1985) *Peat Mining for Energy, Report SIEP-2*, Gland, Switzerland: International Union for the Conservation of Nature.

Maltby, E.D. (1986) *Waterlogged Wealth – Why Waste the World's Wet Places*, London, UK: Earthscan.

Maltby, E. D. (1988) 'Global wetlands – history, current status and future', in D.D. Hook *et al.* (eds) *The Ecology and Management of Wetlands*, London, UK: Croom Helm, pp. 3–15.

Maltby, E.D., Legg, C.J. and Proctor, M.C.F. (1990) 'The ecology of severe moorland fire on the North York Moors: effects of the 1976 fires, and subsequent surface vegetation development', *J. Ecology* 78: 490–518.

Mitsch, W.J. and Gosselink, J.G. (1993) *Wetlands*, New York, USA: Van Nostrand Reinhold.

Murkin, H.R. and Wrubleski, D.A. (1988) 'Aquatic invertebrates of freshwater wetlands: function and ecology', in D.D. Hook *et al.* (eds) *The Ecology and Management of Wetlands, Vol. 1*, London, UK: Croom Helm, pp. 239–249.

OECD (1996) OECD *Development Assistance Committee, Guidelines on Aid and Environment No. 9*, Paris, France: Development Co-operation Directorate, Organization for Economic Co-operation and Development.

Patten, B.C. (ed.) (1990) *Wetlands and Shallow Continental Water Bodies, Vol. 1, Natural and Human Relationships*, The Hague, The Netherlands: SPB Academic Publishing b.v.

Roggeri, H. (1985) *African Dams: Impacts in the Environment – A Case Study of Five Man-made Lakes in Eastern Africa*, Nairobi: Environment Liaison Centre.

Roggeri, H. (1995) *Tropical Freshwater Wetlands. A Guide to Current Knowledge and Sustainable Management*, Dordrecht, The Netherlands: Kluwer Academic Publishers.

Ruddle, K. (1980) 'A preliminary survey of fish culture in ricefields, with special reference to West-Java, Indonesia', *Bull. Nat. Mus. Ethnol.* 5: 801–822.

Sather, J.M. and Smith, R.D. (1984) *An Overview of Major Wetland Functions and Values*, Washington DC, USA: US Fish and Wildlife Service, FWS/OBS-84/18.

Schot, P.P. (1991) 'Solute transport by groundwater flow to wetland ecosystems – the environmental impacts of human activities', Ph.D. thesis, Utrecht, The Netherlands: University of Utrecht.

Schot, P.P. and Blankwaardt, B. (1994) 'Gestion et planification des resources en eau au Burkina Faso', *Communication colloque international 'Resources en eau, Environnement et Development'*, Nouakchott, March 1994 (in French).

Schot, P.P., Barendregt, A. and Wassen, M.J. (1988) 'Hydrology of the Naardermeer; influence of the surrounding area and impact on vegetation', *Agricultural Water Management* 14: 459–470.

Schrader, H.J. (1991) 'Approach of the Ministry of Wildlife, Conservation and Tourism to wetlands in Namibia', *Madoqua*, 17, 2, in R.E. Simmons, C.J. Brown and M. Griffin (eds) (1991) 'The status and conservation of wetlands in Namibia', Special wetlands issue, *Journal of Arid Zone Biology and Nature Conservation Research*, Namibia, pp. 253–254.

Söderland, R. and Svensson, B.H. (1976) 'The global nitrogen cycle', in B.H. Svensson and R. Söderland (eds) *Nitrogen, Phosphorus and Sulphur – Global Cycles*, SCOPE Report No. 7, *Ecological Bulletin* 22, Swedish National Science Research Council, Royal Swedish Academy of Science Stockholm, pp. 23–73.

Sternitzke, H.S. (1976) 'Impact of changing land use on delta hardwood forests', *Journal of Forestry* 74, 1: 25–27.

Tiner, R.W. (1984) *Wetlands of the United States: Current Status and Recent Trends*, Washington DC, USA: US Fish and Wildlife Service.

Turner, R.E. and Boesch, D.F. (1988) 'Aquatic animal production and wetland relationships: insights gleaned following wetland loss or gain', in D.D. Hook *et al.* (eds) *The Ecology and Management of Wetlands, Vol. 1*, London, UK: Croom Helm, pp. 25–39.

Turner, R.E., Forsythe, S.W. and Craig, N.S. (1981) 'Bottomland and hardwood forestland resources of the Southeastern United States', in J.R. Clark and J. Benforado (eds) *Wetlands of Bottomland Hardwood Forests*, New York, USA: Elsevier, pp. 12–28.

UNSO (1990) 'Lakes of grass: regenerating Bourgou in the Inner Delta of the Niger River', in D. Stiles and Technical Support Division of UNSO (eds) *UNSO Technical Publications Series No. 2*, New York, USA: The United Nations Sudano-Sahelian office (UNSO).

US Corps of Engineers (1972) *Charles River Watershed, Massachusetts*, Waltham, Mass, USA: Corps of Engineers, New England Division.

van Liere, L. and Gulatti, R.D. (eds) (1992) 'Restoration and recovery of shallow eutrophic lake ecosystems in the Netherlands', *Developments in Hydrobiology 74*, Reprinted from *Hydrobiologica, 233*, Dordrecht, The Netherlands: Kluwer Academic publishers.

Wassen, M.J., Barendregt, A., Bootsma, M.C. and Schot, P.P. (1989) 'Groundwater chemistry and vegetation gradients from rich to poor fen in the Naardermeer (The Netherlands)', *Vegetatio* 79: 117–132.

Wassen, M.J., Barendregt, A., Schot, P.P. and Beltman, B. (1990) 'Dependency of local mesotrophic fens on a regional groundwater flow system in a poldered river plain in the Netherlands', *Landscape Ecology* 5, 1: 21–38.

Wijkman, A. and Timberlake, L. (1984) *Natural Disasters – Acts of God or Acts of Man?*, London, UK: Earthscan.

Williams, M. (1991) *Wetlands: A Threatened Landscape*, The Institute of British Geographers, Oxford, UK: Basil Blackwell Inc.

Winter, T.C. (1989) 'Hydrologic studies of wetlands in the Northern prairie', in A. van der Valk (ed.) *Northern Prairie Wetlands*, Ia., USA: Iowa State University Press.

Witmer, M.C.H. (1989) 'Integral watermanagement at regional level. An environmental study of the Gooi and Vechtstreek', Ph.D. thesis, Utrecht, The Netherlands: University of Utrecht.

WWF (1989) 'USSR's Aral Sea disappearing fast', in *Wetlands, WWF Special Report No.1*, Gland, Switzerland: World Wide Fund for Nature.

SUGGESTED READING

Denny, P. (ed.) (1985) *The Ecology and Management of African Wetland Vegetation*, Dordrecht, The Netherlands: Junk/Kluwer.

Gore, A.J.P. (1983) *Mires: Swamp, Bog, Fen and Moor, General Studies, Ecosystems of the World 4A*, Amsterdam, The Netherlands: Elsevier.

Hook, D.D. *et al.* (1988) *The Ecology and Management of Wetlands, Volume 1: Wetland Conservation; Volume 2: Wetland Ecology*, London, UK: Croom Helm.

Lugo, A.E., Brinson, M. and Brown, S. (eds) (1990) *Ecosystems of the World 15, Forested Wetlands*, Amsterdam, The Netherlands: Elsevier.

Mitsch, W.J. and Gosselink, J.G. (1986) *Wetlands*. New York, USA: Van Nostrand Reinhold.

Roggeri, H. (1995) *Tropical Freshwater Wetlands. A Guide to Current Knowledge and Sustainable Management*, Dordrecht, The Netherlands: Kluwer Academic Publishers.

Williams, M. (1991) *Wetlands: A Threatened Landscape*, The Institute of British Geographers, Oxford, UK: Basil Blackwell Inc.

SELF-ASSESSMENT QUESTIONS

1   A precise definition of wetlands is necessary in order to:
    (a)   inform the general public on what is a wetland;
    (b)   facilitate classification, inventory and legislation;
    (c)   allow for delineation of wetland boundaries.
2   A universally agreed classification of wetlands is difficult to obtain because:
    (a)   wetlands occur in innumerable varieties all over the globe;
    (b)   the occurrence of seasonal fluctuations in flooding and water availability;
    (c)   people have different perceptions of what a wetland is and how it should be used.
3   The value of a wetland at large is determined by:
    (a)   the goods the wetland provides for humans;
    (b)   the total benefit to society as a result of the functions the wetland performs;
    (c)   the perception of local people of the value of the wetland.
4   Drainage of wetlands:
    (a)   is detrimental to society;
    (b)   is beneficial to society;
    (c)   may be both beneficial and detrimental to society.
5   Participation of local communities in wetland management is necessary because:
    (a)   they have the most rights to the benefits of the wetland;
    (b)   their co-operation is essential to sustainable use of the wetland;
    (c)   they have the most knowledge of the wetland where they live.
6   A landscape-ecological approach pertains to:
    (a)   the relation between wetland functioning and processes in the catchment;
    (b)   the relation on a landscape scale between abiotic and biotic factors, except the human factor;
    (c)   the fact that the landscape determines the ecology of wetlands.
7   Environmental wetland management must be primarily directed at:
    (a)   people;
    (b)   abiotic processes;
    (c)   biotic processes.
8   Environmental wetland management must always have an experimental character up to a certain degree, because:
    (a)   local people are not trained to manage wetlands;
    (b)   interventions in wetlands never occur under exactly the same conditions;
    (c)   we cannot fully oversee all possible effects of human activities on the wetland, the river basin or still larger scales.
9   The principle of Wise Use implies:
    (a)   wetlands must be used for the benefit of humans;
    (b)   wetlands may only be used in a sustainable way keeping the functioning of the ecosystem intact;
    (c)   wetlands must not be used when negative effects may occur.

5

UPLAND AND MOUNTAIN ENVIRONMENTS

Roy W. Tomlinson and Brian Whalley

SUMMARY

This chapter discusses the limitations on, and opportunities for, human activity in upland and mountain environments. Limitations include those of climate, since temperatures are reduced and precipitation and exposure are greater with increasing altitude. These changes affect vegetation and soils which in turn may reduce opportunities for agriculture. People have attempted to modify the environment to suit their agriculture, but in so doing have sometimes damaged the environment. Controlled grazing, burning and reclamation of hill lands are taken as examples of agricultural modifications. Forestry, peat extraction, water storage and tourism and their impacts are also examined. The chapter concludes with a discussion of hazards in high mountain areas.

ACADEMIC OBJECTIVES

On completion of this chapter the reader should be able to appreciate that whether you are considering constraints or opportunities, physical factors of the environment cannot be viewed singly or in isolation from one another. Further, that these physical factors have to be related to the social and economic circumstances of both local residents and the wider world. The reader should be able also to identify some of the methods that might be used in assessing the impacts of various kinds of developments in the uplands.

INTRODUCTION

Definition

The definition of an upland or a mountain area tends to vary with local topography. For example, the Sperrin Mountains of Northern Ireland rise quite abruptly from the surrounding level to a height of over 400 metres and therefore are regarded locally as mountains. Elsewhere in Europe, for example, in the Alpine regions, such areas as the Sperrins might qualify as hills only. Rather than search for a precise definition, which is probably not attainable, the two terms will be used for any area that rises abruptly from the surrounding level and has environments contrasting with the lower ground.

Approach of this chapter

In this chapter the impacts of people in upland and mountain ecosystems are examined mainly through examples taken from Britain and Ireland. The approach adopted is to outline the limitations on, and opportunities for, human activity in these ecosystems. Factors such as temperature and the growing season, increased precipitation, exposure, soil quality and slopes all affect the possibility for human activity. Subsequently, various kinds of uses and the conflicts between these and maintenance of the ecosystems will be examined. Case studies are used throughout. Although the precise problems highlighted will not necessarily occur elsewhere, parallels can be drawn and the examples serve to show the range of considerations necessary in managing human use of these fragile environments.

LIMITATIONS AND OPPORTUNITIES FOR USE

Since the troposphere is heated mainly from below, a decrease in temperature may be expected with height. This is termed the environmental lapse rate. In Britain, temperature generally declines at a rate of 6°C per 1,000 m rise in altitude. However, it must be emphasised that this is a general figure since it may change with season, air mass type, and with variation in topography and vegetation cover. Some of these variations were shown in a classic study by Wagner (1930) where temperatures were recorded from a cable-car in four profiles through the day. In the early morning, when the slopes had cloudy conditions, the decline in temperature with height was only 4°C in 1,700 metres. By mid-morning not only was the general decline in temperature greater, but there was more irregularity with the micro-topography. This irregularity was even more pronounced in mid-afternoon especially over forested areas and a drop in temperature was observed over a hollow on the slopes. In the evening, when the whole slope was in

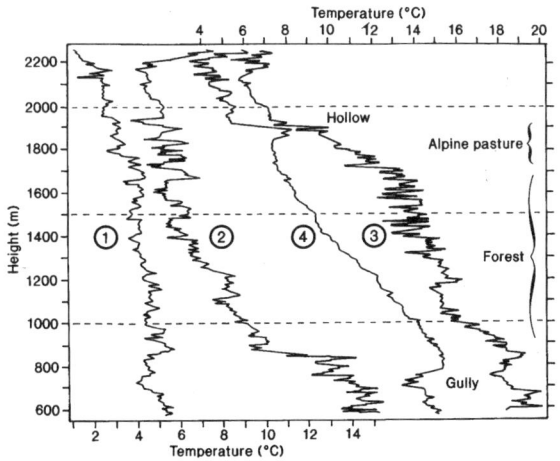

Figure 5.1 *Decline in temperature with altitude near Innsbruck*

Sources: Wagner, 1930; Barry, 1992

shadow, these irregularities disappeared (Figure 5.1). Whereas such micro-variations have implications for land use and developments in local areas, the general values have regional importance. Thus, in Britain, land above 600 m will have minimum temperatures in winter below freezing on most nights. If mean daily temperatures are examined, then the upland margins have around 50 days in the year with a mean of 0°C rising to 100 days in higher areas. Such low temperatures have significance for animal and plant life. For example, in winter deer move off the highest slopes to the lower margins and additional feed may be necessary for some domesticated animals.

The length of the growing season is often defined by climatic, and in particular, temperature conditions. It has been noted that in Britain spring growth does not begin until a mean daily temperature of 6°C is maintained, although occasional days with a lower temperature may occur. In much of lowland southeast and midland England, the growing season may begin in early March, but in the Welsh mountains and in the Pennines spring growth starts 10 to 20 days later. The end of the growing season may be estimated in a similar manner. The length of the growing season, as defined by temperature, in the south-east and the midlands of England is 250–275 days per annum, whereas in the Welsh mountains it is 250 days or less. Such variations in length of growing season may not be as marked in Central Europe. The length of the season is also not constant through time, as may be illustrated by reference to Northern Ireland (Figure 5.2), where variations may be noted according to frequencies of air flows.

Temperatures alone do not limit cultivation; other climatic factors include increased precipitation, humidity and wind speed with altitude. The increase in precipitation is a world-wide feature, but it is neatly summarised in the cross-section through Wales to East Anglia (Figure 5.3). The influence of the Welsh mountains, which provide a barrier to the prevailing moist westerly air streams is shown clearly, but the effects of even minor uplands such as the Leicestershire Wolds and Norfolk Heights are evident. In mountain and upland areas both duration and intensity of precipitation can be higher. In eastern Scotland the average annual precipitation occurs over 500 hours whereas in the north-west Highlands it occurs over 2,000 hours,

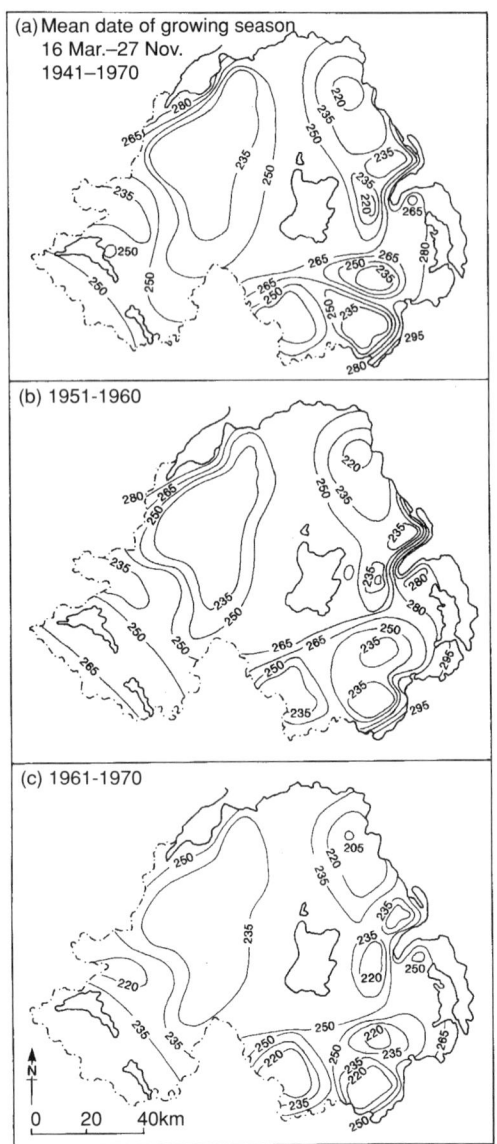

Figure 5.2 Changes in mean length of growing season in Northern Ireland
Source: Betts, 1982

a fourfold increase in time. In amount, however, the north-west Highlands have nine times as much precipitation, a clear indication that intensity must also be much greater.

The effects of wind in limiting land use in mountain areas may be illustrated by a brief discussion of the tree line. This term has a variety of definitions, with the main distinction between latitudinal tree lines (e.g. between the taiga and the tundra in northern Russia) and altitudinal tree lines. In the case of altitudinal tree lines, Central European ecologists distinguish between the timber line, which is the limit of trees of forest density and stature, and the tree line which is the limit of the scattered trees. A considerable amount of research has been conducted as to why trees have an altitudinal limit. Most modern theories suggest that whereas trees may maintain a healthy water balance all year round below the timber line, the shoots of trees above it are often desiccated to a fatal degree in winter. In some areas this desiccation may be brought about by high insolation levels (solar radiation), especially if the preceding summer has been too short to allow new shoots to become 'hardened'. In Northern Europe, bright sunshine is less frequent and desiccation may be related more to persistent, strong winds. Using tatter flags to measure wind exposure, Pears (1967a, 1967b) has shown that exposure is severe at and above the present forest edge in the Cairngorms of Scotland. Such exposure increases water deficits in seedlings through increased transpiration. Combined with the short ripening season desiccation damage would occur, thus limiting the growth and establishment (Figure 5.4).

The increase in severity of climate with altitude offers opportunities as well as limitations. Precipitation may occur as snow, which when it occurs in suitable terrain, provides areas for recreation. Winter sports, especially skiing are now a major source of income for residents in some mountain areas and contribute substantially to national economies, although perhaps at the expense of the ecosystems (see pp. 100–101).

Low temperatures in the uplands, and therefore low rates of evaporation in both winter and summer, mean that the predominant direction of water movement in the soil is downward, producing leaching and soils of low fertility. On the lower slopes there may be quite freely draining brown earths and acid brown earths with a cover of rough grassland, which is possibly derived following clearance of woodland. These soils and the vegetation cover can be improved through liming, use of fertilisers and/or careful

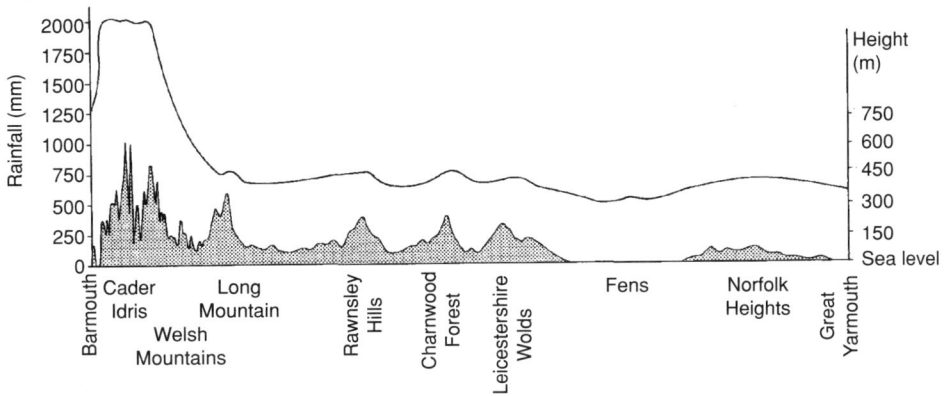

Figure 5.3 Influence of altitude on precipitation – a cross profile from central Wales to East Anglia
Source: MAFF, 1964.

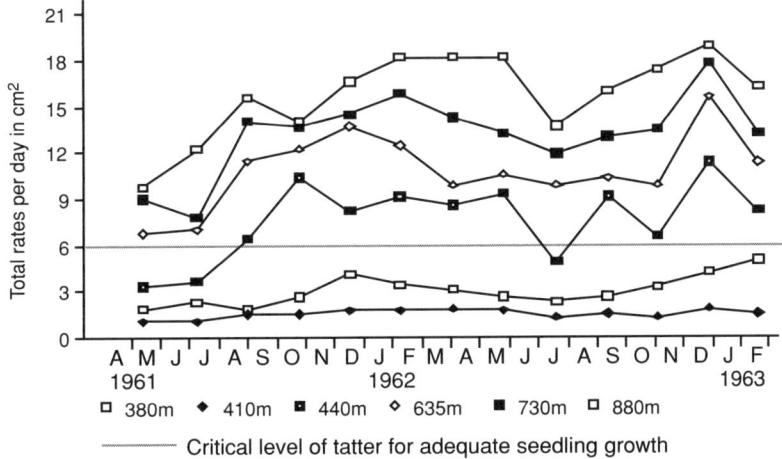

Figure 5.4 Effect of wind exposure on the tattering rate of flags. Note the increase in exposure with height and that site 2, although higher, is sheltered by existing forest.
Source: Pears 1967a

management of stock numbers (see pp. 91–94). Upslope the soil may grade into podsols and peaty podsols. The latter may have an iron pan which restricts water movement, further adding to the difficulties of using those soils for agriculture. Their high acidity, low levels of plant nutrients, very poor quality herbage and possible poor drainage mean that any reclamation methods are quite drastic. On summit areas and steep slopes, soils may be very thin rankers, or non-existent. Of course, slope angle of itself can limit land use and attempts to utilise steep slopes can

result in debris slides and severe vegetation and soil loss (see pp. 101–102).

For alpine mountains, many of the factors discussed above apply, but often to a more extreme extent. This is especially true of the larger mountains where a continental climate generally prevails. Ranges of temperatures, although not necessarily precipitation totals, are frequently much greater than for smaller mountains and hill masses. Precipitation, in the form of snow, presents its own hazards. The very existence of large mountain ranges can produce extreme variations in climate, such as the rainshadow effect, and deflection of storms can lead to an unpredictability of meteorological events.

AGRICULTURE IN THE UPLANDS

The harsher physical environment of the uplands, outlined in the previous section, clearly imposes limits on their agricultural use, but as almost everywhere on earth, people strive to reduce these disadvantages by adapting both the environment to their needs, and their agricultural practices to the environment. These efforts are sometimes misguided and often have ill effects when other forces cause them to be practised too intensely.

Burning

The vegetation of many upland areas in Britain is of low nutritional value, particularly in winter and spring when many grasses are senescent because of low winter temperatures. In summer, low stocking rates cause herbage production to exceed the amount grazed so that dead plant litter accumulates. Other plants, such as *Calluna vulgaris* (common heather), have more woody material and as they age the proportion of woody parts to tender, more nutritious parts increases. To remove much of this grass litter and to produce younger stands of vegetation, burning is a common practice in moorland areas of Scotland, North Yorkshire and the Pennines. Periodic burning, usually carried out in spring, must be done bearing in mind climatic and vegetational characteristics, otherwise severe damage may be done to the ecosystems (Gimingham, 1972; Muirburn Working Party, 1977; Wall, 1993).

Successful burning depends on the amount of heat generated. In some fires temperatures are sufficient only to burn the vegetation lightly. Large amounts of material that can prevent regeneration of plants are left behind. In contrast, if fires are too hot then the underlying peat may also be burnt. The heat sterilises the peat, preventing regeneration of higher plants. If such a burn smoulders for a long period, it causes nutrient loss and eventually erosional activity of water and wind may strip the peat to expose the bedrock. Good practice therefore requires a fire of medium intensity. This is best obtained by burning with a gentle breeze and when the air is relatively dry. Burning in a strong wind not only risks too intense a burn, but may result in very rapid spread of fire so that it passes only through the crowns of plants, especially *Calluna*. This leaves the woody stems unburnt and does not stimulate new shoots from ground level.

Vegetational characteristics affecting the success of burning include the amount of material to be burnt, its composition and moisture content. In older stands of heather, plants are taller and have a higher proportion of woody material and fires tend to be hotter than in younger stands. Vegetation consisting of a mixture of heather, grasses and sedges will tend to be more open and will burn more rapidly and at a higher temperature than stands of grasses and sedges where litter is more densely packed and damp. Therefore, burning of heather should be done before it becomes too old – generally when it has achieved a height of about 20 cm – between about 7 and 15 years of age in the building phase (Watt, 1955) so that a fire of moderate intensity will be achieved. For the damper, dense grass and sedge areas, burning may have to be done when winds are stronger and/or against the wind so as to ensure removal of the litter.

Burning not only needs to be discussed in relation to the dominant vegetation but also with regard to the purpose; whether, for example, it is to produce improved sheep grazing or good grouse moor. In grass dominated moor it is possible to distinguish between the flying bent or purple moor grass (*Molinia caerulea*) dominated moor in the eastern, drier areas of Scotland, for example, and the wetter flying bent and heather moorlands of the west of Scotland. In the east, wavy hair grass (*Deschampsia flexuosa*), sheep's fescue (*Festuca ovina*) and brown bent (*Agrostis canina*) are accompanying species and provide some winter grazing. They are less fire resistant than flying bent, but better able to withstand grazing. If burning is extensive leading to low intensity grazing of new growth, then burning can be counter-productive since this leads to dominance of flying bent. Thus burning should be of small areas, followed by relatively heavy grazing so as to maintain the desirable species.

In the wetter, western moors of flying bent and heather, a decision has to be made as to whether burning should be on the short rotation which encourages new growth of flying bent, or on the longer, heather rotation. If a short rotation is used, then flying bent is encouraged at the expense of heather. Burning should again be of small areas to remove the litter and provide spring–summer grazing without converting large areas from heather to flying bent moorland.

Good grouse moor requires a variety in age and height of heather. Grouse require areas of taller heather for shelter and protection and of young heather with nutritious green shoots for feeding. Variation should be on a small scale and it is suggested that in order to create areas for feeding, burning should be in long, narrow strips up to 30 m wide so that the protective cover of the older, taller heather is nearby.

Burning of moorland vegetation is carried out to produce enhanced grazing and it may be expected that species variety will be reduced. Trees and shrubs are prevented from colonising and only those plant species with some resistance to fire may survive. Nevertheless, these areas are ecosystems with sites for particular species of plants and animals and of landscape importance; for example, the much admired North Yorkshire Moors with their expanse of well-maintained heather. The combination of human activities for economic gain and the physical environment in this case has produced a valued ecosystem, but one that requires very careful management. Sadly, there are examples where accidental fires or bad practice have led to the destruction of the ecosystem.

Upland reclamation

Many upland areas have changed in the recent past as demand for agricultural products has increased. In Britain, over most of the post-World War II period, moorland has been lost through conversion or reclamation to improved grazings and pastures. A high proportion of conversion or reclamation has been at the moorland edge and is secondary, that is reclamation of land that has reverted after an earlier period of improvement. For example, Parry *et al.* (1982) found in the Yorkshire Dales National Park that much of the reclamation was of land that had reverted in the depression years of the 1930s. They

showed that many reclaimed or improved areas on the margins of moorland are small and subject to recurrent changes, fluctuating between improvement and reversion, depending on their agricultural profitability. The plant communities are finely balanced with regard to both their physical location between lowland and upland environments, and the economics of agriculture at different periods. When returns from grazing are poor, stocking rates may be reduced and the vegetation reverts in this example to the coarser, more vigorous *Agrostis–Festuca* grasses.

The overall trend in loss of moorland has raised concern in Britain, particularly in Exmoor where between 1947 and 1979 over 17 per cent of the heathland was reclaimed and there was an annual loss of 160 ha of moorland (Porchester, 1977).

Moorland reclamation takes several forms. It may involve only controlled grazing or it may be by vegetation removal (by fire, herbicides) followed by fertiliser application. On wetter types of moorland deep ploughing may be necessary to break up any iron pan and to mix mor humus or peat with underlying mineral soil. Drainage is often required, as well as the use of herbicides to prevent regrowth of rushes, bracken or heather. Liming and fertiliser application are needed to ensure successful reseeding with ryegrass mixtures.

An example of this most extreme type of reclamation may be observed on the south side of the Mourne Area of Outstanding Natural Beauty in south-east Northern Ireland. Here, below the steep slopes of the Mourne Mountains, there is a gradual decrease in height towards the raised coastal terrace which consists of glacial drift and out-wash sands and gravels. Over the last 200 years this land has been enclosed. Huge boulders within the drift were used to build the retaining walls. The enclosed areas were subject to minimal reclamation and some land has reverted subsequently towards its 'natural state'. Recently there has been increased reclamation in these poorly reclaimed or reverted fields and reclamation has extended up the slopes of the mountains (Figure 5.5). This was encouraged by government grants which allowed up to 50 per cent of costs to be recovered. Since the area is within the Less Favoured Area (LFA), higher stocking rates possible

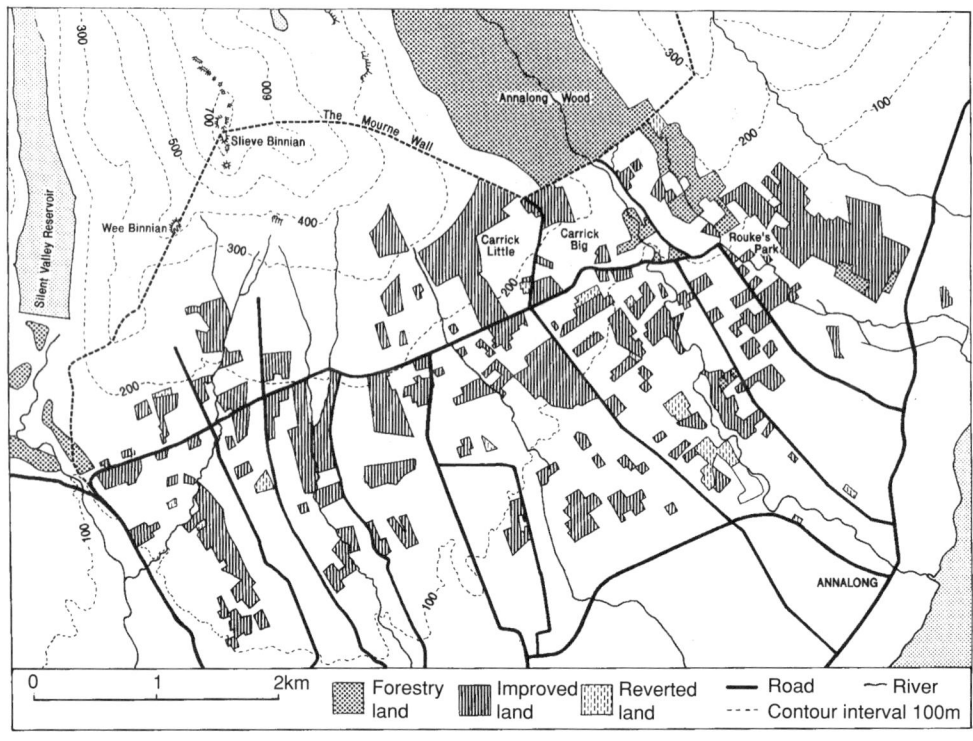

Figure 5.5 Reclamation of moorland and hill margins on the south side of the Mourne Mountains, Northern Ireland

on the new lands allowed increased income through headage payments. Such grants were devised to assist farmers to remain in these areas and thereby to help conserve the countryside, but often the grants seem to have been administered in such a way as to disregard the future appearance of the landscape. Thus a change towards larger fields is occurring, with a consequent loss of vegetation communities and a reduction of the patchwork of fields of varying quality which was of importance both to wildlife and the scenic value (Greer, 1986).

Less intensive reclamation will change the species composition, as indeed it is designed to do. On hill margins there may be areas of better grazing, 'green areas' with *Agrostis tenuis* and *Festuca* spp., surrounded by the coarser, less palatable *Molinia caerulea–Nardus stricta* dominated grasses. Without fencing the green areas may be overgrazed. By controlling the grazing, not only can these be protected but, in the absence of an alternative, the

coarser grasses have to be grazed. This, and trampling, reduces sward height and litter, and allows more palatable species to enter. Floate *et al.* (1973) showed that, although much better results could be obtained through more intensive treatments, such as the use of fertilisers, lime and reseeding, controlled grazing alone did reduce the less desirable species. *Molinia caerulea* was more sensitive than *Nardus stricta* which required very heavy grazing to reduce its content in the sward.

Whereas sheep and cattle may be used to improve the quality of grazing, there is evidence that too high stocking rates can decrease the grazing quality. An examination of Exmoor shows a great contrast between vegetation at the centre, where *Nardus stricta* grassland is dominant, together with *Molinia caerulea*, and at the periphery where *Calluna vulgaris* is dominant. Study of its history reveals that the central, enclosed part of the moor was a Royal Forest until 1819 and that towards the end of that period, land was rented out to grazing with the fee levied per head of livestock. Heavy stocking rates resulted so that the combination of high sheep numbers, cattle and ponies prevented regeneration of heather and dominance of coarse, unpalatable grass species. Outside the enclosure, land was held in common, stocking rates were lower and the *Calluna* could survive.

Other studies have shown in contrast that reduction in grazing intensity may lead to reversion and not only at the margins. Ball *et al.* (1982) summarised findings from field studies concentrated in 12 areas of England and Wales that had mountain and hill farming. While recognising that upland vegetation is highly variable in space and over time, the authors nevertheless divided it into four main groups: improved pastures, rough pastures, grassy heaths and shrubby heaths. They proposed general changes that could result from a decrease in intensity of management. For example, they suggested that over a period of 20–50 years, and with a decline from 4–7 sheep per hectare to 2–4 sheep per hectare, rough pastures initially dominated by *Agrostis–Juncus* or *Festuca–Juncus* would change to *Festuca–Nardus–Molinia*

heath, and there could be an overall decline in species diversity.

Rawes (1981) showed that in plots protected against sheep grazing in the northern Pennines, the number of pin contacts per quadrat increased significantly at most sites, especially in the lower levels of the vegetation (Figure 5.6). However, the study showed also that there was a very marked reduction in species diversity. An earlier study (Welch and Rawes, 1964) on three sites in the same area, which had been protected from grazing for eight years, showed that bryophytes, lichens and flowering plants (except grasses) decreased. *Nardus stricta* and *Juncus squarrosus* declined markedly and were replaced by *Deschampsia flexuosa*.

Clearly, the management of upland grazing involves a delicate balance between stocking levels and environmental conditions at particular sites if vegetation communities, associated fauna and scenery are to be retained (Macgillivray, 1994). It should be remembered, though, that most of these upland moors have been created by human activity and that many of their attributes are of relatively recent origin. Widespread travel since the 1950s has made people

Stratum 1, > 30 cm above ground level; 2, 20–30 cm/ 3, 10–20 cm; 4, 0–10cm

Figure 5.6 Total number of pin contacts of vascular plants in four strata of grazed and protected grasslands
Source: Rawes, 1981

more conscious of these areas, but they tend to believe that the moors are natural. Whether the land should be managed so as to maintain the moors is open to question because future generations may prefer a different landscape. Had our predecessors prevented burning, some moors would have been colonised by birchwood or scrub and their present appearance could not now be enjoyed.

Agriculture in the uplands has impacts on the soil as well as on the vegetation. Clearance of the vegetation cover, combined with rainfall of high intensity, especially on steep slopes, can produce soil erosion and associated problems. Soil conservation strategies are therefore an important part of the management of upland areas under agriculture (Staver *et al.*, 1991). Not only is the soil resource lost, but the resulting sediment, together with that from construction sites and landslides, is washed into rivers. Subsequently, the riverborne sediment can cause silting of reservoirs and damage to riparian ecosystems.

FORESTRY

The British Forestry Commission was established immediately after World War I, with the aim of increasing national timber, pulp and paper supplies. Since then, political decisions have been made to expand forestry. However, with government subsidies to upland farmers, acquisition of land has been difficult and forestry has been confined largely to the poorest lands. These are often peat covered, of low nutrient status, have a harsh climate and are waterlogged. New techniques have been adopted for forestry in these areas and non-native species have been introduced, including *Picea sitchensis* and *Pinus contorta*. The latter is usually grown in the wettest, poorest plantable areas. Techniques employed to overcome physical disadvantages and to reduce costs include draining, planting in straight lines, aerial application of fertilisers, line thinning and clearfelling. All of these have ecological and/or landscape consequences.

The effects of peatland drainage have long been a matter of dispute. The argument centres on whether or not, in view of the low hydraulic conductivity of peat, the drain has much effect in lowering the

water table beyond a decrease in a narrow band at either side of the drain. Many studies have shown that most water movement is through the root zone (acrotelm) (Chapman, 1964; Tallis, 1973; Tomlinson, 1979). Therefore drains will have consequences for plants adapted to high moisture contents, particularly *Sphagnum* species which cannot withstand long periods of desiccation. *Sphagnum* species are often dominant so that there will be a change in the vegetation composition unless the drains themselves become colonised and no longer function. Removal of water from the acrotelm also results in loss of features of the micro-topography, such as pools and hollows.

Once an area has been planted, then the effects of drainage are subsumed by other direct effects. It should be borne in mind, though, that large areas may undergo drainage and remain unplanted for many years so that the effects of afforestation extend beyond the area actually planted. Whether peatlands around the forest are affected by large drains within the forest is more difficult to determine. Some conservationists believe that there is a general drawdown of the water table and therefore change in the species composition and structure of the peatland. This conflicts with findings that there is limited water movement in a bog, except in the acrotelm. Perhaps there should be more focus on changes in precipitation arising from different patterns of air turbulence induced by surrounding forests.

Water drained from forests eventually enters natural stream channels and there is evidence that conifer plantations on base-poor, upland soils increase the acidity and the concentrations of metals in surface waters. In Wales, the pH of streams draining forested areas has been found to be 0.5 to 1.0 units lower than similar streams draining unplanted moorland and the aluminium concentrations to be 0.1 to 0.4 mg/l higher (Stoner *et al.*, 1984). Increases in acidity and aluminium concentrations may limit plant species in the streams and affect the diversity of invertebrates, fish populations and, indirectly, bird populations along rivers. Amongst the evidence presented is the decline in catches or numbers of salmonid fish in the river Fleet in Galloway, S.W. Scotland, on Plynlimon in Wales, and the higher loss of salmon eggs and young trout

BOX 5.1 MAINTAINING RURAL POPULATIONS AND PROBLEMS OF OVER-GRAZING IN NORTH-WEST IRELAND

The north-western coastal counties of Ireland are popular with tourists of many nationalities. The combination of sea and coast with windswept mountains and open peatlands appeals to many people, especially with the low population densities. These low densities are no accident, but reflect a history of emigration brought about in part by the physical environment. While wild windswept landscapes may have attractions for the holiday visitor, they are a very different matter for the permanent resident. Precipitation is more than 1,200 mm a year, increasing to over 2,800 mm on the mountains of northern Galway and southern Mayo, and there are over 200 days a year with 1 mm or more of rain. Evapotranspiration is low and there is considerable excess of precipitation over evaporation, reaching around 700 mm a year. The widespread occurrence of blanket peatland, even at sea level, is no surprise. These physical limitations of climate and soils have led to large parts being classed by the EC as 'disadvantaged areas' so that subsidies have been paid to farmers to support incomes and for general improvement of agriculture.

Private forestry has been encouraged through various grant schemes and state planting has been extended, but this can have deleterious impacts. Low altitude and montane peatland is lost to the planting and streams may have 'flashier' regimes as a result of the drainage necessary for tree planting. The acidity of the streams may be increased and their fauna and flora changed (see p. 94). More significant loss of peatland may result from agriculture. In an attempt to maintain farming and retain rural population in these disadvantaged regions, EU-funded livestock headage payments have been made to farmers. This has encouraged them to overstock; because the payments are by headage there has been a large increase in sheep numbers. Since 1980 the sheep population in Ireland has more than doubled to a total in 1991 of 8.9 million. Extensive overgrazing has occurred in the uplands and other blanket peatlands; upland areas in Mayo have been particularly badly affected, along with similar country in Donegal, Galway and Kerry. The increased sheep numbers have led to selective grazing of plant species, loss of vegetation cover, and consequent poaching of the surface. This is followed by weathering and erosion of the peat. As with forestry, deposition of peat particles in streams can damage nursery beds of salmonid fish and raise the acidity levels. In the longer term this accelerated peat erosion will reduce the scenic value and thus the tourist industry which is such an important part of the economy of these western areas. A change in emphasis is required so that farmers are paid to maintain these fragile peatland landscapes. There is a need to increase payments for reducing stock numbers and for farming in an extensive, sustainable manner. That would maintain both the rural population and the landscape, which itself is a valuable resource.

in plantation streams as compared with moorland streams (Nature Conservancy Council, 1986). Nisbet (1992) has pointed out, however, that non-forested catchments have also become acidified and he does not regard the evidence that forests have added to this acidification as conclusive.

Application of fertilisers to forest plantations can result also in stream pollution, either directly through drift of aerially applied fertilisers or indirectly, over a longer term, through leaching of forest soils. In addition to polluting streams, aerial applications of fertilisers can affect surrounding vegetation. Peatland vegetation is adapted to low nutrient conditions and fertilisers could, especially if water tables have also fallen, create an environment unsuitable for the survival of peatland plants. Forest planting can also increase the sediment load and the 'flashiness' of the stream hydrograph.

Planting in even rows, each receiving similar amounts of fertiliser and enjoying similar environmental conditions, will produce a well-developed, dense canopy beneath which light intensity is a fraction of that outside. Vegetation within the plantation is therefore scarce – perhaps only a few strands of heather or wavy hair grass and some mosses, lichens or liverworts. The type and density of vegetation depends on the age of the plantation, the original vegetation and the tree species planted. In a stand of Sitka spruce, vascular plants decreased rapidly after planting and were only 50 per cent of ground cover after 15 years and almost zero per cent after 25 years. Bryophytes increased in the first 15 years to about 10 per cent cover and this level was maintained subsequently (Hill, 1979). Goldsmith and Wood (1983) showed also that upland plants have a gradation in tolerance to shade so that:

BOX 5.2   INFLUENCE OF UPLANDS ON LAND DEGRADATION IN ZIMBABWE

This chapter has been written largely around examples from Britain and Ireland, but parallels can be drawn in other parts of the world. For example, in this second case study, the effects of overgrazing and land clearance in a particular upland environment of Zimbabwe are illustrated. Whitlow (1988) estimated that there were 1.8 million hectares of eroded land in Zimbabwe, or 4.7 per cent of the country. Most, 1.5 million hectares, was in the Communal Lands where population densities were high and traditional agricultural practices were followed. These included late planting (so that a cover is not established before the onset of the rainy season), shallow ploughing (only the uppermost soil layer with 'looser' soil particles is turned) and limited application of fertilisers and manure (crop growth and hence cover is poor and lack of manure means that soil particles are not bound together).

Erosion was found in the high mountains along the eastern border of the country but, because large areas are in land protected for wildlife and scenery, tended to be localised. The most acute erosion was in areas of granitic domes or bornhardts. These intrusive granite domes are found especially in a broad arc which stretches from the Matopos hills, south of Bulawayo, to the Mutoko region, north-east of Harare. In this arc, the domes may occupy up to one-third of the land area and because of their poorly vegetated, steep, often convex slopes, shed considerable amounts of runoff on to surrounding land. Where the land has not been cleared by grazing or for agriculture (see Figure 1, below), the savanna vegetation reduces the potential for erosion since the soil surface is covered. This slows the flow of water and allows greater infiltration into the soil. There is less bare earth and soil particles to be transported by sheetwash from the dome or by water channelled into rills and streams. In contrast, when clearance for agriculture or overgrazing has taken place up to the foot of the dome (see Figure 2, below), sheetwash erosion and gullying is common (an idealised plan view is shown in part 3 of the figure below) (based on Whitlow, 1988).

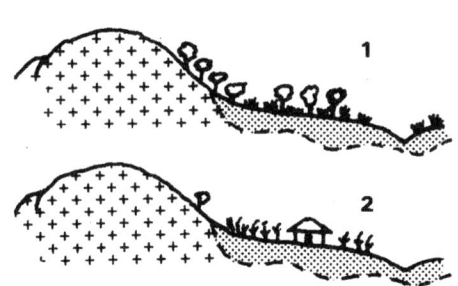
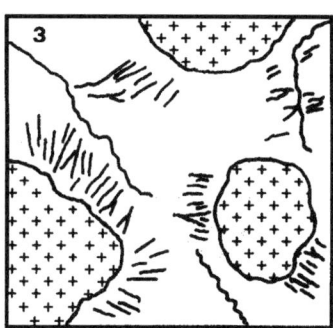

This example shows that uplands not only have environments of their own, but affect surrounding areas. It demonstrates also the combination of factors that create specific environments. For example, the seasonality of rainfall and the high intensity of the early rains interact with the stage of growth and cover of crops to affect the erodiblity of soils. The steepness and cover of slopes also affect whether and to what degree sheetwash and gully erosion occur. The mechanical composition of the soil is important; soils that are sandy and have a low organic matter content may have poorly developed aggregates so that they disintegrate when ploughed and when submitted to the impact of large raindrops. All of these factors are interrelated, but also interact with human influences. The Communal Lands have been and are under considerable pressure as the human population has increased. The greater demands for food must be met as far as possible from national production and, for the rural inhabitants, from their lands. Cultivation is extended into less suitable areas and grazing intensity increased, so that soil erosion is greater – in this example at the foot of the upland domes. As with the first case study (Box 5.1) there is a need for sustainable agriculture, but it may be even more difficult to attain in developing societies and almost certainly not without external funding and co-operation between neighbouring states. The 'common market' of southern Africa offers some solutions to problems of food shortage and the environment, but requires assistance from its European counterpart and elsewhere.

More shade tolerant   ⟶   Shade intolerant

More shade tolerant			Shade intolerant
Calluna vulgaris	*Vaccinium myrtillus*	*Deschampsia flexuosa*	*Rubus fruticosus*	*Epilobium angusti-*
	Galium saxatile	*Agrostis canina*	*Pteridium aquilinum*	*folium*
		Dryopteris dilitata		*Betula* spp.

Upland forestry may also affect fauna, particularly bird life. In the harsh upland environment many species are large and require extensive territories. Populations of such birds as hen harrier (*Circus cyaneus*), merlin (*Falco columbarius*) and golden eagle (*Aquila chrysaètos*) are likely to be affected by any dissection of open moorland by patches of forestry. Populations of the largely summer visiting waders such as curlew (*Numerius arquata*), golden plover (*Charadrius apricarius*) and dunlin (*Calidris alpiria*), which rely on invertebrate prey, may also be reduced. A comparison was made of the populations of dunlin, golden plover and greenshanks (*Tringa rubularia*) for the Caithness and Sutherland peatlands at sites before and after plantation. It was estimated that there has been an actual or predictable loss of 19 per cent of golden plovers, 17 per cent of dunlins and 17 per cent of greenshanks (Nature Conservancy Council, 1988). It is worth noting that this area has 20 per cent of black-throated divers (*Gavia arctica*), 35 per cent of dunlin and 66 per cent of greenshank breeding in the EU.

Clearly, the ecological consequences of forestry in the uplands can be severe and there is a conflict between the desire to produce more timber, pulp and jobs and the demands of the environment. Balancing economic interests and protecting the environment is made more difficult when government policies such as tax incentives and grants make ordinarily unprofitable, poor land profitable.

In the Flow Country of Caithness and Sutherland, the Forestry Commission has reduced its planting in recent years, but from about 1980 licensed a private forest company to plant trees. The company owns around twice the hectarage of the Forestry Commission. Cheap land, tax incentives and government grants made forestry an attractive investment for the company's clients. Through those tax incentives and grants it was possible to obtain a forest for less than one-third of its true cost. After ten years the forest estate could be sold with minimal capital gains tax and even at a low asking price would yield a substantial profit to the investor. The new owner would acquire the young forest at a bargain price. Changes in the 1988 Budget reduced the tax advantages, but extensive private forestry is still occurring in upland areas. The question remains as to whether the timber produced, replacing foreign imports, the jobs created and the return on public money from both public and private forestry can be weighed against the environmental damage (Anglia Television, 1988). It should be noted also that although planting of conifers may help to sequester carbon dioxide from the atmosphere and thereby help to reduce global warming, drainage of the peat prior to planting may release even more carbon dioxide and methane.

OTHER ECONOMIC ACTIVITY

Peat extraction

Ireland and Western Scotland are the type area of upland blanket peatland (Moore and Bellamy, 1974). However, in addition to pressure from reclamation and forestry these areas have had renewed exploitation for fuel. This is an excellent example of conflict in the uplands between concern for the environment and economic need of the resident population. In Northern Ireland the fuel price rises of the 1970s first encouraged more hand-cutting, but development by the early 1980s of small, lightweight machines powered by a farm tractor led to rapid expansion of machine-cutting of fuel peat. A survey in 1990 (Cruickshank and Tomlinson, 1991a, 1991b) showed that the distribution of machine-cutting was widespread, but the 1,021 sites were predominantly small – 64 per cent less than 1.0 ha – and occupied only 2.6 per cent of the blanket peat area. Almost three-quarters of the area was on peatland which had formerly been hand-cut and only 14 per cent occurred on intact peatland. Nevertheless, since this peatland type is relatively rare in Northern Ireland, comprising 10 per cent of the blanket peatland, concern was expressed about this exploitation. Research into the impacts on the peatland has not been extensive and

Table 5.1 Effects of extrusion cutting on vegetation cover and species composition of a lowland blanket bog. Visual estimates of percentage cover

Species	Number of cuts		
	0	*2*	*4–5*
Calluna vulgaris	60	35	0
Eriophorum vaginatum	15	15	3
E. angustifolium	2	+	+
Trichophorum cespitosum	10	10	2
Molinia caerulea	2	+	0
Narthecium ossifragum	+	0	+
Sphagnum spp.	25	10	0
Hypnum cupressiforme	10	5	0
Campylopus introflexus	0	0	+
Polytrichum spp.	5	+	+
Bare ground	+	15	95

+: the plant is present but of low cover value
Source: Bayfield *et al.*, 1991

views are based largely on visual inspection. Bayfield *et al.* (1991) list a number of impacts of extrusion cutting including crushing and abrasion of the plants by wheels; changes to the surface topography, drainage and peat structure; burial of plants by the extruded peat; and changes to the physical structure of the subsurface peat chiefly through aeration and drainage caused by the slits. They also present a table indicating changes in plant species composition following different numbers of cuts (Table 5.1).

Laverty (1988) has also shown the effects of extrusion cutting on the peatland flora. On sites cut in successive years the cover built up quite quickly so that by the fourth year there was little bare ground. However, he noted that *Eriophorum* species and *Trichophorum cespitosum* dominated any plant cover in the early years after cutting and even after four years their percentage cover was higher than in the uncut sample (Figure 5.7).

Impacts are often more severe in wet seasons. For example, in the wet summers of 1985 and 1986 much of the extruded peat was not harvested and the vegetation was buried for a long period. Observations at sites have shown that it has taken several years for the extruded peat rolls to break up and for a vegetation cover to re-establish. *Eriophorum vaginatum* dominated this cover for the first few years of re-colonisation and even today its occurrence in rows may be seen. Future expansion of this form of extraction will, if these observations are confirmed by experimental

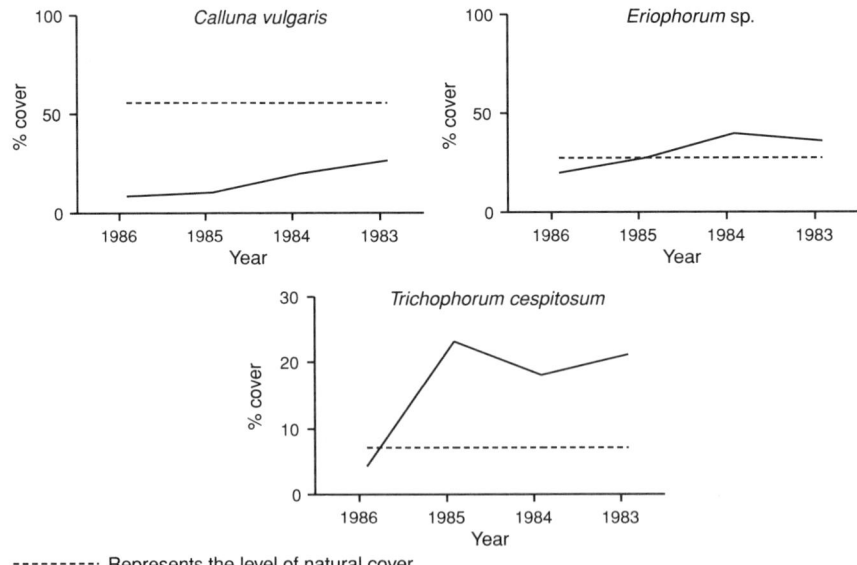

Figure 5.7 Percentage cover of dominant bog species in 1987 on peat areas machine-cut between 1983 and 1986
Source: Laverty, 1988

results, bring about significant change to a valued vegetation and habitat type. Impacts on animals and invertebrates have received little attention until recently and there are few results that may be quoted. It should be noted, however, that the margins of upland blanket bog are areas for migrant waders and any increase in the extent of machine-cutting will disturb these birds (see also p. 97).

Loss of plant species has economic as well as biological consequences. Some farmers have claimed successfully in the courts that peat extraction has damaged their livelihood. They allege that their incomes have been reduced because the quality of grazing has been reduced, adversely affecting the growth and finish of their sheep. A recent survey of attitudes to machine-cutting has shown that 65 per cent of graziers interviewed believed that machine-cutting damaged peatland.

Blanket peatland is an important element in the attractive landscapes of Northern Ireland; indeed many uplands have been designated as Areas of Outstanding Natural Beauty and yet they contain a high proportion of existing extraction sites which, at least temporarily, scar the landscape. Evidence suggesting that machine-cutting is regarded as deleterious to landscape quality is limited, but the possible implications for countryside conservation, tourism and associated industries should not be ignored (Cruickshank et al., 1995).

Effects of peat extraction are not restricted to the peatland. Since most cutting is on small sites and non-commercial, effective drainage works and associated settling ponds are not constructed so that peat particles enter stream headwaters. If these particles are deposited on the gravel spawning beds and nursery areas of salmonid fish, fish stocks are reduced and the quality of angling jeopardised. Angling as a whole contributes £8–10 million per annum to the Northern Ireland economy and while only part of that comes from rivers affected by peat extraction, there is still potential for both increased loss of species and higher monetary costs. Water quality for domestic and industrial consumption may be affected, imposing greater filtration costs.

In view of all the attendant disadvantages, it may well be asked why machine-cutting has not been prohibited. The disadvantages must be set in the context of the social environment. In general, areas affected have high unemployment rates and lie within the Severely Disadvantaged Areas where the majority of farm businesses are too small to support families at a reasonable standard of living, even with subsidies. Poverty, unemployment and under-employment have to be relieved by part- or full-time jobs outside farming or by emigration. Other employment opportunities are few and the physical environment restricts farm diversification. One way of increasing income is to buy a machine to either contract cut, or cut fuel peat for sale. Since the annual fuel needs of a household may be cut in one or two hours, the number of people owning a machine will be limited. If income cannot be expanded, an alternative is to reduce household expenditure. It has been estimated (Cruickshank and Tomlinson, 1991b) that heating costs can be halved by having the peat cut for fuel.

How is this conflict between economic need of residents and the needs of conservation to be resolved? Planning restrictions are presently of little help since they apply only to overtly commercial operations. Extension of planning laws is impractical in the present circumstances of land and peat-cutting rights and would require an army of monitors since only the largest of sites can be detected by satellites. Even were it possible to introduce new restrictions, there would be strong opposition from the resident population who would point out their disadvantaged position and that peat cutting is a tradition – only the technology has changed. Rather than enforcement, compromise is required so that the best peatland sites can be preserved and the remainder cut according to agreed guidelines to minimise damage. This could make extraction more expensive, especially if access roads and drains had to be constructed. Nor would such guidelines necessarily serve the demands of some conservationists. For instance, ill-informed peat conservation campaigns based on information from England, where peatland is rare, might not be mollified. Those concerned for birds recognise that they require expanses of peat rather than sites. It may be necessary, therefore, to provide a mechanism similar to the Environmentally Sensitive Areas (ESAs), but instead of paying farmers to practise traditional, non-intensive agriculture, payment could be made for hand-cutting or a low intensity machine-cutting.

Water storage

Water extraction and storage present both advantages and problems for mountain areas. The creation of artificial reservoirs may lead to local disputes and public inquiries. Not only is the scenery changed, but villages may be destroyed. The Derwent Reservoir in the Peak District and Thirlmere Reservoir in the Lake District (both in northern England) are good examples. These were constructed early this century to supply drinking water to the cities of the East Midlands and Manchester respectively. Although flooding of villages is unlikely in the UK today, there are still sites of significant importance which need to be protected. When new reservoirs are built there is usually an attempt to make them into tourist attractions, but these may have attendant effects on scenery, vegetation and soils of the surrounding countryside. Control of the watershed may also present problems. Planting trees is one way to control runoff, but blocks of alien tree species (usually conifers) are now considered a visual blight and more scenically sympathetic planting is required.

In the high alpine mountains, water storage schemes for hydroelectricity generation play a nationally important economic role. Dam construction provides jobs, but also hazards. In Switzerland, the Mattmark disaster of 1965 killed 40 Italian workers when a section of the glacier fell on their camp (Vivian, 1966). A knowledge of snow hydrology and glaciology therefore is an important part of designing these storage schemes. The capital costs of these schemes are high, but the balance between development and conservation is significant because hydro-electricity is essentially a renewable resource.

RECREATION

Disturbance and damage to plants and animals

Increased affluence and leisure time have brought new conflicts into the countryside and the uplands and mountains have not escaped. Vegetation has been damaged by trampling, soil erosion has increased, animals have been disturbed and whole areas changed completely by planned removal of vegetation, redesign of slopes and construction of ski resorts.

One of the upland areas in Britain that demonstrates the conflict between conservation and recreation is the Cairngorms in north-central Scotland. As long ago as 1967, the Cairngorm Area Report (Scottish Development Department, 1967) drew attention to the high number of nature reserves and sites of special scientific interest and the desirability of conservation of the area. At the same time, there was acknowledgement of the growing demand for leisure facilities, particularly walking and skiing, which provided a boost to the economy of this relatively isolated region with few natural advantages. The plan proposed developments for recreation even though it was noted that use of the hills was causing damage, even extending into local nature reserves (Watson, 1967).

Bayfield (1971) attributed most damage to construction of ski-lifts and associated facilities and to the use of tracked vehicles, and noted adverse impacts of trampling through walking and skiing. He monitored the actual impact of walking and skiing on areas close to visitor locations and also conducted a series of experiments which simulated the human activities. The simulations are particularly interesting since they imposed a measure of control so that effects on individual plant species could be demonstrated. In a small area, near a car park by a ski-tow, which was visited infrequently, there was a typical blanket peat vegetation including *Trichophorum cespitosum*, *Calluna vulgaris*, *Eriophorum vaginatum*, *E. angustifolium*, *Erica tetralix* and *Sphagnum* species. Plots were arranged in randomised blocks and subjected to different trampling treatments. Upon completion of the trampling, the plots were assessed for damage and monitored subsequently. This showed that 34 per cent of *Trichophorum* plots were damaged under maximum impact, but recovery was quite rapid, so that after only 12 months, only 12 per cent of the plot showed damage. In the case of *Calluna*, damage was more serious and recovery slow. At the start of the experiment the plot to undergo maximum trampling had 35 per cent cover of live heather, but even 23 months after completion of the experiment, there was still only 18 per cent cover. *Sphagnum* species also showed very little recovery after 23 months.

In a subsequent study, Bayfield (1979) established three groups of species according to their susceptibility to damage (Table 5.2). The most susceptible may be related to their position in the micro-topography. They tend to be species of the drier, hummock sites where snow cover would be thinner and therefore protection less. Studies elsewhere have shown that heather is susceptible to trampling (Bayfield and Brookes, 1979; Harrison, 1981) and also that *Sphagnum* species are susceptible and slow to recover (Slater and Agnew, 1977). Great care has to be taken when conducting research on peat bogs since periodic visits to the same monitoring points can damage the surface quite severely and it is often necessary to construct a raised path for the research period. More permanent raised paths have to be established where long-distance walks cross susceptible areas. A long-distance trail, the Lyke Wake Walk, which traverses the North Yorkshire Moors, passes through Fen Bogs. Several thousand walkers each year have eroded the path on the surrounding moorland so that it is now down to bedrock in places. Were it not for a raised 'bridge', irreparable damage to this Grade 1 Site of Special Scientific Interest would have resulted. The huge number of walkers along the Pennine Way, another long-distance footpath, has caused similar problems, but on a larger scale, necessitating not only control measures but also re-establishment of a vegetation cover (Phillips *et al.*, 1981; Tallis and Yalden, 1983).

The Mourne Wall Walk in Northern Ireland (35 km around the periphery of the Water Services catchment area) has been suspended because of footpath erosion caused during the walk and the many 'practice runs' (Lowther and Smith, 1988). Research has shown that here, as on other long-distance paths, the extent of erosion depends on slope angle and soil and vegetation type. Peat is particularly vulnerable.

Evidence of disturbance of animals is more difficult to obtain. Watson (1979) showed an increase of reindeer (*Rangifer tarandus*) near the Cairngorm ski areas, possibly because of improved grazing from revegetated areas, and that populations of red grouse (*Lagopus lagopus scoticus*) and ptarmigan (*Lagopus mutes*) had not declined despite a loss of heather cover. The higher nutritive value of remaining heather and other new vegetation compensated for this loss. Populations of scavenging birds increased and had the undesirable side effect of altering the bird populations of nearby nature reserves.

Tourism in high mountains

Many mountain communities have traditionally relied on transhumance – the migration of people with flocks or herds to higher shelters for summer grazing. Although this produced grazing pressures, the cyclic system was in equilibrium with the environment. With the development of tourism, this practice has largely died out in Europe and has been replaced by tourist villages and restaurants with associated roads or paths (Bätzing, 1991). As a consequence, many more people live in high regions throughout the year. New apartment-related villages have been built to cater almost entirely for the winter and spring skier (Knafou, 1978). It is not yet known what the full impact of this will be, but there are disturbing signs of enhanced erosion, greater runoff and river incision. There is evidence that the creation of pistes for skiing is detrimental to the flora of these areas in a similar way to those in Scotland, as well as increasing slope failure potential. Even in relatively remote regions of Europe there may be anthropogenic pressures which alter, if not destroy, the ecology. Mirek (1992) describes such changes in the Tatra Mountains of Poland. The general development of mountainous areas, which has usually gone hand-in-

Table 5.2 Susceptibility of species to trampling in an upland heath in the Cairngorms

Most susceptible	Initial damage high and recovery poor – Calluna, Empetrum, Sphagnum rubellum, Arctostaphylos
Moderately susceptible	Moderate to high initial damage recovery fair – Racomitrium lanuginosum, Trichophorum cespitosum, Vaccinium myrtillus, lichens
Less susceptibility	Low to moderate initial damage, increase in relative cover – Erica spp., Eriophorum spp., Carex bigelowii

Source: Bayfield, 1979

hand with enhancement of tourism and the facilities required by 'global tourism', has produced a variety of consequences in upland and mountain areas. Certain areas of the Himalaya-Karakorum chain have seen an unprecedented increase of tourism in some areas. The approach routes to mountains such as Anapurna and Chogolungma (Everest) are especially pressured. Aspects of the development in these areas are discussed in, for example, Ives and Messerli (1989) and Hewitt (1997).

The UNESCO Man and the Biosphere Project (MAB-6), 'Study of the Impact of Human Activities on Mountain and Tundra Ecosystems', has provided much useful information about the problems of development in mountain areas. Detailed studies at Obergurgl (Austria) highlight this complexity, particularly with respect to recreational demand, population and economic development, farming and ecological change, and land use and development control (Moser and Moser, 1986). In the very highest mountains, tourist trekking presents its own pressures on traditionally subsistence agricultural communities, leading to greater economic dependence on the world beyond the mountains (Bjønness, 1983). New routeways also bring in fresh impacts to traditional communities. Kreutzmann (1991) discusses some of these problems associated with the Karakorum Highway (KKH) from the plains in Pakistan to the Chinese border at the Kunjerab Pass.

HAZARDS

As hills give way to the mountains of an area, so the potential energy difference between peak and summit increases. Further, the range of climate, soil and vegetation types also increases in number and perhaps size. These differences between hills and mountains are especially important with respect to the hazards involved.

The formation of large water storage reservoirs has been mentioned in a previous section. Once in place, the lakes may be promoted as scenic tourist attractions as well as the dams themselves, especially gravity arch dams. Such tourism in the summer, coupled with winter sports, has placed a considerable stress on the road systems in alpine areas. Enlarging roads to accommodate the extra traffic itself increases the

problems where rock walls have to be blasted. In extreme cases, such as the Brenner Pass between Italy and Austria, completely new roads have to be engineered along and across the valley. Yet rerouting of old roads is costly and may also present new dangers or place travellers in new dangers. The development of the Karakorum Highway is one such zone, where sudden rains bring landslides, mudflows, rockfalls and associated hazards. These, and other hazard-high risk areas and phenomena are discussed in Hewitt (1992, 1997) and Schroder (1989) with other examples in the books by Gerrard (1990), Price (1981) and Allan *et al.* (1988). These books show the often complex interrelationships between development and consequences. The sections below mention several specific hazard phenomena.

Glaciers

The presence of glaciers produces a very special hazard to mountain areas. In the past, glacier advances, such as during the Little Ice Age of the sixteenth to nieteenth centuries, produced considerable economic problems in certain alpine areas. The advances and associated geomorphologic events in western Norway have been examined by Grove (1972) using old tax records. In many cases, fields were covered by advancing glaciers or their moraines. A comprehensive review is given by Tufnell (1984). Although there is a tendency to think of a current 'global warming' with a concomitant recession of glaciers, this is not always the case. Of 109 Swiss glaciers monitored in 1982–1983, some 46 were advancing and 57 retreating. However, in 1988–1989, of 105 glaciers, 19 were advancing and 83 retreating and in 1990–1991, of 109 glaciers, only 8 were advancing and 100 retreating. For places such as Switzerland, these changes have little human impact, but snout movements associated with glacier lakes still do have important consequences. Grubengletscher in the Swiss Canton of Wallis has a small ice dammed lake which drained subglacially (usually called jökulhlaups or self-dumping lakes) in the 1970s. A sudden surge of water carved out bank and bed material which was carried rapidly down to the valley. Fortunately, although the debris flow produced passed through the centre of the village, there was no loss of life. A similar event occurred in 1980 at Gupis in the

Ghizar valley in north-west frontier province, Pakistan. That neither directly involved loss of life was probably because these occurrences were common in the historical past (indeed some jökulhlaups are annual events) and so a 'freeway' was left for them. Unfortunately, two men were killed in trying to cross the flow, but this was after the event and before the mud had fully de-watered so the surface was not yet stable. At Grubengletscher, there was enough money available to provide a permanent solution in the form of a subglacial drain which means that the lake will never be allowed to fill completely. It is rarely possible to provide such a permanent solution. Ice-dammed or moraine-dammed lakes have been reported in many parts of the world and any of these may undergo catastrophic failure and consequent flooding, landsliding and debris flows (Hewitt, 1982; Tufnell, 1984; Haeberli *et al.*, 1989).

Rockfalls

Rockfalls are an almost everyday hazard in the mountains, but there have been large falls of a million cubic metres or more in recent history. The most famous of these were at Elm, Switzerland and Hope, Canada. Both examples became airborne on a cushion of air and thus travelled further than they would have if they had slowly settled to the valley floor. This type of event is often referred to as a Berghlaup or Sturzström. Although extremely rare and unlikely to be predicted, they are significant events that need to be taken into account when making long-term plans. More recently, a section of Mt Cook in New Zealand fell away and, most catastrophically, the face of Mt Huascaran in Peru. The latter was particularly significant, as a mud and debris flow was initiated as well. The loss of life was chiefly by the mudflow through the streets of a town (Morales, 1966). Discussion of a variety of slope failures, their mechanisms and effects can be found in Brunsden and Prior (1984).

Avalanches

Avalanches are considered a hazard, particularly by skiers, but they are the normal way in which mountain slopes shed their snow load. They only become a hazard when a road is placed in an avalanche track.

Indeed, tracks of avalanches are generally well known and roads and railways can be protected. This is usually done with sheds or with more costly tunnels. However, when the full costs of employing snow-ploughs in hazardous locations are considered, tunnels may be more cost-effective. Avalanches differ markedly in their type according to the season and type of snow. This is a very complex field, and local knowledge is important in making predictions; for instance, closing a piste costs money. Attempts are now being made to use expert computer systems to help with analysing avalanche paths (Buisson and Charlier, 1989).

Avalanches not only occur on very steep slopes, but also present a particular problem in gullies. High angle gullies in mountaineering areas are obviously risky but hunters or cross-country skiers can cross avalanche terrain just above the timber line. Because of the channelling effect, even small avalanches have increased potential and may be lethal. Once submerged by snow, victims rarely survive.

Snowmelt and debris flows

Rapid snowmelt can produce similar effects as jokulhlaups, that is scoured stream sides and mudflows. At the very least, the floodwater will be fast and contain mud and boulders, giving it a high density. The increased density itself leads to more erosion and a higher kinetic energy. As with avalanches, the paths of streams are usually well known. However, the power of high density streams that are created rapidly in spring is not always appreciated, especially in areas that have been newly developed for tourism or industry. The effects of forests on stream hydrographs have been discussed above, but they also create particular hazards related to snowmelt and landsliding (Furbish and Rice, 1983).

Even relatively low-relief areas may create problems. One of the best known of these was the Lynmouth floods in 1952. Runoff from an intense storm in a 'flashy' upland catchment on the northern margin of Exmoor, England, was channelled down through the seaside village. Seven people were killed and much property destroyed (Kidson, 1953). More recent occurrences have been noted from the French and Italian Alps. Another well-known example is that

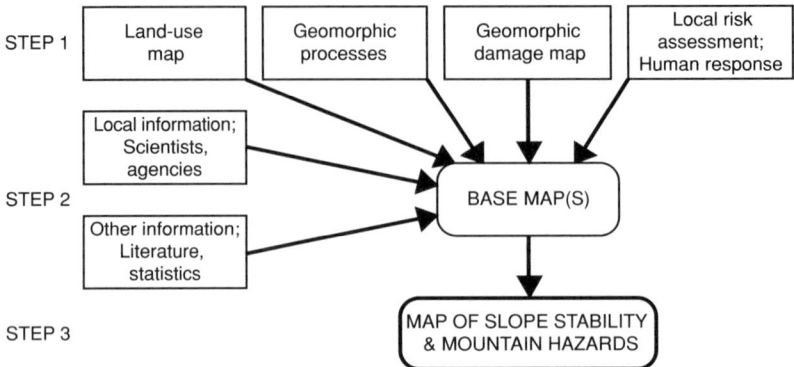

Figure 5.8 Plan of research design for the UNESCO-MAB project in the Kathmandu-Kakani area, Nepal
Source: Kienholz *et al.,* 1983

at Wrightwood on the eastern slopes of the Sierra Nevada, California (Sharp and Nobles, 1953). Here rapidly melting snows produced streams which then became debris flows with high erosion potential.

Slopes and hazard mapping

The potential difference in height between peaks and valley floors, together with steep mountain slopes, means that size and intensity of failures increase with an increase in these factors. However, there are other considerations, such as rainfall intensity and particularly the relationship between vegetation, soils and geology. One way in which progress has been made in evaluating some of the potential hazards is by providing a map or GIS-related database. A good example of the technique and illustration of the complexities involved is given by Kienholz *et al.* (1983) in a report on the UNESCO-MAB project in the Kathmandu-Kakani area of Nepal. Figure 5.8 shows a plan for the research design of this project. The landslide hazard mapping in Tuscany by Brunori *et al.* (1996) is an example of the way in which GIS and statistical methods can be employed in evaluation. An engineering-oriented approach is given by Gostelow *et al.* (1997) in their study of slope instability associated with hilltop towns in southern Italy. A traditional, mapping-based approach to mountain environments and landforms can be seen in the work of Kotarba (1992) in the Tatra Mountains (Poland). Geomorphological features have been used as indicators of environmental impact by

Cavallin *et al.* (1994) and Rivas *et al.* (1997) in a methodological approach in which impact is expressed as magnitudes or as fractions of some maximum theoretical value.

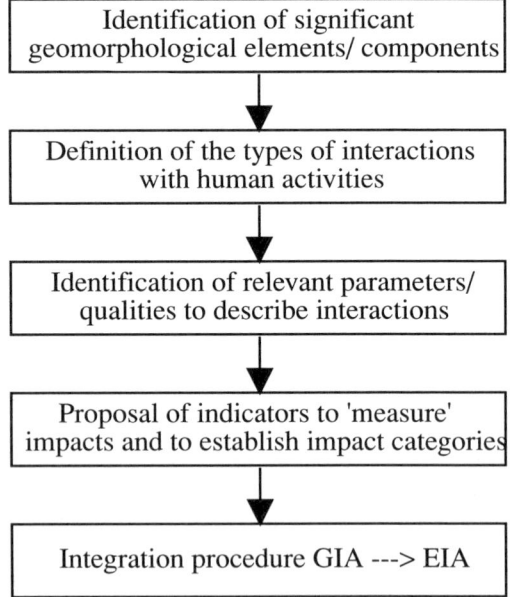

Figure 5.9 Methodological flow diagram for the production of an environmental impact assessment (EIA) from geomorphological components through geomorphological impact assessment (GIA)
Source: Rivas *et al.,* 1997

CONCLUSIONS

All the examples given in this chapter have shown that physical factors of the environment cannot be viewed in isolation and that human activities have interrelated impacts. For example, overgrazing may be manifest in changing plant species and in plant cover, but may also lead to erosion and to acidification of streams and changes in their flora and fauna. Equally, afforestation, peat cutting, burning of heather moorland and increased recreational activities have diverse impacts. Many of the examples have also shown that there should be no conflict between conserving the environment and its use by humans; their welfare and prosperity rely on using the uplands in a sustainable manner. Keeping more sheep on the hills can bring greater short-term rewards in headage payments, but if this leads to loss of plant cover and to soil erosion, there is no long-term future for upland grazing.

> Sustainability is not just about tackling global warming and the hole in the ozone layer. Closer to home, it is also about making sure that the quality of the country-side – its beauty and diversity – can be conserved and enhanced for the enjoyment of this and future generations.
>
> (Editorial, *Countryside* 63, 1993)

To which we should add that the needs of the residents are important and that policies for sustainability need to be worked out with them rather than for them.

REFERENCES

Allan, N.J.R., Knapp, G.W. and Stadel, C. (eds) (1988) *Human Impact on Mountains*, Totawa, NJ, USA: Rowman and Littlefield.

Anglia Television (1988) *Paradise Ploughed: A Survival Special*, Broadcast 9 February 1988. Accompanying brochure of same title.

Ball, D.F., Dale, J., Shaeil, J.E. and Heal, O.W. (1982) *Vegetation Change in Upland Landscapes*, Cambridge, UK: Institute of Terrestrial Ecology, NERC.

Barry, R,G, (1992) *Mountain Weather and Climate*, London, UK: Routledge.

Bätzing, W. (1991) *Die Alpen. Entstehung und Geföhrdung einer europäischen Kulturlandscaft*, Munich, Germany: C.H. Beck (in German).

Bayfield, N.G. (1971) 'Some effects of walking and skiing on vegetation at Cairngorm', in E. Duffy and A.S. Watt *The scientific Management of Animal and Plant Communities for Conservation*, Oxford, UK: Blackwell, pp. 469–486.

Bayfield, N.G. (1979) 'Recovery of four montane heath communities on Cairngorm, Scotland, from disturbance by trampling', *Biological Conservation* 15: 165–179.

Bayfield, N.G. and Brookes, B.S. (1979) 'Effects of repeated use of an area of heather, *Calluna vulgaris* (L) Hull, moor at Kindrogan, Scotland, for teaching purposes', *Biological Conservation* 16: 31–41.

Bayfield, N.G., Picozzi, N., Staines, B.W., Crisp, D.T., Tipping, E., Carling, P., Robinson, M., Gustard, A. and Shipman, P. (1991) *Ecological Impacts of Blanket Peat Extraction in Northern Ireland: Summary*, Belfast, Northern Ireland: Report to Wildlife and Countryside Branch, Department of the Environment.

Betts, N.L. (1982) 'Climate', in J.G. Cruickshank and D.N. Wilcock (eds) *Northern Ireland – Environment and Natural Resources*, Queen's University, Belfast, Northern Ireland.

Bjønness, I.-M. (1983) 'External economic dependency and changing human adjustments to marginal environment in the high Himalaya, Nepal', *Mountain Research and Development* 3: 263–272.

Brunori, F., Casagli, N. and Fiaschi, S. (1996) 'Landslide hazard mapping in Tuscany, Italy: an example of automatic evaluation', in O. Slaymaker *Geomorphic Hazards*, Chichester, UK: Wiley.

Brunsden, D. and Prior, D.B. (1984) *Slope Instability*, Chichester, UK: Wiley.

Buisson, L. and Charlier, C. (1989) 'Avalanche starting-zone analysis by use of a knowledge-based system', *Annals of Glaciology* 13: 27–30.

Cavallin, A., Marchetti, M., Panizza, M. and Soldati, M. (1994) 'The role of geomorphology in environmental impact assessment', *Geomorphology* 9: 143–153.

Chapman, S.B. (1964) 'The ecology of Coom Rigg Moss, Northumberland; I. Stratigraphy and present vegetation', *J. Ecol.* 52: 299–313.

Cruickshank, M.M. and Tomlinson, R.W. (1990) 'Peatland in Northern Ireland: Inventory and prospect', *Irish Geography* 23: 17–30.

Cruickshank, M.M. and Tomlinson, R.W. (eds) (1991a) *Survey of the Scale, Extent and Rate of Peat Extraction from Blanket Bogs in Northern Ireland. Summary*, Northern Ireland: Report to Countryside and Wildlife Branch, Department of the Environment.

Cruickshank, M.M. and Tomlinson, R.W. (eds) (1991b) *Socio-economic Factors Affecting Peat Extraction Practices in Northern Ireland and an Assessment of Likely Future Trends. Summary*, Northern Ireland: Report to Countryside and Wildlife Branch, Department of the Environment.

Cruickshank, M.M., Tomlinson, R.W., Bond, D., Devine, P.M. and Edwards, C.J.W. (1995) 'Peat extraction, conservation and the rural economy in Northern Ireland', *Applied Geography* 15: 365–383.

Floate, M.J.S., Eadie, J., Black, J.S. and Nicholson, A. (1973) 'The improvement of *Nardus* dominant hill pasture by grazing control and fertiliser treatment and its economic assessment', *Potassium Institute Ltd Colloquium Proceeding* 3: 33–39.

Furbish, D.J. and Rice, R.M. (1983) 'Predicting landslides related to clear-cut logging, north-western California, USA', *Mountain Research and Development* 3: 253–259.

Gerrard, A. (1990) *Mountain Environments: An Examination of the Physical Geography of Mountains*, London, UK: Belhaven Press.

Gimingham, C.H. (1972) *Ecology of Heathlands*, London, UK: Chapman and Hall.

Goldsmith, F.B. and Wood, J.B. (1983) 'Ecological effects of upland afforestation', in A. Warren and F.B. Goldsmith *Conservation in Perspective*, Chichester, UK: Wiley, pp. 287–311.

Gostelow, T.P., Del Prete, M. and Simoni, M. (1997) 'Slope instability in historic hilltop towns of Basilicata, southern Italy', *Quarterly Journal of Engineering Geology* 30: 3–26.

Greer, J. (1986) 'Land use change 1938–85 in an area of the south-eastern margins of the Mourne Mountains, Co. Down', Unpubl. BA Dissertation, Belfast, Northern Ireland: Dept of Geography, Queen's University of Belfast.

Grove, J.M. (1972) 'The incidence of landslides, avalanches and floods in western Norway during the Little Ice Age', *Arctic and Alpine Research* 4: 131–138.

Grove, J.M. (1989) *The Little Ice Age*, London, UK: Methuen.

Haeberli, W., Alean, J.-C., Miller, P. and Funk, M. (1989) 'Assessing risks from glacier hazards in high mountain regions: some experiences from the Swiss Alps', *Annals of Glaciology* 13: 96–102.

Harrison, C. (1981) 'Recovery of lowland grassland and heathland in southern England from disturbance by seasonal trampling', *Biological Conservation* 19: 119–80.

Hewitt, K. (1982) *Natural Dams and Outburst Floods of the Karakoram Himalaya; Hydrological Aspects of High Mountain Areas*, Publication 138, International Association of Scientific Hydrology, pp. 259–269.

Hewitt, K. (1992) 'Mountain hazards', *GeoJournal* 27: 47–60.

Hewitt, K. (1997) *Regions of Risk. A Geographical Introduction of Disasters*, London, UK: Longman.

Hill, M.D. (1979) 'The development of a flora in even-aged plantations', in E.D. Ford, D.C. Malcolm and J. Atterson *The Ecology of Even-aged Forest Plantation*, Cambridge, UK: ITE, pp. 175–192.

Ives, J. and Messerli, B. (1989) *The Himalayan Dilemma: Reconciling Development and Conservation*, London, UK: Routledge.

Kidson. C. (1953) 'The Exmoor storm and Lynmouth floods', *Geography* 38: 1–9.

Kienholz, H., Hafner, H., Schneider, G. and Tamrakar, R. (1983) 'Mountain hazards mapping in Nepal's Middle Mountains with maps of land use and geomorphic damages (Kathmandu-Kakani area)', *Mountain Research and Development* 3: 195–220.

Knafou, R. (1978) *Les stations intégrées de sports d'hiver des Alpes françaises*, Paris, France: Masson (in French).

Kotarba, A. (1992) 'Natural environment and landform dynamics of the Tatra Mountains', *Mountain Research and Development* 12: 105–129.

Kreutzmann, H. (1991) 'The Karakoram Highway: the impact of road construction on mountain societies', *Modern Asian Studies* 13: 19–39.

Laverty, A. (1988) 'An investigation to find how mechanical cutting of peat in the Glendun area has affected the peat and its associated flora', Unpubl. B.Sc. Dissertation, Belfast, Northern Ireland: Dept of Geography, Queen's University of Belfast.

Lowther, K.A. and Smith, B.J. (1988) 'The environmental impact of recreation in upland areas: a case study of footpath erosion in the High Mourne Mountains, Co. Down', in W.I. Mongomery, J.H. McAdam and B.J. Smith *The High Country*, Northern Ireland: Institute of Biology and Ireland: Geographical Society, pp. 56–71.

Macgillivray, N. (1994) 'Cairnsmore of Fleet', *Enact – Managing Land for Wildlife* 2: 7–8.

Ministry of Agriculture, Fisheries and Food (MAFF) (1964) *The Farmer's Weather*, London, UK: HMSO.

Mirek, Z. (1992) 'Threats to the natural environment in the Polish Tatra Mountains', *Mountain Research and Development* 12: 193–203.

Moore, P.D. and Bellamy, D.J. (1974) *Peatlands*, London, UK: Elek. Science.

Morales, B. (1966) 'The Huascaran avalanche in the Santa valley, Peru', in *International Symposium on the Scientific Aspects of Snow and Ice Avalanches*, Davos: International Association of Scientific Hydrology, Publication 69, pp. 304–15.

Moser, P. and Moser, W. (1986) 'Reflections of the MAB-6 Obergugl Project and tourism in an alpine environment', *Mountain Research and Development* 6: 101–118.

Muirburn Working Party (1977) *A Guide to Good Muirburn Practice*, Edinburgh, Scotland, UK: HMSO.

Nature Conservancy Council (1986) *Nature Conservation and Afforestation in Britain*, Peterborough, UK: NCC.

Nature Conservancy Council (1988) *The Flow Country*, Peterborough, UK: NCC.

Nisbet, T.R. (1992) *Forests and Surface Water Acidification*, Forestry Commission Bulletin 86. London, UK: HMSO.

Parry, M.L., Bruce, A. and Harkness, C.E. (1982) *Changes in the Extent of Moorland and Roughland in the Yorkshire Dales National Park*, Report 9 of the Moorland Change Project, Birmingham, UK: University of Birmingham.

Pears, N. (1967a) 'Present tree lines of the Cairngorm Mountains, Scotland', *J. Ecol.* 55: 815–830.

Pears, N. (1967b) 'Wind as a factor in mountain ecology: some data from the Cairngorm Mountains, Scotland', *Scott. Geog. Mag.* 83: 118–124.

Phillips, J., Yalden, D. and Tallis, J. (1981) *Peak District Moorland Erosion Study. Phase 1 Report*, Bakewell, Derbyshire, UK: Peak Park Joint Planning Board.

Porchester H.G. (1977) *A Study of Exmoor*, London, UK: HMSO.

Price, L.W. (1981) *Mountains and Men*, Los Angeles and Berkeley, Calif., USA: University of California Press.

Rawes, M. (1981) 'Further results of excluding sheep from high-level grasslands in the north Pennines', *J. Ecol.* 69: 651–670.

Rivas, V., Rix, K., Francés, E., Cendredo, A. and Brundsen, D. (1997) 'Geomorphological indicators for environmental impact assessment: consumable and non-consumable geomorphological resources', *Geomorphology* 18: 169–182.

Schroder, J.F. Jr. (1989) 'Hazards of the Himalaya', *American Scientist* 77: 564–572.

Scottish Development Department (1967) *Cairngorm Area. Report of the Technical Group on the Cairngorm Area of the Eastern Highlands of Scotland*, Edinburgh, Scotland, UK: HMSO.

Sharp, R.P. and Nobles, L.H. (1953) 'Mudflow of 1941 at Wrightwood, southern California', *Geological Society of America Bulletin* 64: 547–560.

Slater, F.H. and Agnew, A.D.Q. (1977) 'Observations on a peat bog's ability to withstand increasing public pressure', *Biological Conservation* 11: 21–27.

Staver, C.P., Byers, A.C., Ravelo, A. and Dickinson, J.C. (1991) 'Refining soil conservation strategies in the mountain environment – the climatic factor: an Ecuadorian case study', *Mountain Research and Development* 11: 127–144.

Stoner, J.H., Gee, A.S. and Wade, K.R. (1984) 'The effects of acidification on the ecology of streams in the upper Tywi catchment in west Wales', *Environmental Pollution A* 35: 125–157.

Tallis, J.H. (1973) 'Studies on southern Pennine peats V. Direct observations on peat erosion and peat hydrology at Featherbed Moss', *J. Ecol.* 61: 1–22.

Tallis, J.H. and Yalden, D.W. (1983) *Peak District Moorland Erosion Study. Phase 2 Report: Re-vegetation Trials*, Bakewell, Derbyshire, UK: Peak Park Joint Planning Board.

Tomlinson, R.W. (1979) 'Water levels in peatlands and some implications for runoff and erosional processes', in A. Pitty, *Geographical Approaches to Fluvial Processes*, Norwich, UK: Geobooks, pp. 149–162.

Tufnell, L. (1984) *Glacier Hazards*, London, UK: Longman.

Vivian, R. (1966) 'La catastrophe du glacier Allalin', *Revue de geographie alpine* 54: 97–112 (in French).

Wagner, A. (1930) 'Über die Feinstruktur des Temperaturgradienten langs Berghängen', *Zeit. Geophys.* 6: 310–318 (in German).

Wall, T. (1993) 'Burning issues', *Enact – Managing Land for Wildlife* 1: 12–14.

Watson, A. (1967) 'Public pressures on soils, plants and animals near ski lifts in the Cairngorms', in E. Duffy (ed.) *The Biotic Effects of Public Pressures on the Environment*, Peterborough, UK: NERC.

Watson, A. (1979) 'Bird and animal populations in relation to human impact at ski lifts on Scottish hills', *J. Ecol.* 16: 753–764.

Watt, A.S. (1955) 'Pattern and process in the plant community', *J. Ecol.* 43: 490–506.

Welch, D. and Rawes, M. (1964) 'The early effects of excluding sheep from high-level grasslands in the northern Pennines', *J. Appl. Ecol.* 1: 281–300.

Whitlow, J.R. (1988) *Land Degradation in Zimbabwe*, Zimbabwe: Department of Natural Resources.

SUGGESTED READING

Enact – Managing Land for Wildlife is published by English Nature four times a year and gives details of management for sites of nature conservation interest. A good source for case examples.

Goudie, A.S., Brunsden, D., Collins, D.N., Derbyshire, E., Ferguson, R.I., Hashmet, Z., Jones, D.K.C., Perrot, F.A., Said, M., Waters, R.S. and Whalley, W.B. (1984) 'The geomorphology of the Hunza valley, Karakoram Mountains, Pakistan', in K.J. Miller (ed.) *The International Karakoram Project*, Vol. 2, Cambridge, UK: Cambridge University Press, pp. 359–410.

Perry, A.H. (1981) *Environmental Hazards in the British Isles*, London, UK: Allen and Unwin.

SELF-ASSESSMENT QUESTIONS

1 In parts of Britain, heathland is managed for sheep grazing and for grouse moor by the use of fire. Which of the following practices should *normally* be followed if heathland vegetation is to be preserved?
   (a) Fires should be of (i) high intensity (ii) low intensity (iii) moderate intensity.
   (b) Burning should be in (i) high winds (ii) gentle winds.
   (c) Burning should be carried out on (i) old heather (ii) young heather (iii) heather in the building phase.
2 Show, by examples, that high stocking levels can
   (a) improve the quality of grazing;
   (b) reduce species diversity and grazing quality.

3 Show, by examples, that low stocking levels can
   (a) lead to reversion of marginal grasslands;
   (b) lead to improved grass cover but lower species
        diversity.
4 Attempt a construction of a flow diagram that shows
   how planting of conifers on an upland peat-covered
   hill mass can damage the environment.
5 How may recreational activities damage plant
   communities in upland and mountain areas?
6 On Figure 5.10 examine the potential hazards carefully and list them according to the following criteria:
   (a) Frequency of occurrence.
   (b) Potential magnitude.
   (c) Ability to prevent them.
   (d) Ability to control or mitigate their effects.
   (e) Imagine that this area is to be developed as a
        ski and tourist resort with hotels, car parks,
        shops, ski-tows, ski slopes and summer walking
        areas. Take your list of 'hazard potentials' and
        explain how you would develop the area with
        regard to safety and cost to the environment.

Figure 5.10 Question 6 illustration

6

SAVANNAS

R. Laurie Robbins and Thomas A. Eddy

SUMMARY

Savannas are variously described. Some limit the term to grasslands with scattered broadleaf and evergreen trees bordering tropical rain forests. In the broadest sense, savannas include both temperate and tropical grasslands with scattered woody vegetation, with transitions into deserts at one extreme and forests at the other. In either definition, they receive (250) 500–1,500 (2,000) mm rainfall per year, and support perennial grasses, often on relatively fertile soils. These characteristics have long made this system highly desirable for grazing domestic animals and for cultivation. The open nature of savannas, especially in areas with woody vegetation, is maintained by fire.

With increasing human settlement, *natural* fires have decreased, and adjacent woody vegetation has encroached into areas not converted to crop production. Those areas often represent prime agricultural land for production of annual grains and legumes. Yet other areas are under severe grazing pressure. These human intrusions, in combination with increasing urbanisation and with normal, fluctuating weather patterns, result in systematic degradation of the land and loss of biodiversity. Important issues to consider in effective management and recovery of these lands include political and practical issues (grazing methods, cropping systems, exotic species), goal-setting, planning, research, inventory and restoration plans.

ACADEMIC OBJECTIVES

In this chapter, the reader will gain an understanding of the basic dynamics of the diverse continental savannas as well as insights into their management. After careful examination of the contents, the following concepts, principles and management processes should be evident:

1 The role of the primary determinants of fire and herbivory in shaping the formation and maintenance of savannas.
2 The historical stages of human interaction with savannas and how humans have impacted the ecology of the system (cropping, grazing and browsing, fire frequency and preservation).
3 The use of modern management strategies to improve the productivity of savannas.
4 The process of planning and carrying out the restoration of degraded savannas.
5 The effect of exotics (plants and animals) on the savanna ecosystem and the management of these aliens.

INTRODUCTION

The term 'savanna' has been used variously in the literature. Definitions can be broadly based on either general appearance (grassland with scattered shrubs and trees) or ecology (tropical, subtropical, or temperate areas intermediate between deserts and forests; seasonality; and patterns and amounts of water availability). Both definitions include temperate as well as subtropical and tropical environments. For a thorough discussion of the historical and proposed future use of the term 'savanna', refer to Eiten (1986). When examining literature regarding a particular geographic area, biome or ecosystem, the reader will find that the authors invariably provide a definition of the area under discussion. For a summary of various savanna/woodland classification schemes based on percentage of canopy cover, see Figure 6.1.

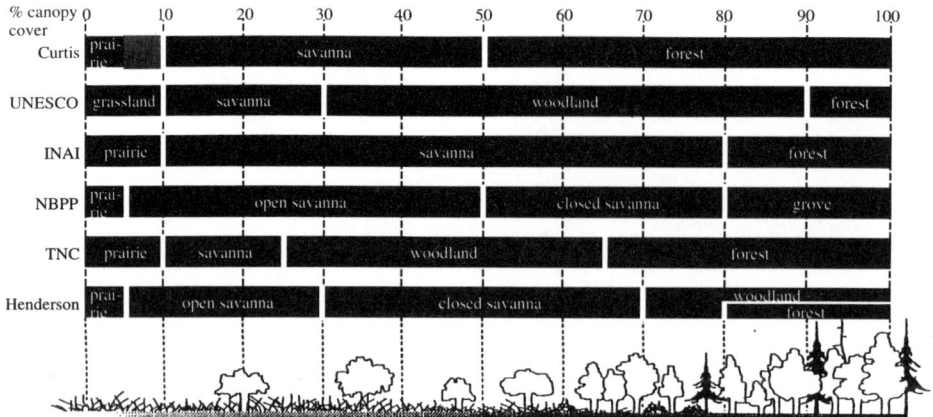

Figure 6.1 Savanna/woodland classification systems
Forests: multilayered with well-developed woody understoreys; prevalence of spring ephemerals and fire-sensitive herbs; absence or reduced importance of summer- and fall-blooming groundlayer plants; infrequent to rare fire.
Woodlands: two-layered, with little or no woody shrub or understorey layer; summer- and fall-blooming groundlayer plants prevalent and productive; frequent, but low intensity, fire.
Source: Botts et al., 1994

Figure 6.1 appears in *Midwest Oak Ecosystems Recovery Plan: A Call to Action* (1995). It is adapted from Attachment 1 in Botts *et al.* (1994), and was originally based on a sketch by Don Faber-Langendoen. Curtis (1959) did not use per cent cover to distinguish prairie (grassland) from savanna; instead he indicated that a savanna has a minimum of one mature tree per acre, and a prairie has less than one mature tree per acre. Other system citations are UNESCO (1973); INIA, Illinois (USA) Natural Areas Survey (White and Madany, 1978); NBPP, North Branch Prairie Project, Illinois (USA) (no citation); TNC (Faber-Langendoen, 1994); Henderson (in Botts *et al.*, 1994).

In defining savannas, most authors agree that climate, soils and fire, and to a lesser extent, herbivory, are the major determinants of this particular ecosystem. Walker (1985) groups these into 'fixed determinants' – rainfall regime and soils – and 'variable determinants' – fire and herbivory. These systems show a pronounced seasonality, with warm/hot summers and cool/cold winters. Most estimates of annual precipitation range from 500–1,500 mm per year with most occurring in the warm/hot summer months in tropical systems and the opposite rainfall distribution for the temperate systems. Some authors

designate subclassifications such as core, wet and dry savannas (Johnson and Tothill, 1985); savanna woodlands, savanna parkland, savanna grasslands, low tree and scrub savanna, and thicket and scrub (Cole, 1963); ecophysiological classification based on soil–water relationships including semi-seasonal, seasonal, hyperseasonal savannas and seasonal swamps (Sarmiento and Monasterio, 1975).

For the purposes of this chapter, 'savanna' will be broadly circumscribed to include any environment, temperate, subtropical or tropical, that is dominated by perennial grasses and herbaceous vegetation, and supports from 5–80 per cent shrub and tree cover in a heterogeneous pattern. This definition allows a broader view of such ecosystems, and provides the reader with a glimpse of the grasslands, which are not presently covered in this volume. Generally speaking, then, the savannas are situated between grasslands and desert steppes at the arid extreme, and the forested woodlands on the more mesic margins. Floral composition varies from one system to another, and it should be noted that in tropical systems C4 grasses predominate; in temperate systems C3 grasses are more common.

Slightly less than one-third of the Earth's land surface is covered by tropical savannas; over half the area of Africa and Australia and about 45 per cent of South America, and approximately 10 per cent of India and South-East Asia are considered savannas (Werner *et al.*, 1991). Adding temperate savannas increases the total percentage of land in savannas to well over one-third of the Earth's land surface. This ecosystem boasts both a large and growing human population as well as having the majority of the world's rangelands for livestock. Thus demographic and economic changes are significant factors that influence the functioning of this ecosystem; impact is being seen in degradation of both soils and vegetation. The projected increase in population and human use as well as the spectre of global warming and consequent climatic changes would suggest that this ecosystem be watched and managed very carefully in the future. Because of the importance of this ecosystem to human agriculture and animal husbandry, there is considerable information available to the student interested in further examining this ecosystem. Books, research articles and symposium volumes abound, providing the interested individual with a wealth of information ranging from wildlife management, agriculture use, soils, microbial physiology, nutrient cycling and ecology to socio-political issues and all aspects of savanna biology and management. (See, for example, Tothill and Mott, 1985.)

DETERMINING ELEMENTS OF SAVANNAS

Any system such as a savanna, however described, should be regarded as an integrated entity, including a consideration of both the abiotic and biotic factors, as well as the ways these elements interact. Within the broad context of this discussion, then, we can look at the following broad categories that serve to define a savanna: soils, climate, fire, herbivory, human interactions and other processes. Clearly, most of these components are held in common with other ecosystems as common determinants in the nature of the system. It is, however, the unique combinations that help to define each of these different systems. Thus, a brief overview of the contribution of each to the overall make-up of a savanna follows. The reader should be aware at this point that comparative studies between and among savannas on different continents are difficult because, although all fit the broad definition, there can be vast differences within each of the determining factors. For example, see Table 6.1.

Soils

Soils are developed over a parent material, whether it be bedrock of igneous, sedimentary or metamorphic nature or wind-blown or water-deposited organic

Table 6.1 A preliminary synopsis of dominant features and the relative strength of processes in savannas of three continents

Process	Africa	South America	Australia
Primary productivity	High	Intermediate	Low
Herbivory	Large vertebrates	Smaller vertebrates	Invertebrates
Predation	Large vertebrates	Medium vertebrates	Small vertebrates
Nutrient acquisition	Easy	Intermediate	Difficult
Mutualism	Uncommon	Common	Common
Decomposition	Rapid	Intermediate	Slow
Pollination	Specialised	Very specialised	Unspecialised
Plant dispersal	Large vertebrates	Smaller vertebrates	Invertebrates
Animal migration	Major	Intermediate	Minor
Individual animal metabolism	Medium	High	Low
Animal population productivity	High	Medium	Low
Competition	Widespread	Patchy	Patchy
Fire accommodation	Medium	Low	High

Source: Braithwaite, 1991

or inorganic sediment. The nature of this parent material dictates much of the character of that soil, including its chemistry (pH, unique assemblage of elements, 'fertility', etc.), structure, depth, water-holding capabilities and other such physical characteristics. These, in turn, combined with topography and climatic factors, dictate, to a large extent, the organisms that associate with a particular soil, depending on their own ecological requirements. These organisms include, but are not limited to, micro-organisms that provide the bulk of nutrient recycling, the vegetation, the micorrhizal fungi that form symbiotic associations with the plants, ants, termites and other insects, the burrowing and grazing animals that depend upon the vegetation as a food source, and the myriad other organisms that make the soil either directly or indirectly part of their own habitat. Each organism in turn has an impact on the soil, affecting such physical characteristics as structure, organic materials (and hence fertility), water-holding and draining capacities, and even chemistry. Events range from micro-events to macro-events such as geomorphological changes (land-shifts from earthquakes and slumping of sloping topographical features), for example. In most of these interactions, savannas are no different from any other ecosystem.

Climate

Climatic influence involves light (duration, intensity and quality) and associated temperature, and rainfall, both amount and distribution throughout the year, which largely dictates atmospheric humidity and thus temperature fluctuations. As mentioned previously, tropical systems tend to receive the bulk of their moisture during summer months, whereas temperate systems receive most of their moisture during the winter months. It is difficult to give average, mean and maximum for 'savannas'; rather, one must examine each system individually. For example, African tropical savannas have a high mean temperature range (20–30°C), with extremes ranging from 4 to 44°C (Okigbo, 1985). More mesic areas have less extreme temperatures than xeric areas because of heat-holding capacity of atmospheric humidity. Additionally, variations in latitude and altitude affect not only average temperatures, but also minimum and maximum values. Each savanna has a characteristic set of environmental conditions that must be taken into consideration when studying or examining the ecology of that particular system. All of these elements in turn influence wind, flooding and other weather-related phenomena. The strong seasonality of available moisture in savannas determines the productivity of the system as well as the species composition of organisms that live there.

Fire

Although fire is a component of nearly all ecosystems, it is of special import to the savannas – it is the force that maintains the open nature of the savanna. Without fire, woody vegetation can encroach, turning a patchy environment of fire-tolerant species and scattered woody vegetation into a woodland or, depending upon other factors, a forest. All ecosystems grade into adjacent systems; a savanna gradually gives way to more woody habitats as precipitation increases, and on the other precipitation extreme, to grasslands or deserts. Causes of fires can range from natural (lightning) to human-initiated. Frequency of fires can significantly affect the amount of litter that accumulates between fires, and thus the speed, temperature and intensity of the fire influences the vegetation. Fire controls the build-up of litter (which can decrease primary productivity and thus influence species composition), stimulates vegetative growth of fire-adapted species and controls the spread (and presence) of non-fire-adapted species (Old, 1969; Peet et al., 1975; Rice and Parenti, 1978). In a savanna in which the cause of fire is largely natural, the margin between the savanna and the adjacent ecosystem fluctuates over time, depending on climate, intensity and frequency of fire, etc. In human fire-managed savannas, however, encroachment of woody vegetation is usually held at bay, if not driven further back to expand the savanna itself. Despite concern of urban dwellers who become alarmed at human-initiated burning in savannas and grasslands, it is a necessary maintenance process that was nature-initiated prior to human presence. For example, paleobotanical studies reveal pollen associated with regularly deposited layers of charcoal in regions where savannas have long existed (Watts, 1980; Winkler et al., 1986; Abrams, 1992).

Herbivory

It is difficult to draw parallels between and among the world's savannas in this respect, because, depending on the continent, the type and size of herbivores vary widely. As seen in Table 6.1, herbivore sizes and types are quite different on the three continents compared in the table.

Savannas have a high ratio of plant productivity to plant biomass; as a result, the forage potential of these areas is high. Palatable vegetation is readily available to herbivores who don't have to climb trees for food (Huston, 1994). Types of herbivory will be examined more closely in the next sections. Although herbivores and fires might compete for the same resources, fire stimulates new, nitrogen-rich, highly palatable growth for herbivores. Negative impacts on the herbivores (carnivores, disease, etc.) can also affect the plant/herbivore interactions (McNaughton, 1985).

Issues related to outright herbivory should also include pollination, seed dispersal and seed predation; all have significant effects on the overall make-up of a savanna.

Human impact

The range of human impact is great; the following list ranges from the most primitive (hunting-gathering) and having the least impact to the most 'advanced'

and having the greatest impact: nomadic and semi-nomadic pastoralism, permanent pastoralism, subsistence cropping, cash cropping (ranch, estate and plantation farming), urbanisation, mining and war. Also to be considered are recreational use and mis-management of human activities such as fire.

As a direct result of many of these activities, erosion becomes an important aspect in the rate and degree of degradation of the savannas. Overgrazing and cultivation, in combination with natural drought cycles, have the greatest impact, baring the soil to wind and water erosion. The nature of the soil also influences the degree of erosion under these conditions. As pointed out by McTanish (1985), this problem becomes most evident to the world-wide public during drought years.

Extent of human impact on savannas varies from continent to continent. Historical population levels and land use have dictated the length and extent of use and resultant modification of these areas. However, some elements have been repeated in most savannas. For example, Table 6.2 traces the development of farming methods in tropical Africa.

Other processes

As mentioned in Table 6.1, decomposition rates vary from system to system, and are influenced by species of decomposers, organic materials to be decomposed

Table 6.2 Changes in animal production systems in savanna and arid areas in relation to association with crop production

Stages in evolution of crop-associated farming systems	Remarks and extent of livestock association with crops
1 Hunting, fishing and collecting	No settlements.
2 Total nomadism	No settlements, no cultivation, foodstuffs purchased from cultivators. Tents and makeshift dwellings pitched when necessary.
3 Semi-nomadism	Permanent place of residence. Supplementary cultivation often practised. Some might plant short-duration crops and migrate, then return later to harvest and graze.
4 Transhumance	Permanent place of residence. Herds of livestock managed by herdsmen who migrate to distant grazing grounds. Owner of cattle might or might not be farmer. Might graze crop residues in dry season in farmers' fields.
5 Partial nomadism	Permanent settlement; herds remain in vicinity of settlement and mixed farming is practised. Farmer keeps a few livestock under controlled grazing.
6 Stationary animal husbandry	Animals remain in holding throughout year and settlement permanent. Movement of livestock not far. Usually mixed farmers who also grow crops.
7 Non-farm occupations	Pastoral peoples or their descendants are also involved in trading, crafts, technical and other non-agricultural occupations.

Source: modified from Okigbo, 1985

and climatic factors such as temperature and available moisture which contribute to the overall habitat requirements of the decomposers themselves. According to Solbrig *et al.* (1992), cycling of nutrients in savannas is only partially understood. Loss of nutrients from the system can occur by means of fire, cropping or intensive animal harvesting. Restoration of nutrients to the system via rainfall, nitrogen fixation and decomposition do not occur steadily throughout the season, but respond to seasonal climatic variables such as rainfall and temperature. Nitrogen input with rainfall clearly occurs in pulses during the wet season, whereas temperature and available moisture dictate rates of nitrogen fixation and decomposition by the organisms' optimal functioning environments. How well those nutrients are maintained in the savanna results in part from the nature of the soils. For example, dystrophic soils are low in weatherable nutrients, primarily because of the nature of the parent material. Conversely, eutrophic soils are high in weatherable nutrients.

In pastoral situations, degree of grazing provides selective pressures for and against plant species, and in turn their associated organisms (pollinators, predators, etc.). Adjacent ecosystems encroach with varying amounts of success at the edges of the savannas, depending upon human activities and natural climatic events. More radical encroachment by humans comes in the form of urbanisation. This process not only occupies physical space (homes, car parks, businesses, artificial habitats, roads and streets, etc.), but also fragments existing ecosystems, often to the extreme detriment of that system. Habitat fragmentation is a significant issue in ecosystem management that is at long last being addressed.

Preservation is an issue that can significantly impact an ecosystem, positively or even negatively if undertaken in a good-spirited, but uninformed manner. For example, preservation without benefit of natural processes such as fire in a savanna ultimately degrades the very savanna the effort was intended to protect. If protected rigorously from such natural events, species composition significantly changes, thus changing the savanna. Furthermore, size of preserved habitat is critical. Questions are finally being raised regarding the smallest effective size one can preserve and still have an adequate fragment that represents the larger ecosystem.

Although fire, floods and herbivory have been mentioned, other stressors (factors that impact specific elements within the system) cannot be ignored. These include drought, disease, introduction of exotic species (including livestock), and regionally important events such as tornadoes and ice storms (in temperate savannas). Stressors affect different species in different ways, and thus have complex effects on species diversity of *all* organisms, not just plants. With respect to plants, species-specific characteristics such as thickness of bark, resistance to pathogen invasion after fire or other injury, germination ability after stress, and nature of the seed beds available to the plants after a stressor are but a few of the factors important in the survival of a particular species. Additionally, the size of the plant, the intensity and duration of the stress and the season during which it occurs are also important in maintaining the balance of species in the system.

The more mesic areas of savannas abut forests and grasslands and suffer continual encroachment of those ecosystems. At the other extreme, drought often plays a significant part in determining the nature of that part of the savanna and can, in fact, reduce the size of the savanna just as dramatically as do the forests at the other moisture extreme. Other plant issues also come into play such as morphological (deep roots, xeromorphic leaves) and physiological characteristics (low water potential thresholds for stomatal closure, ability to adjust osmotically) (Abrams, 1990).

USE AND MANAGEMENT OF SAVANNAS, PAST AND PRESENT

Savannas have a long history of human occupation and use, with activities ranging from simple hunting-gathering to extensive cash cropping to urbanisation. Obstacles to improvement of this ecosystem vary from region to region, and include limitations to forage and cropping by available water, natural pests, poor soils and poor drainage. Numerous measures have been taken to augment the productivity of the savannas: mechanical or physical work on soil or vegetation; planting, seeding or reseeding; application of chemicals; altering the timing and use of land by livestock; and adjusting numbers of animals and the resultant pressure from grazing (Hadley, 1985).

In total, the characteristics that affect management of the savannas are similar to those encountered in any other ecosystem: the relatively unchanging physical characteristics, socio-economic and political realities, and technical and legislative interventions.

Co-adaptations between vegetation and herbivores

Current studies to unravel the evolutionary history of plants, animals and human cultures in savannas can improve insight into the co-adaptation among these entities and provide a basis for more rational management of the elements of the ecosystem. Special adaptation by plants (e.g. spines and toxins) reduce consumption by herbivores, whereas counteradaptations by herbivores (e.g. tongue modifications, feeding strategies, specialised digestive floras) allow use of otherwise unpalatable vegetation. Herbivory, if not excessive, apparently has a stimulating effect on the growth of the plant (McNaughton *et al.*, 1983).

Wild herbivores show characteristic patterns of grazing and browsing as documented by Okigbo (1985) and shown in Table 6.3. For example, giraffe and elephants browse higher than eland and impalas, whereas zebras, antelopes and the like are primarily grass eaters.

Livestock also have specialised feeding activities (Strange, 1980). Cattle consume the upper level of understorey plants, but prefer grasses. Grazing can create a patchwork of lightly or heavily grazed areas in the savanna. Sheep are highly competitive with cattle as they select the most desirable plant parts and graze close to the soil surface, thus encouraging a short, compact turf. Goats primarily use shrubs and understorey vegetation that are unpalatable to sheep and cattle. Overbrowsing in forested areas can severely interrupt replacement of trees and reproduction of understorey species. Camels consume desert browse species that are not palatable to other herbivores. Donkeys and horses are feeders that nip leaves and stems with a bite from the upper and lower incisors. If confined over too long a period, they can cause severe damage to the vegetation (Okigbo, 1985).

Use of African and Asian savannas by domestic herbivores has altered the structure and composition of vegetation developed by a long history of co-adaptation with wild herbivores, whereas the Australian and American savannas, only recently occupied by domestic herds, reflect adaptive patterns from pressures from native herbivores (Nyerges, 1982). However, in some African savannas, presence of the tsetse fly has long discouraged the encroachment of man and his domesticated animals, thus maintaining the original pattern of co-evolved animals and land use (Adkins, 1978).

Further complexity is demonstrated by numerous savannas being characterised by disjunct patterns of

Table 6.3 Feeding specialisations of savanna herbivores, taking into consideration vegetation types and broad habitat differences and food preferences

Specialisations related to:	Examples of specialisations
*Habitats where overlaps occur	Zebras, antelopes, gazelles and giraffe prefer open grassland with scattered trees
	Reedbucks prefer dense mats of grass and reeds
	Rhinoceros and elephants prefer a mix of forest and grassland
	Kudus eat a great variety of grass, shrubs and wild fruit
*Different grazing heights/levels	Giraffe eat tops
	Elephants feed high on trees, but lower than giraffe
	Eland and impalas eat low levels of trees
	Steenbuck feed on lowest grasses and forbs
	White gazelles eat new grass shoots
Different dry season refuge	Distributes pressure on dormant vegetation
Use of same food sources at different times	Migrating herds arrive at different times of forage development

Source: *modified from Okigbo, 1985

fertile and infertile patches that result from soil differences, grazing or browsing behaviours of animals, and from impacts of human activities. Both wild and domestic herbivores use both fertile and infertile patches to balance their nutrient needs and thus more efficiently use the vegetation of the savanna. This also maintains the essential pattern of the savanna.

Over more extensive savanna regions, especially fertile areas (e.g. Sweetveld in South Africa), animal stocking rate is determined by the amount of grass produced annually, whereas in infertile savannas the number of grazers per unit is related to the quality of grass acceptable to the grazing animals. In either case, much effort must be made to stock animals to match forage production for that year or season. Stocking rates of wild browsers are difficult to determine. The browse available in the late dry season is the limiting factor for most regions. Multi-seasonal browse use could be substantially increased if supplemental feeding were employed. Data obtained from African savannas by Owen-Smith (1985) indicate that natural density of browsers is 10–20 per cent of that recommended as stocking rates for cattle. Further, it was noted that animal production from savanna use could be increased by stocking a variety of grazers and browsers that could use different parts, species and levels of the vegetation.

The increase of animal production in savannas can be accomplished by including wild browsers in ranges presently dominated by domestic grazers. The success of this strategy is to be determined by the food choice of the browsers and if they will exert sufficient control on certain woody species, especially in infertile savannas. These specialisations are examined in Table 6.3.

Comparisons between wild and domestic herbivores give insights into the productivity of savannas for each type of herbivore. The standing biomass of wildlife is five times that of domestic animals. Greater production of wildlife is attributed to more tolerance to disease and heat stress and their earlier maturity, shorter gestation periods and more rapid growth (Knight, 1976).

Management of grazing systems with livestock in Australia is carried out on a fenced, ranch-style, commercial basis. However, in other savannas (especially in Africa) livestock grazing is practised on communal lands on a subsistence level. Modern Australian savanna management attempts are aimed at optimising the beneficial aspects of animal science within the limitations of economics, place, organisational structure and labour. Improvements in nutrition, within the constraints of economics and time, must be planned to upgrade production of livestock herds. The overseeding of legumes on more fertile savannas (e.g. *Medicago sativa*) to improve nitrogen fixation and quality of foraging is promising, but might concentrate grazing on the existing grasses and increase invasion of low-quality woody species.

Current inquiry has progressed beyond the study of animal stocking rates to direct determinations of quantities of plant defoliation. These studies should determine the time of foliage loss as well as mechanical and physiological responses of the plant to the loss. This approach provides a useful aspect to the study of herbivory.

Crops and grazing integrations

Australian savannas are most successfully cropped in northern areas where the soils are both nutrient- and water-rich. However, Leslie (1984) has cautioned that these soils can be damaged unless they are carefully managed. Mid-grass savannas in Australia also have some potential for crop production but cultivation removes the protective vegetative cover and damages the soil by sealing (preventing water infiltration) and thus increasing soil erosion. Early economic benefits of recently cropped soil encourage further expansion of cropping into marginal savanna pasture and cropland to sustain economic survival through periods of low rainfall and low markets (Leslie, 1984). Unfortunately, sustained use of these areas is difficult, if not impossible, and degradation continues at an increased pace.

At the other extreme, South American poorly drained savannas might be developed further in the production of wetland rice as the marketing systems improve (Cochrane *et al.*, 1985). Well-drained savannas of Brazil have considerable potential for crop production that will be further increased with the development of cultivation, soil management and irrigation (Cochrane *et al.*, 1985).

Despite the interventions of mixed farms, large-scale mechanised farms, extensive irrigation projects, livestock ranching and horticultural enterprises of improving agriculture, these efforts have generally failed or have been only marginally successful in Africa. These innovations have not achieved the 4 per cent annual increase in agricultural production to meet the growing needs of food for the human population (Okigbo, 1985).

Failures of these projects were attributable to inadequacies of the ecosystem being managed, weaknesses in planning and various socio-economic factors. Strategies employed by native Africans prior to European intrusion to use savannas often involved practices that preserved, maintained or improved environmental qualities that sustained the people's way of life.

Increased production from existing savannas can be realised by two approaches: (1) the expansion of current crops or grazing into suitable but unexploited environments, and (2) identifying, through research, new approaches to improving production on existing lands. The second approach of improving existing pastoral and crop productivity requires substantial investments in research programmes to identify new technologies, developing higher yielding genotypes and implementing on-farm research to improve efficiency of the farm unit.

RESEARCH

The majority of current research is oriented towards use of the savannas rather than conservation *per se*. Associated studies examine small research projects within a small area, a specific taxon, a physiological or soil parameter or the like. Relatively little is being done regarding overall preservation of the savanna as a functional limit.

Henzell (1985) discusses two significant shortcomings of recent savanna research: (1) inadequate communication to take into account the pastoralists' and farmers' needs and wishes, and (2) inadequate communication of potential benefits of scientific studies if they were adopted by savanna dwellers and their governments.

Putting a segment of any ecosystem under governmental protection for observation, research or recreational use by humanity at large has almost always resulted in loud, sustained and organised opposition by those segments of the population living in and deriving their primary income from this particular environment. Examples in North America include the much-discussed (and much-cursed) 'Buffalo Commons' proposed by Frank and Deborah Popper and the Z-Bar Ranch National Tallgrass Prairie Park proposed in Kansas in the USA.

In the Buffalo Commons issue, Frank Popper (an urban planner at Rutgers University) and Deborah Popper (who teaches in New York) proposed that the Great Plains 'wasteland' of central North America is essentially unfit for human habitation and should be returned, at least in part, to the pre-European residents, the buffalo (*Bos bison*) and the native American Indians. Predictably, those residing in the vast central 'breadbasket' of the continent have protested loudly, arguing that properly managed, this area is one of the most productive in the world, especially if managed carefully within the limitations of the various parts of the area. For example, some areas are more suited to grazing than cropping, and other heavily cropped, irrigated areas should logically be returned to grazing before below-ground stores of water (aquifers) are exhausted. It is clear to many that there *is* a carrying capacity for the area and in recent years, as the population reversed, this part of the continent is much more sensibly populated. No-till agriculture is rapidly becoming a possibility in much of this intensively cultivated land. Much of this region ranges from temperate savannas to grasslands.

In a related issue, the Land Institute, a non-profit research group, has the philosophy that perennial polycultures, consisting of high-yielding grasses and legumes, can at once provide humans and domesticated animals with needed sustenance, but also halt the erosion of topsoil associated with cultivation. Even this increasingly well-documented management scheme meets resistance from the people now engaged in deriving their livelihood from this area. (See, for example, Jackson, 1980; Soule and Piper, 1992.)

The second issue mentioned above was the establishment of a National Tallgrass Prairie Park. It was done by means of a legislative procedure at the national level, with voluminous amounts of argument and discussion. Issues involved here included concern about diseases introduced by the native buffalo, to loss of livelihood, to loss of the tax base used to support local functions (roads, schools, etc.). This issue is hotly debated, and is likely to remain so for years to come.

Social considerations are critical to the success of most savanna management programmes. Non-traditional uses (i.e. development) of savannas can inflict tragic consequences on the native peoples of an area. The survival of aboriginal people in their traditional cultures continues in northern Australia and in parts of Africa, in spite of European invasions. Biernoff (1985a) notes that these cultures are experimenting with modern land uses, technologies and political means to adapt to the economic and rapidly changing world without disrupting the continuity of their ways of life.

The development of aboriginal lands poses significant consequences for those peoples not actively participating in the change by hastening the collapse of their economic and social structure (Biernoff, 1985a). Biernoff suggests that developers be asked these five questions before affecting change:

1   Development and improvements for whom?
2   Who makes the decisions and who benefits?
3   Who determines and chooses the solutions?
4   Where does the power to effect change lie, internationally, nationally or locally?
5   Who will be most disadvantaged by the change?

The questions, honestly answered, might help avoid many failures in the health and stability of native cultures.

Independent aboriginal communities in northern Australia continue to exist with impetus from the Aboriginal Land Rights (Northern Territory) Act of 1976. This Act permits native people to return to their ancestral homes and obtain title to land (Biernoff, 1985b). The aboriginals are striving for self-sufficiency, which they believe will lead to independence and vitality of their people.

Goal-oriented projects and comparative technical studies are suggested by Hadley (1985, Table 6.4). The former projects are usually land-use oriented and are committed to understanding the socio-cultural and political aspects of development. The latter are projects that seek solutions through scientifically derived answers to practical problems of soil fertility, plant growth dynamics, etc.

According to Okigbo (1985), research organisations concerned with the future of savannas have identified strategies that show promise for improving productivity and management, including:

- genetic improvement of crops adapted to savanna ecosystems;
- establish the scientific soundness of intercropping to increase yield;
- make others more aware of known potentials for savanna reforestation;
- study and attempt to increase potentials for water harvesting in savanna areas where moisture is the greatest constraint to increased production;
- examine the possibility of eliminating fallow areas through alley cropping, thus facilitating maintenance of soil fertility and fuel wood production, especially in more humid savannas;
- demonstrate that improved tillage techniques (e.g. tie-ridging) can increase crop yields in certain circumstances;
- aim for significantly higher maize yields in the savanna as compared with the humid tropics;
- effectively use timing and intensity of fire to manage vegetation;
- identify physical, biological and socio-economic constraints to increased human population;
- make further progress in the study of savanna ecosystems and development of management techniques;
- further explore game ranching as an efficient method of managing wildlife that is compatible with various objectives.

RESTORATION, MANAGEMENT AND RECOVERY GOALS – A MODEL FOR TEMPERATE NORTH AMERICAN OAK SAVANNAS

This model, for a temperate savanna, used with permission, is derived from Leach and Ross (1995). Elements to be considered are:

Restoration and management. Restoration activities need to be customised for the unique qualities and purposes of each site.

Purposes and goals. What are the purposes of the restoration? What ecological models will be used as

Table 6.4 Identifying unifying scientific hypotheses for improved management in savanna areas

General management problem	Scientific hypothesis/observation	Management implications
Whether or not to eliminate all woody plants in order to achieve maximum grass yield.	Vegetation structure determined by soil moisture conditions (cf. floristic composition determined by soil nutrient conditions) (Cole, 1982). Light, high tree canopy leads to greater production of grasses (in Nigerian Guinea savanna) than either full exposure or dense canopy (Sanford *et al.*, 1982).	Major changes in vegetation structure and composition effected by relatively minor manipulations in soil moisture conditions (Tinley, 1982, based on data on southern African savannas) and in soil nutrient content. Desirability of maintaining low density of large trees, preferably legumes.
How to improve production while protecting against soil erosion and loss in soil fertility.	Soil organic carbon (humus) is a key factor in ensuring continued production as it is positively related to water-holding capacity, soil aeration, cation exchange capacity and content of nitrogen; significant positive correlation between soil carbon content and tree density (see Sanford *et al.*, 1983).	Maximum humus formation favoured by management techniques that favour maximum biomass production. On slopes, maintenance of both tree and grass cover important for control of wind and water erosion.
Whether and how to manage ungulate populations in extensive savanna areas set aside as national parks.	Strong interactions exist between a few components of savanna ecosystems, e.g. in the Serengeti, between the dominant herbivore (wildebeest) and perennial grasses, between giraffe, elephant and trees. Natural negative feedback mechanisms operate between some components of system. Provided negative feedbacks strong enough, system able to absorb disturbance and perturbation (Sinclair and Norton-Griffiths, 1979).	Changes in a few major components can have far-reaching effects on other components (e.g. removal of wildebeest would probably result in major changes in all trophic levels in contrast to likely minimal effects on system of disappearance of impala or buffalo). System should be maintained in as natural a state as possible to fulfil function as national park and ecological baseline area.
Presence, timing, frequency and intensity of burning, and its relation to soil fertility and production.	Effects of annual burning vary with geographic position, climate and vegetation; great caution required in interpreting data. In Nigerian Guinea savanna, burning may reduce tree canopy; with less canopy more grass produced, except in drier, warmer regions (Sanford *et al.*, 1982).	Burning should be controlled as to time (intensity), which varies with vegetation, climate, soil and management goal. For highest primary production, early (light) burning beneficial in drier savannas; in moist savannas, late (intense) burning probably more satisfactory.

Source: Hadley, 1985

a guide? Is there a commitment to long-term monitoring and management?

Site conditions. What were the past land uses? What are the present conditions? What is the structure and composition of current vegetation? Has succession from oak savanna or woodland to, for example, sugar maple forest progressed to the point where restoration might be difficult and expensive? Will there need to be remedial planting? What kinds of animals use the site? What are the abiotic features?

Models. What kinds of plant and animal communities previously occupied the site? What was the physiognomy of that vegetation as deduced from soils, topography, the surrounding landscape and historical records? What was the fire regime of those communities? What are the present (and future) biotic and abiotic conditions that affect the choice of future communities? Are there nearby and affordable sources for plant and animal reintroduction? Will the chosen model emphasise composition, structure or processes (e.g. fire regime)?

Size. Size might be the primary factor when considering how many species and ecosystem properties the site can support. Adjoining land use also affects those properties, especially as a source of invading species, nutrient inputs and alteration of hydrology. Most restorations would benefit from being nested within buffer areas that could be semi-restored areas which limit undesired inputs and increase the effective size of the restoration.

Fire. What actions need to be taken to ensure that prescribed fire regimes can be realised legally, safely, efficiently and without causing problems with neighbours?

Unwanted species. Which undesirable species (plants and animals) are present? How will each of these be dealt with? How will new invaders be identified and dealt with?

Defensibility. How will the restored ecosystem be protected from human abuse? What agency or agencies will serve as stewards for the sites?

Use. How will people use the site? How will decisions be made about what uses are and are not appropriate? What kinds of research will be encouraged or permitted? Who owns the land, private or public agency?

Viability. What is the outlook for long-term viability of the restoration? Who will provide stewardship in 20, 50 or 100 years?

The following policies have been adopted by the Board of Directors of the Society for Ecological Restoration (USA). They are included here for consideration during the restoration planning process.

Restoration plans

The Society for Ecological Restoration advises that plans for restoration projects should contain, at a minimum, the following items.

(a) A baseline ecological description of the kind of ecosystem designated for restoration, which accounts for the regional expression of that ecosystem in terms of the biota and poignant features of the abiotic environment.

(b) An evaluation of how the proposed restoration will integrate with other components of the regional landscape, especially those aspects of the landscape that might affect the long-term sustainability of the restored ecosystem.

(c) Explicit plans and schedules for all on-site preparation and installation activities, including plans for contingencies.

(d) Well-developed and explicitly stated performance standards, by which the project can be evaluated objectively.

(e) Monitoring protocols by which the performance standards can be measured.

(f) Provision for the procurement of suitable planting stocks and for supervision to guarantee their proper installation.

(g) Procedures to expedite promptly any needed post-installation activity.

Exotic species at restoration sites

An exotic species (or lesser taxon) of plant or animal is one that was introduced, either intentionally or unintentionally, by human endeavour into a locality where it previously did not occur. The invasiveness of exotic species of plants and animals challenges a basic goal of ecological restorationists to re-create environments like those that existed prior to widespread human disturbance. Ideally, a restoration project should consist entirely of indigenous species. In order to meet this goal for virtually all restoration projects, the control of exotic species will require ongoing management, monitoring and evaluation.

To that end, the Society for Ecological Restoration recommends that the following principles be followed during the planning, implementation, and evaluation of restoration projects and programmes. These can be modified to fit restorations made in other savannas.

1 The control of exotic species (including livestock) should be an integral component of all restoration projects and programmes.

2 Monitoring of exotics and periodic reassessment of their control should be integrated into all restoration plans and programmes.

3 Highest priority should be given to the control of those species that pose the greatest threats, namely:
- exotics that replace indigenous key species;
- exotics that substantially reduce indigenous species diversity, particularly with respect to the species richness and abundance of species with stable populations;
- exotics that significantly alter ecosystem or community structure or function;
- exotics that persist indefinitely as sizeable sexually reproducing or clonally spreading populations;
- exotics that are very mobile and/or expanding locally.

4 Restoration plans and management programmes should include contingencies for removing exotics as they first appear and for implementing new control methods as they become available.

5 Control programmes should cause the least possible disturbance to indigenous species and communities and, for this reason, can be phased in over time.

6 The restoration and management programme must, of necessity, be strategic. Protection of indigenous habitats, levels of infestation, appropriate resource allocation and knowledge of control methods should be integrated into the monitoring and management programme.

7 Exotic species should not be introduced to the site in the restoration plan.

8 Native species should also be evaluated for their potential threat to indigenous communities. Weedy native species should be avoided in restoration plans as well as native planting stocks representing non-indigenous ecotypes.

Recovery goals

1 Achieve broad consensus for a regional oak ecosystems recovery effort.

2 Establish a networked system of reserves that captures the full array of oak ecosystem species, communities and processes, and that conserves viable populations of all plants and animals known to inhabit them.

3 Complete inventory and mapping of oak savannas.

4 Establish a network of pilot restoration projects with intensive monitoring.

5 Foster the development of influential local stewardship groups and education networks.

6 Develop incentives and information in order to involve and engage the private sector (e.g. nurseries, timber harvesters, farmers, land developers, recreation landowners, etc.).

7 Develop a regional co-operative burning policy.

8 Develop a research agenda.

9 Establish coalitions and partnerships dedicated to conservation of oak ecosystems both locally and regionally.

10 Investigate the role of native and domestic herbivores in managing and restoring functioning savanna ecosystems.

RECENT DEVELOPMENTS

There are a number of recent developments that should be noted in the management of savannas and grasslands. Holistic resource management is a system increasingly employed by ranchers to simulate the intensive but relatively short-term grazing pressures of large roaming herds of wild or native herbivores. By moving cattle into paddocks for short but intensive periods of grazing, it is being found that long-term impact on soil and vegetation is less profound than that of a traditional pasturing system. For more information, see Savory (1988).

We are seeing an increase in computer-aided planning of grazing. For example, RSPM (resource systems planning model) has been developed by the Ranching Systems Group at Texas A&M University (USA) for nation-wide distribution by the United States Department of Agriculture–Natural Resources Conservation Field Service in the United States (Stuth et al., 1991). In this model, a decision support system is used to assist the landowner in making informed decisions about his/her operation. There are three levels of decision-making: strategic (assessment of management goals, inventory of forage and animal resources, enterprise analysis and long-term investment of selected technologies); tactical (allocation of resources for a prescribed amount of time); and operational (adjustment of forage and animal resources to weather and market events). Another decision support system is the CSIRO RANGE-PACK, an Australian project involving pastoral land management on microcomputer (Smith and Foran, 1991).

SHRUBKILL, an advisory programme for prescribed burning to control shrubs is another computer-based tool available (Ludwig, 1991).

BOX 6.1   GROUP RANCHING AND MANAGEMENT OF COMMUNAL GRAZING IN BOTSWANA

Botswana in southern Africa is semi-arid, with limited opportunities for arable agriculture; thus traditional livestock husbandry is an important aspect of this country. The greatest rainfall, soil fertility, available groundwater, and most of the human population are located on the eastern side of the country, although increasing population pressure is expanding grazing areas into the more arid regions of the country, namely into the Kalahari sands. Superimposed on these determining factors is a complex web of social, political and economic factors that surround ownership of cattle.

Several items were of major concern to the government, when writing a National Policy on Tribal Grazing:

1 Sustain or increase beef protection.
2 Realistic grazing control and better range management.
3 Safeguard the interests of all people, whether they have few or no cattle.
4 Preserve the right of every tribesman to have as much land as necessary to sustain himself and his family.

Towards this end, after considerable consultation and publicity, the Tribal Grazing Land Policy (TGLP) was instituted in 1976, designating land into one of four zones, which are briefly outlined below. First, *Commercial Areas* with leasing rights granted for ranching effectively resettled larger cattle operations with fewer operators into these areas, thus alleviating pressure on the communal areas. These ranches were made available to individuals and to groups, with an eye towards operations large enough to provide satisfactory net farm income and return on capital invested. To assist in this, subsidies in development costs from the National Development Bank were provided to encourage groups of small cattle owners to commercialise their production.

The second zone established was the *Communal Areas*, which were designated to maintain communal access. The first effort was to form Communal Grazing Cells within the overgrazed areas surrounding villages. These units were smaller than those established in the commercial areas, aiming for approximately 300 head of young livestock. Through these cells, once organised, funds were available, primarily for educating participants how to better manage their available resources. Several items were provided: peripheral fencing, paddock subdivision, central watering point and handling facilities. This programme met with less success than that in the commercial areas; however, one cell is functioning well. A major handicap for this phase was finding usable water. Because the early attempt was less successful than anticipated, another approach was initiated in 1980. Emphasis was placed on agricultural production combined with resource conservation, with special focus on land-use planning and strengthening local institutions.

Third, *Wildlife Management Areas* were designated where primary purpose is wildlife use, but where domestic animals are permitted, and, finally, *Reserved Areas* were set aside for possible future needs and uses.

As a result of this programme, experience has revealed several very important principles in introducing new management strategies to any group of people.

1 Government cannot successfully impose or enforce solutions on a society unless it has become an institutional force from within the community itself.
2 Initiatives are more successful if employed in a 'bottom-up', rather than a 'top-down' manner.
3 Issues such as grazing rights and livestock control can be tackled only after more pressing issues within the society are addressed.
4 An essential requirement for voluntary stock control must involve boundary recognition; cohesive groups make this easier to accept.
5 Where possible, development programmes should work through local institutional frameworks, in logical steps, and with reasonable time-scales.

Source: Sweet and Addy, 1985

BOX 6.2 RESTORING MESQUITE SAVANNA IN WESTERN TEXAS, USA THROUGH BRUSH AND CACTI MANAGEMENT

Much of western Texas was historically populated by a savanna with warm-season grasses and scattered honey mesquite (*Prosopis juliflora* var. *glandulosa*). The mesquite was more dense along watercourses. In the past century, likely as a result of overgrazing, cessation of fires and scatter of seed by livestock, these open savannas have become extremely dense, mesquite–prickly-pear cactus (*Opuntia* sp.) thickets. Prickly-pear causes mechanical damage to livestock mouths and digestive distress resulting from seed compaction in the rumen. Mechanical brush control methods such as root-ploughing, chaining and root-raking generally increases cactus density. Removal by hand is effective, but is labour intensive and expensive.

Traditional attitudes and governmental incentives have led to a 'brush eradication' mentality, with total removal of all brush in an attempt to obtain a 'pure' grassland, thus ignoring the essential savanna-like nature of the area. The author supports the notion that this area is a woody-shrub climax, and that naturally occurring fires, combined with selective methods of shrub and cactus control, can return this area to its 'normal' or natural state.

In a traditional management scheme, designed to return these lands to grasslands, there are three approaches: fire, herbicides and mechanical removal.

- *Fire.* Mesquite is vulnerable to fire, depending on the size and age of the plant, presence or absence of dead stems, and factors regulating the intensity of the fire (Wright, 1972, 1974; Wright *et al.*, 1976). Prescribed burning early in the spring has been found to be effective for controlling mesquite.
- *Herbicides.* Mesquite and prickly-pear cacti have been most widely controlled by the aerial application of liquid herbicides, with varying results. Some chemicals provided significant top kill, but resprouting can occur quickly. Others provided more significant reduction of vegetation. Attention needs to be paid to timing of application of foliar herbicides, both with respect to biotic (physiological) and abiotic factors.
- *Mechanical.* Most mechanical methods can result in production of denser stands unless supplemented with herbicide treatments.

It is proposed by this author that these west Texas rangelands be reconsidered as savannas rather than pure grasslands, and that a holistic management scheme reflect this changed philosophy. With a judicious use of all three methods, the land can be returned to a more stable state with a number of benefits. These benefits would include economic gains, ecological stability, diversity of land use, increased livestock performance and increased aesthetic quality. The cost of maintenance through selective burning and individual plant treatment should be less. This increased ecological diversity would also provide livestock with shade in summer and protection in winter. Additionally, this management method would increase wildlife habitat, with an increase in hunting potential, another source of income for landowners. In total, it is felt that this restructured management method would require more management input by landowners and a more conservative approach to grazing management. This then would extend the use of the land while at the same time allowing sufficient fuel build-up for periodic burning. It is likely that in the future, economic realities will require this or a similar management scheme for these lands.

Source: Jacoby, 1985

CONCLUSIONS

Savannas represent highly variable ecosystems, represented on five continents. Their complexity is evidenced by historical co-evolutionary processes with wild herbivores (invertebrates to vertebrates) to more recent pressure from domesticated animals. Throughout their history of human occupation, they have suffered extensive degradation; in the most extreme cases in Africa and India, production of consumable resources from savannas is inadequate to meet present demand, despite recent innovations in agriculture and management. There is a clear need for extensive research to better understand both the ecology and the socio-economic realities that exist on these highly desirable lands. Additionally, restoration efforts, improved methods of management and realistic and thoughtful planning are imperative if these ecosystems are to survive.

REFERENCES

Abrams, M.D. (1990) 'Adaptations and responses to drought in Quercus species of North America', *Tree Physiology* 7: 227–238.

Abrams, M.D. (1992) 'Fire and the development of oak forests', *BioScience* 42, 5: 346–353.

Adkins, M.D. (1978) *Insects in Perspective*, New York, USA: Macmillan.

Biernoff, D.C. (1985a) 'Integration or conflicts in human use of savannas: introductory comments', in J.C. Tothill and J.J. Mott (eds) *Ecology and Management of the World's Savannas*, Brunswick, Victoria, Australia: Australian Academy of Sciences, pp. 173–174.

Biernoff, D.C. (1985b) 'Perspectives for aboriginal community development in northern Australia', in J.C. Tothill and J.J. Mott (eds) *Ecology and Management of the World's Savannas*, Brunswick, Victoria, Australia: Australian Academy of Sciences, pp. 181–184.

Botts, P., Haney, A., Holland, K. and Packard, S. (eds) (1994) *Midwest Oak Ecosystems Recovery Plan* (Draft September 1994), Chicago, Ill., USA: Nature Conservancy, Illinois Field Office.

Braithwaite, R.W. (1991) 'Australia's unique biota: implications for ecological processes', in P.A. Werner (ed.) *Savanna Ecology and Management: Australian Perspectives and Intercontinental Comparisons*, London, UK: Blackwell Scientific Publications, pp. 3–10.

Cochrane, T.T., de Azevedo, L.G., Thomas, D., Madeira Netto, J., Adamoli, J. and Verdesio, J.J. (1985) 'Land use and productive potential of American savannas', in J.C. Tothill and J.J. Mott (eds) *Ecology and Management of the World's Savannas*, Brunswick, Victoria, Australia: Australian Academy of Sciences, pp. 114–124.

Cole, M.M. (1963) 'Vegetation nomenclature and classification with particular reference to the savannas', *South African Geographical Journal* 45: 3–14.

Cole, M.M. (1982) 'The influence of soils, geomorphology and geology on the distribution of plant communities in savanna ecosystems', in B.J. Huntley and B.H. Walker (eds) *Ecology of Tropical Savannas, Ecological Studies 42*, New York, USA: Springer-Verlag, pp. 145–174.

Curtis, J.T. (1959) *The Vegetation of Wisconsin*, Madison, Wis., USA: University of Wisconsin Press.

Eiten, G. (1986) 'The use of the term "savanna"', *Tropical Ecology* 27: 10–23.

Faber-Langendoen, D. (1994) 'A proposed classification for savannas and woodlands in the Midwest', in P.A. Botts, A. Haney, K. Holland and S. Packard (eds) *Midwest Oak Ecosystems Recovery Plan* (Draft September 1994), Attachment 2, Chicago, Ill., USA: Nature Conservancy, Illinois Field Office, pp. 65–74.

Hadley, M. (1985) 'Comparative aspects of land use and resource management in savanna environments', in J.C. Tothill and J.J. Mott (eds) *Ecology and Management of the World's Savannas*, Brunswick, Victoria, Australia: Australian Academy of Sciences, pp. 142–158.

Henzell, E.F. (1985) 'Summative address', in J.C. Tothill and J.J. Mott (eds) *Ecology and Management of the World's Savannas*, Brunswick, Victoria, Australia: Australian Academy of Sciences, pp. 367–369.

Huston, M.A. (1994) *Biological Diversity*, Cambridge, UK: Cambridge University Press.

Jackson, W. (1980) *New Roots for Agriculture*, Lincoln, Nebr., USA: University of Nebraska Press.

Jacoby, P.W. Jr. (1985) 'Restoring mesquite savanna in western Texas, USA through brush and cacti management', in J.C. Tothill and J.J. Mott (eds) *Ecology and Management of the World's Savannas*, Brunswick, Victoria, Australia: Australian Academy of Sciences, pp. 223–228.

Johnson, R.W. and Tothill, J.C. (1985) 'Definitions and broad geographic outline of savanna lands', in J.C. Tothill and J.J. Mott (eds) *Ecology and Management of the World's Savannas*, Brunswick, Victoria, Australia: Australian Academy of Sciences, pp. 1–13.

Knight, C.G. (1976) 'Wildlife', in C.G. Knight and J.L. Newman (eds) *Contemporary Africa*, Englewood Cliffs, NJ, USA: Prentice-Hall, pp. 159–168.

Leach, M.K. and Ross, L. (eds) (1995) *Midwest Oak Ecosystems Recovery Plan: A Call to Action*, Chicago, Ill., USA: The Nature Conservancy Illinois Field Office, 27 September.

Leslie, J.K. (1984) 'Soil and crop husbandry: recent trends in production and research', in *ACIAR (Australian Centre for International Agricultural Research) Proceedings, Eastern Africa – ACIAR Consultation on Agricultural Research, Nairobi, Kenya*, Kenya: National Council for Science and Technology and Canberra, Australia: ACIAR, pp. 97–107.

Ludwig, J.A. (1991) 'SHRUBKILL: a decision support system for management burns in Australian savannas', in P.A. Werner (ed.) *Savanna Ecology and Management: Australian Perspectives and Intercontinental Comparisons*, London, UK: Blackwell Scientific Publications, pp. 203–206.

McNaughton, S.J. (1985) 'Ecology of a grazing ecosystem: the Serengeti', *Ecological Monographs* 55: 259–294.

McNaughton, S.J., Wallace, L.L. and Coughenous, M.B. (1983) 'Plant adaptation in an ecosystem context: effects of defoliation, nitrogen, and water on growth of an African C_4 sedge', *Ecology* 64, 2: 307–318.

McTanish, G. (1985) 'The role of aeolian processes in the savannas of northern Nigeria', in J.C. Tothill and J.J. Mott (eds) *Ecology and Management of the World's Savannas*, Brunswick, Victoria, Australia: Australian Academy of Sciences, pp. 197–199.

Nyerges, A.E. (1982) 'Pastoralists, flocks and vegetation: processes of co-adaptation', in B. Spooner and H.S. Mann (eds) *Desertification and Development: Dryland Ecology in Social Perspective*, London, UK: Academic Press, pp. 217–247.

Okigbo, B.N. (1985) 'Land use and production potentials of African savanna', in J.C. Tothill and J.J. Mott (eds) *Ecology and Management of the World's Savannas*, Brunswick, Victoria, Australia: Australian Academy of Sciences, pp. 95–106.

Old, S. (1969) 'Microclimate, fire and plant production in an Illinois Prairie', *Ecological Monographs* 39: 355–384.

Owen-Smith, N. (1985) 'The ecology of browsing by African wild ungulates', in J.C. Tothill and J.J. Mott (eds) *Ecology and Management of the World's Savannas*, Brunswick, Victoria, Australia: Australian Academy of Sciences, pp. 345–349.

Peet, M., Anderson, R. and Adams, M. (1975) 'Effects of fire on big bluestem production', *American Midland Naturalist* 94: 15–26.

Rice, E.L. and Parenti, R.L. (1978) 'Causes of decreases in productivity in undisturbed tall grass prairie', *American Journal of Botany* 65: 1091–1097.

Sanford, S. (1982) 'Pastoral strategies and desertification: opportunism and conservatism in dry lands', in B. Spooner and H.S. Mann (eds) *Desertification and Development: Dryland Ecology in Social Perspective*, London, UK: Academic Press, pp. 61–81.

Sanford, W.W., Usman, S., Obot, E.O., Isichei, A.O. and Wari, M. (1982) 'Relationship of woody plants to herbaceous production in Nigerian savanna', *Tropical Agriculture (Trinidad)* 59, 4: 315–318.

Sanford, W.W., Wangari, E. and di Castri, F. (1983) 'Problems of the African savanna: soil fertility and efficient utilisation', paper presented to international symposium on savanna and woodland ecosystems in tropical America and Africa, University of Brasilia, Brazil, October 1983.

Sarmiento, G. and Monasterio, M. (1975) 'A critical consideration of the environmental conditions associated with the occurrence of savanna ecosystems in tropical America', in F.B. Golley and E. Medina (eds) *Tropical Ecological Systems*, New York, USA: Springer-Verlag, pp. 223–250.

Savory, A. (1988) *Holistic Resource Management*, Covelo, Calif., USA: Island Press.

Sinclair, A.R.E. and Norton-Griffiths, M. (eds) (1979) *Serengeti: Dynamics of an Ecosystem*, Chicago, Ill., USA: Chicago University Press.

Smith, D.M.S. and Foran, B.D. (1991) 'RANGEPACK: the philosophy underlying the development of a micro-computer-based decision support system for pastoral land management', in P.A. Werner (ed.) *Savanna Ecology and Management: Australian Perspectives and Intercontinental Comparisons*, London, UK: Blackwell Scientific Publications, pp. 197–202.

Solbrig, O.T., Goldstein, G., Medina, E., Sarmiento, G. and Silva, J. (1992) 'Responses of tropical savannas to stress and disturbance: a research approach', in M.K. Wali (ed.) *Ecosystem Rehabilitation: Preamble to Sustainable Development, Vol. 2: Ecosystem Analysis and Synthesis*, The Hague, The Netherlands: SPB Academic Publishing bv., pp. 63–73.

Soule, J.D. and Piper, J.K. (1992) *Farming in Nature's Image: An Ecological Approach to Agriculture*, Washington DC, USA: Island Press.

Strange, L.R.N. (1980) *An Introduction to African*

Pastureland Ecology, Nos. 5 and 6, Rome, Italy: FAO.

Stuth, J.W., Conner, J.R., Hamilton, W.T., Riegel, D.A., Lyons, B.G., Myrick, B.R. and Couch, M.J. (1991) 'RSPM: a resource systems planning model for integrated resource management', in P.A. Werner (ed.) *Savanna Ecology and Management: Australian Perspectives and Intercontinental Comparisons*, London, UK: Blackwell Scientific Publications, pp. 187–196.

Sweet, R.J. and Addy, B.L. (1985) 'An approach to group ranching and the management of communal grazing in Botswana', in J.C. Tothill and J.J. Mott (eds) *Ecology and Management of the World's Savannas*, Brunswick, Victoria, Australia: Australian Academy of Sciences, pp. 178–180.

Tinley, K.L. (1982) 'The influence of soil moisture balance on ecosystem patterns in southern Africa', in B.J. Huntley and B.H. Walker (eds) *Ecology of Tropical Savannas, Ecological Studies 42*, New York, USA: Springer-Verlag, pp. 175–192.

Tothill, J.C. and Mott, J.J. (eds) (1985) *Ecology and Management of the World's Savannas*, Brunswick, Victoria, Australia: Australian Academy of Sciences.

United Nations Educational, Scientific and Cultural Organization (UNESCO) (1973) *International Classification and Mapping of Vegetation*, Series 6, Ecology and Conservation, Paris, France: UNESCO.

Walker, B.H. (1985) 'Structure and function of savannas: an overview', in J.C. Tothill and J.J. Mott (eds) *Ecology and Management of the World's Savannas*, Brunswick, Victoria, Australia: Australian Academy of Sciences, pp. 83–91.

Watts, W.A. (1980) 'Late Quaternary vegetation of the central Appalachian and the New Jersey coastal plain', *Ecological Monographs* 49: 427–469.

Werner, P.A., Walker, B.H. and Stott, P.A. (1991) 'Introduction', in P.A. Werner (ed.) *Savanna Ecology and Management: Australian Perspectives and Intercontinental Comparisons*, London, UK: Blackwell Scientific Publications, pp. xi–xii.

White, J. and Madany, M. (1978) *Classification of Natural Communities in Illinois*, Illinois Natural Areas Inventory Technical Report, Illinois, USA: Illinois Department of Conservation.

Winkler, J.G., Swain, A.M. and Kutzback, J.E. (1986) 'Middle Holocene dry period in the northern midwestern United States: lake levels and pollen stratigraphy', *Quaternary Research* 25: 235–250.

Wright, H.A. (1972) 'Fire as a tool to manage tobosa grasslands', *Proceedings, Tall Timbers Fire Ecology Conference* 12: 153–167.

Wright, H.A. (1974) 'Range burning', *Journal of Range Management* 27: 5–11.

Wright, H.A., Bunting, S.C. and Neuenschwander, L.F. (1976) 'Effect of fire on honey mesquite', *Journal of Range Management* 29: 467–471.

SUGGESTED READING

The following are symposium volumes that contain vast amounts of information should the reader wish to explore further the topic of savannas.

Leach, M.K. and Ross, L. (eds) (1995) *Midwest Oak Ecosystems Recovery Plan: A Call to Action*, 27 September 1995.

Sarmiento, G. (1984) *The Ecology of Neotropical Savannas*, Cambridge, MA, USA: Harvard University Press.

Tothill, J.C. and J.J. Mott (eds) (1985) *Ecology and Management of the World's Savannas*, Brunswick, Victoria, Australia: Australian Academy of Sciences.

Werner, P.A. (ed.) (1991) *Savanna Ecology and Management: Australian Perspectives and Intercontinental Comparisons*, reprinted from *The Journal of Biogeography* 17, 4/5, London, UK: Blackwell Scientific Publications.

SELF-ASSESSMENT QUESTIONS

1 Stocking rates of domestic animals on savannas attempt to:
   (a) maximise use of available forage;
   (b) insure highest income from forage to stimulate local economy;
   (c) balance livestock numbers with available forage without damaging vegetative resources;
   (d) balance wet and dry season forage resources with the number of livestock owned by area pastoralists;
   (e) estimate the number of livestock produced in the region.

2 African savannas do not produce adequate crops and livestock to meet the growing needs of the region's human population because of:
   (a) abandonment of traditional land-use practices that prevailed prior to pre-European intrusion;
   (b) errors in planning for use of land resources;
   (c) population growth and economic problems;
   (d) relatively low productivity of soils and cyclic droughts;
   (e) all of the above.

3 To produce lasting positive benefits, savanna research must:
   (a) abandon the 'old ways' of resource use and identify high technology solutions;
   (b) concentrate on successful savanna management in other parts of the world;
   (c) aim at extracting maximum income from all savanna resources;
   (d) consider the socio-economic situation of the farmers and ranchers and adequately communicate benefits of research to the affected public;
   (e) all of the above.

4 Which is the least important concern in the management of exotic species in a savanna?
   (a) Competition with native species.
   (b) Population monitoring and control of problem exotic species.
   (c) Interspecific competition among exotic species.
   (d) Strategies to minimise damage to native species in exotic control efforts.
   (e) Identification of highly mobile exotic species.

5 The primary stressors that developed and maintain savanna vegetation are:
   (a) fire and herbivory;
   (b) bacterial and fungal diseases;
   (c) weather variations and climate changes;
   (d) greenhouse effects;
   (e) changes in soil structure and nutrient availability.

6 A phrase that best characterises a savanna is:
   (a) grass and trees;
   (b) browsing and grazing large herbivores;
   (c) seasonally variable climate;
   (d) vegetative stability;
   (e) a region where there is an interaction of the above ecological components.

7 Restoration of savannas is accomplished most efficiently by:
   (a) a planned and monitored restructuring of the previous system's organisation;
   (b) letting natural recovery proceed with no intervention;

   (c) nurturing the surviving species and eliminating competition for food and space;

   (d) promotion of studies that evaluate the current ecology of the system;

   (e) use of known recovery patterns in other savannas in similar climates.

8 Continuance of a productive savanna is dependent upon:

   (a) the elimination of fire as an element in vegetation management;

   (b) absence of human activities;

   (c) total soil management to reduce variation in landscape and vegetative structure;

   (d) elimination of grazing as land-use practice;

   (e) allowance of basic natural processes to proceed unchecked by artificial disturbances.

9 Holistic resource management

   (a) has little potential in natural savannas;

   (b) promotes traditional vegetative management techniques;

   (c) would not be effective on restored savannas;

   (d) attempts to mimic natural herbivore–forage relationships;

   (e) is succeeding because of reduced management inputs.

10 The use of savannas by early humans:

   (a) left only minimal impact on the savanna's ecology and productivity;

   (b) eliminated many of the large herbivores;

   (c) exploited the land resources and interfered with natural processes;

   (d) produced no known impacts;

   (e) cannot be evaluated because of lack of knowledge of early human history.

7

DESERT MARGINS

The Problem of Desertification

Bernard J. Smith

SUMMARY

Desertification is a global problem encompassing social, political and economic changes which, when superimposed upon agricultural marginality and climatic variability, result in progressive environmental degradation in dryland areas. In this chapter the different aspects of desertification are first defined, to illustrate the extent of the problem, its inter-disciplinary character and its relationship with different forms of drought. The underlying causes of desertification are investigated in the context of environmental marginality and the cumulative pressures of overpopulation, and the social, economic and political changes consequent upon development. The cumulative effects of these pressures can instigate desertification, but more frequently it is triggered by sudden additional stresses associated with conflict or drought. Drought as a trigger mechanism is examined through a study of the West African Sahel, where it has been superimposed upon overcultivation, overgrazing, deforestation and failed irrigation. Approaches to the prevention of desertification and rehabilitation of areas already affected are briefly investigated through policy requirements as laid down by the UN, the choices available to policy-makers, and techniques available for soil erosion control. The chapter concludes with a review of recent research that questions previously held ideas on the spread and intensity of desertification and has highlighted the resilience of indigenous communities.

ACADEMIC OBJECTIVES

Desertification is a problem that transcends disciplinary boundaries, affects millions of people and, with the exception of Antarctica, is found on every continent. It is impossible therefore to cover all aspects of the subject in one chapter. The best one can hope for is to establish a framework into which ideas and information can be fitted, to point the reader in the direction of more detailed material and hopefully to stimulate thought on possible future actions. This chapter is based upon the premise that understanding the nature of a problem is the first and essential step towards its successful management. There is thus an emphasis upon the precise definition of the problem and especially its causes. It will be shown that not only are there many factors that contribute to desertification but that they act in concert to destroy the balance between soil stability, vegetation, landscape, climate and human activities. In terms of management, space does not permit explanations of every strategy available for tackling the problem at global, regional, national and local levels. Some indications are given as to required actions, but the principal objective is to educate managers so that they can choose appropriate solutions.

NATURE OF THE PROBLEM

Desertification and drought

Desertification is a term that is widely used, but remains an area of considerable contention among those concerned with detailed studies of environmental degradation in dryland regions. It was first used as long ago as 1949 by Aubréville to describe replacement of forest by savannas in Africa (Grainger, 1990). But it came to prominence in the 1970s following widespread environmental degradation, drought and famine in the Sahel zone of West Africa. As Verstraete (1986) pointed out in his review of desertification, this coincided with growing international concern over environmental matters in general and in 1974 the UN General Assembly instructed the UN Environmental Programmeme (UNEP) to organise an International Conference on Desertification. This conference (UNCOD) finally took place in Nairobi in 1977. At the conference and in the studies leading up to it, it became clear that there was dispute over the precise definition of desertification and doubt about whether the term should be used at all. Le Houérou (1959) had, for example, proposed the term *desertization* to refer specifically to degradation of semi-desert areas on the margins of climatic deserts, and took exception to the use of *desertification* when applied to environments far removed from deserts. None the less, the definition finally agreed by UNCOD did not include any geographic limitations and stated that:

> Desertification is the diminution or destruction of the biological potential of the land, and can lead ultimately to desert-like conditions. It is an aspect of the widespread deterioration of ecosystems and has diminished the biological potential of the land, i.e. plant and animal production, for multiple use purposes at a time when increased productivity is needed to support growing populations in quest of development.

This definition was later modified by UNEP to include reference to:

> The combined pressures of adverse and fluctuating climate and excessive exploitation.

As Verstraete (1986) notes, however, this addition is frequently omitted from studies and the relationship between drought and desertification is often confused. The role of drought was recently discussed by Agnew and O'Connor (1991) who stressed that environmental degradation in continents such as Africa is multivariate and that by concentrating on the importance of drought 'we may be guilty of diverting attention from other immediate causes and from relevant underlying structures – and also from the possibility of long-term environmental deterioration on which more reliable information is surely needed' (p. 4). They further point out, by reference to the work of Farmer and Wigley (1985), that drought itself is a complex phenomenon which can take several forms. Most people when considering drought tend to identify it with changes in the weather patterns of an area. *Meteorological drought*

can therefore be considered as a protracted departure from normal water availability, resulting in a water deficit that exists for long enough to cause hardship. Clearly, however, there are other drought conditions which do not relate to overall rainfall amount, but to specific water needs. *Agricultural drought*, for example, occurs when there is insufficient water to meet crop requirements. This may reflect a decrease in total rainfall but frequently it can be caused by a change in rainfall pattern, whereby seasonal rains start too late, finish too early or contain prolonged dry spells. Similarly, a *hydrological drought* may occur if there is insufficient water to sustain required levels of river flow or groundwater supply, while *economic drought* can result from water shortages that cause a shortfall in industrial production. In each case the stage at which drought occurs or can be declared is somewhat arbitrary. Agnew and O'Connor (1991) note that anything between a 20–40 per cent departure from the mean is frequently used, but such figures do not, for example, take into account the possible cumulative effects of several years of persistent, less extreme shortfall. Neither do these definitions incorporate changes in water need and it must not be forgotten that water shortages may result from increased demand as well as decreased supply.

Irrespective of the specific local causes of desertification, a common characteristic is that the end product is the deterioration of ecosystems. This is manifested in vegetation destruction and soil degradation, and writers such as Mortimore (1987) have considered it sufficient to define desertification as simply the 'diminution or destruction of the biological potential of the land'. He also makes a specific link between desertification and dryland areas that was reinforced by Mendoza (1994, quoted in Thomas and Middleton, 1994) who saw desertification as 'land degradation in arid, semi-arid and dry sub-humid areas resulting mainly from adverse human impact'. This is very similar to the definition adopted at the Earth Summit at Rio de Janeiro in 1992, and contained within the Agenda 21 document. This states that desertification is 'land degradation in arid, semi-arid and dry sub-humid areas resulting from various factors including climatic variations and human activities'. Both definitions reflect a growing consensus that the term should be restricted to dryland environments and human-induced degradation that imparts changes in the environment that reduce the potential for sustainable development (Thomas, 1993).

Soil degradation

Soil degradation is often interpreted as soil erosion but while loss of soil material is the most obvious and arguably the commonest form of degradation there are numerous other means by which the growth potential of soils can be reduced. The factors that influence degradation and the processes responsible have been summarised by Rapp (1986) and Lal *et al.* (1989) and can be seen in Table 7.1. Rapp and Lal recognise three process categories, namely physical, chemical and biological, but it is evident that the distinctions are not independent; most of the processes mentioned operate in conjunction with, or as a consequence of, others. Chemical leaching and reduction in organic matter, for example, are often key factors in the deterioration of soil structure, while increased toxicity or waterlogging of soils can reduce and/or modify soil fauna. Likewise, the factors controlling

degradation rarely operate in complete isolation and the range of agricultural activities listed may change as an area is cleared of forest, cultivated and then abandoned to grazing. Alternatively, within one area different sites will experience various types and intensities of land use, producing a patchwork of land-use types and degrees of degradation. None the less, Table 7.1 does illustrate the great variety of and potential for soil degradation. It also highlights the range of processes in addition to erosion, together with non-agricultural factors, that can lead to soil degradation and in extreme cases desertification.

Vegetation destruction

Vegetation removal is a crucial element in soil degradation. A stable and complete vegetation cover plays many roles in ensuring the stability of an ecosystem including the following:

1 Interception of rainfall, which generally reduces kinetic energy of rain reaching the soil, thereby reducing crusting and encouraging infiltration.
2 Maintenance of soil structure through root action and organic matter supply.

Table 7.1 Factors and processes of soil degradation

Factors influencing degradation		
Agricultural	*Industrial*	*Urban*
Deforestation	Waste disposal	City waste
Inappropriate ploughing	Acid deposition	Conversion to non-agricultural use
Intensive cropping and monoculture		
Indiscriminate/excessive use of chemicals		

Processes of degradation		
Physical	*Chemical*	*Biological*
Structure deterioration	Leaching	Biomass carbon decline
Compaction	Fertility depletion	Reduced organic matter
Crusting	Alkalinisation	Reduced soil fauna and flora
Accelerated erosion	Salinisation	
Hard setting	Lateritisation	Switch to unfavourable trends in biological processes
Waterlogging	Toxification, especially Al^{3+}, Mn^{+3}, heavy metals	
Drought		
Temperature extremes (high or low)		

Source: After Lal *et al.,* 1989

3 Recycling of nutrients which would otherwise be leached from the soil.
4 Reduction of soil surface temperatures and maintenance of high ground level relative humidity.
5 Supply of transpired moisture to the atmosphere which would otherwise run off and may therefore contribute to higher rainfall totals.
6 Reduction of wind speed near the ground and provision of ground cover thereby reducing the possibility of wind erosion.

Destruction of the vegetation cover frequently leads to more rainfall reaching the soil surface, less infiltration, lower soil moisture contents and increased surface runoff. This leads to erosion, often preferentially of fine material including organic matter on to which nutrients are bonded. At the same time, the ground level microclimate becomes harsher and this, combined with the loss of soil mass, reduces the likelihood of vegetation regeneration. In this way, 'desertification is a self-accelerating process, feeding on itself, and as it advances, rehabilitation costs rise exponentially' (Mortimore, 1988: 61). This process is most rapid when vegetation is completely removed, if not permanently then at least for substantial parts of the year.

Desertification: extent and severity

Grainger (1990) pointed out that desertification is a world-wide problem that can occur in any dryland area. As such, the distribution of lands affected by desertification correlates very closely with areas that are naturally semi-arid or at least marginal to desert areas. Because of this there is a need to distinguish between those areas that are natural, climatic deserts and desertified areas where conditions are the consequence of poor land use. Naturally arid areas were first comprehensively classified by Meigs (1953) into hyperarid, arid and semi-arid zones. In hyperarid and most arid areas the risk of desertification is slight as the environment is already marginal. Instead various studies, beginning with the map of desertification hazard commissioned for the UNCOD meeting of 1977, have shown that the areas of greatest risk are the semi-arid and steppe on the fringes of arid zones. This is echoed in Figure 7.1 which shows areas of actual desertification as identified by Dregne (1983). In this he recognises four degrees of desertification which also provide an insight into the relevant criteria for identifying the problem. The classes are as follows:

(a) *Slight* – little or no deterioration in vegetation cover or soil due to human activities. Most of the world's deserts automatically fall in to this category because vegetation is so sparse in the natural state.

(b) *Moderate* – where the level of environmental deterioration is such that the plant cover has been degraded to the point where grazing conditions are only fair and 26–50 per cent of the plant community consists of climax species. Water and wind erosion is accelerated producing small dunes or small gullies with 25–75 per cent of topsoil lost, and erosion and possibly salinisation result in a 10–50 per cent reduction in crops. Moderate damage is significant because it represents a loss of productivity that will worsen unless management practices are improved.

(c) *Severe* – when undesirable broad-leaved herbs and shrubs have replaced desirable grasses and shrubs or have spread to the point where they dominate the flora with only 10–25 per cent of climax species remaining. Wind and water erosion has largely removed the topsoil and largely denuded the land of vegetation. Crop yields are reduced by at least 50 per cent or salinity has made the land unsuitable for sustained cultivation.

(d) *Very severe* – irreversible desertification with less than 10 per cent of climax species in the vegetation community. Gullies are large and numerous, there may be large shifting sand dunes or salt crusts. Small areas may be reclaimable at moderate cost, but total reclamation is generally uneconomic, especially where there are mobile sand dunes. World-wide there are few extensive areas of severe degradation, but within an area there may be numerous small areas.

There are equivalent classifications for rangeland and Mabbutt (1984) proposed that areas be classified as moderately desertified if livestock-carrying capacity were reduced by up to 25 per cent, severely desertified if reduced by 25–50 per cent and very severely desertified if reduced by more than 50 per cent.

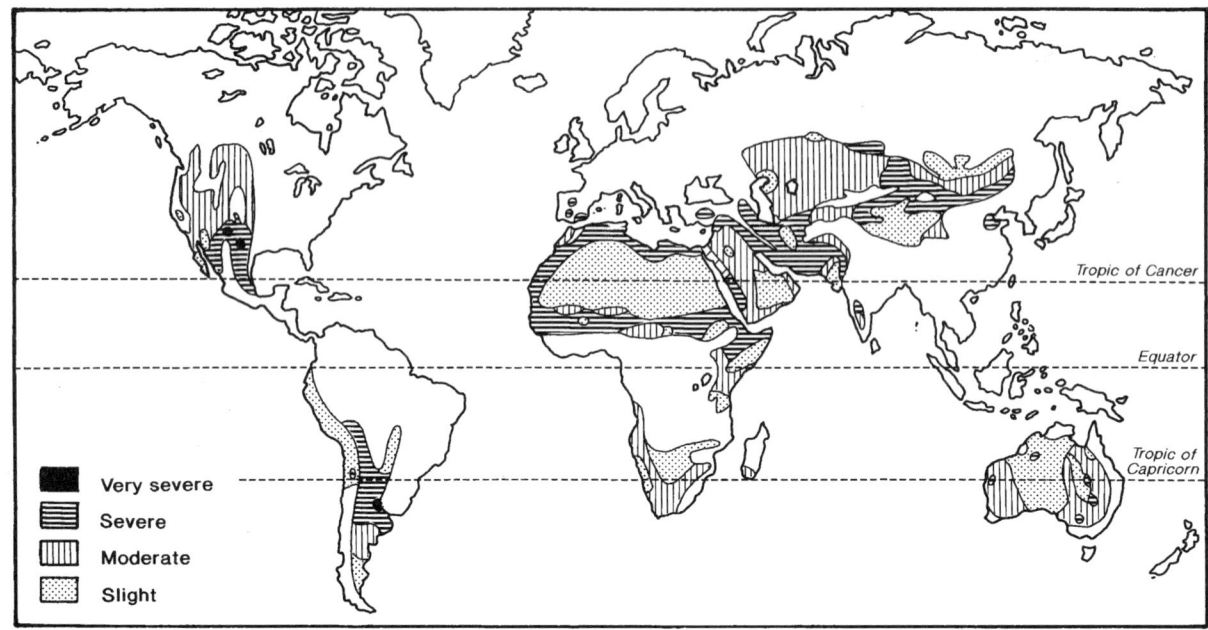

Figure 7.1 Map showing areas of desertification
Source: redrawn from Dregne, 1983

The extent of areas affected is a subject of considerable debate, reflecting the paucity of reliable information, from what are generally the world's poorest countries. Dregne (1983) did, however, estimate that globally some 1.731 million km² of rainfed cropland was at least moderately affected, 0.271 million km² of irrigated cropland and 30.486 million km² of rangeland. As can be seen from Table 7.2, these areas are overwhelmingly concentrated in the developing countries of Africa and Asia.

The figures given in Table 7.2 represent a snapshot of a continuing process, and Dregne (1983) estimated annual rates of increase of desertification to be 177,000 km² for rangeland, 2,000 km² for rainfed agriculture and 5,460 km² for irrigated agriculture. It should not be thought, however, that desertification is solely a modern phenomenon. It has been present for at least the last 1,000 years, but has undoubtedly accelerated in recent years as increasing populations have sought to support themselves on a resource of finite extent (Dregne, 1977). Similarly,

it is misleading to think of increasing desertification as the inexorable and uniform outward migration of desert margins. There are some studies that have claimed to have plotted gradual desert expansion; Rapp *et al.* (1976), for example, measured the southwards movement of the Sahara Desert in the Sudan

Table 7.2 Areas in different continents (thousand km²) that experience at least moderate desertification

	Rangeland	Rainfed agriculture	Irrigated agriculture
Africa	10,678	396	14
Asia	12,007	912	206
Australia	3,087	15	2
Europe [Spain]	206	42	9
N. America & Mexico	3,185	247	28
S. America	3,325	119	12
Total	32,488	1,731	271

Source: Dregne, 1983

between 1957 and 1975 at an average rate of 9 km per year. But in most areas desertification is a more subtle process, which mainly occurs away from the desert fringe in response to particular land uses and ground conditions.

SOWING THE SEEDS

The origins of desertification fall into two broad categories. First, there are those factors that are essentially passive and represent marginality and the natural susceptibility of certain areas to vegetation and soil degradation. Second, there are dynamic factors associated with human activities which propel environments towards change. Many factors operate indirectly to increase pressures on the natural environment, whereas other activities are directly responsible for degradation. In this section the marginality of areas susceptible to desertification will be discussed, together with those indirect factors that increase risk.

Environmental marginality

Climate

Desertification predominates in drylands with low annual rainfall, and because rainfall variability is inversely proportional to amount one neither expects nor finds a high correlation from year to year. Rainfall is often concentrated within a wet season whose beginning and end also varies from year to year. The length of this wet season decreases with proximity to the desert margin, and to ensure a successful yield, land must be cultivated in advance of the rains. Failure accurately to predict the rains and/or a false start of the rains can reduce yields or necessitate replanting of crops. The rain itself is often intense, especially at the beginning of the rainy season in areas such as West Africa, and frequently falls as isolated storms giving considerable spatial as well as temporal variability. Relative humidities during the dry season are often very low, high winds are common, both regionally and locally in response to convection and downdraughts from storms. These factors, combined with high temperatures, produce high potential evapotranspiration as well as conditions ideal for wind erosion of loose unprotected topsoil.

Vegetation

Dryland vegetation is strongly influenced by the seasonal and often sporadic nature of the rainfall, and is dominated by species adapted to prolonged seasonal drought. Some species flower and set seed before the rains begin and others survive by exploiting groundwater resources. Most vegetation, however, either sheds its leaves after the rains cease, dying back to root stock or a tuber, or complete their life cycles during the rains and see out the dry season as seeds. As a consequence, vegetation cover decreases dramatically during the dry season making most areas naturally prone to wind erosion. There is also a time-lag between the commencement of the rains and the development of a new ground cover, especially of grasses, during which topsoil is exposed to potentially high rates of erosion by rainsplash and surface runoff. Even during the rains, the degree of cover afforded by vegetation is variable and as annual rainfall decreases vegetation communities degenerate into isolated individuals separated by bare ground.

Soils

Dryland soils, because of limited chemical alteration, are commonly coarse textured with low clay and organic matter content. This is frequently exacerbated by the fact that soils are developed on stabilised sand dunes relict from previous, climatically driven episodes of desert expansion. Such sandy soils have low nutrient and moisture-holding capacities, are of low fertility (especially low in nitrogen and phosphorous) and are prone to wind erosion. High soil temperatures encourage evaporation and rapid decomposition of organic matter aided by termite activity. Loamy soils are prone to capping which reduces infiltration and encourages overland flow and erosion by rainsplash, sheetwash, rills and gullies.

Geomorphology and hydrology

Runoff, in response to climate, is highly seasonal and many tributary streams have extremely 'flashy' regimes. Flow only occurs during and immediately after storms and peak discharges can be very high, leading to flooding and damage to roads, bridges, etc. High transmission losses mean, however, that flows can disappear before storage areas such as dams are reached. Much of the world's drylands, especially on deltaic areas and the stable shields of Africa,

Australia and India have a flat topography. This can make water storage and distribution through gravity difficult, and restricts adequate drainage of irrigated areas. Rapid surface runoff and long dry seasons combine to restrict the availability of shallow groundwater away from river courses.

Biological factors

High temperatures encourage rapid growth of weeds once rains come as well as a high incidence of pests and human and animal diseases. The widespread occurrence of diseases such as *Trypanosomiasis* (sleeping sickness) can prevent integrated livestock and arable farming in wetter areas, which removes a source of power for cultivation and of nutrients in the form of animal wastes (Harrison, 1987). Numerous chronic human diseases, invariably water related, water washed or carried by water-related vectors (see Pacey (1977) for a review) result in low labour productivity and through diseases such as river blindness (*Onchocerciasis*) can create large dependent populations.

The pressure of population

Many countries that border or include desert areas have very high birth rates. In common with many other developing nations they are experiencing a demographic transition from conditions of high birth rate, high death rate and low, stable population levels to a situation where death rates have declined but birth rates remain high and population rapidly increases. It is projected that as the need for high birth rates – to ensure survival of some offspring – disappears and education and availability of contraception increases, a new stability will be reached of low birth and death rates and a new higher, stable population. At present, however, passage through this transition has left many countries at risk from desertification with large, youthful populations which, as they reach reproductive age, will continue to fuel population growth even if births per family are reduced.

One consequence of rapid population growth has been substantial rural to urban migration. For example, the urban population of West Africa rose from 12,425,000 (12.8 per cent of total population) in 1965, to 30,500,000 (21 per cent of the population) in 1980 (Lewis and Berry, 1988: 341). This has created large urban populations that can only be fed by importation of food, intensified production from existing farmland or extension of agriculture to more marginal lands.

Political and economic pressures

Desertification is overwhelmingly a problem of developing countries and the political and economic consequences of development have indirectly placed considerable pressures on their environments. As previously indicated, one consequence has been the growth of large urban populations. This now includes educated urban elites that, together with the concentrated populations of the cities, constitute the most organised power base and possible threat to the government of the day. These urban populations are thus invariably indulged by governments to ensure their support. One means of placating the urban population is through keeping the price of food artificially low. This is possible only at the expense of the agricultural sector which is often excessively taxed, starved of resources from central government and yet expected to produce higher and higher yields. Taxes may take the form of a proportion of the crop, but invariably farmers are required to grow cash crops to generate an income that can be taxed. Continuous monoculture of crops such as cotton and groundnuts has a deleterious effect upon soil fertility which can only be remedied by expensive inputs of artificial fertiliser that are rarely available. Monocultivation also increases susceptibility to pests. As more land is allocated to cash crops the possibility of self-sufficiency decreases, especially when rains fail and cultivation is extended to more and more marginal land.

In addition to subsidising the needs and lifestyles of large urban populations, farmers in developing countries are increasingly forced to service the national debts of their countries which borrowed heavily during the 1960s and 1970s. Endemic corruption and massive expansion of armed forces meant that little of this money percolated down to individual farmers. In some countries, for example, Nigeria, large amounts of money were put into agricultural development, but invariably into large-scale projects designed either for cash crops such as cotton and sugar or crops such as wheat for import substitution. In many cases these schemes, often based upon large dams, have, through mismanagement, environmental and technological inappropriateness plus the vagaries of world markets and climate, failed to live up to expectations (for discussions see Adams, 1986, 1991 and Adams and Grove, 1984). These projects often take already cultivated land out of food production rather than bringing new land into cultivation. This, together with increased mechanisation, added to urban migration as well as encouraging rural populations to expand into land that was previously considered too marginal for cultivation (Smith and Baillie, 1985). Similar pressures were placed on pastoralists who were confined to grazing reservations with the consequent threat of overgrazing. Alternatively, they had to abandon traditional migratory routes, often in favour of areas with less assured rainfall. Such changes also interrupted the symbiotic relationships between sedentary farmers such as the Hausa in West Africa and pastoralists such as the Fulani who were traditionally allowed to graze their herds on stubble in return for the manure that is left behind.

The most dramatic disruptions of rural communities and environments have come about, however, not from economic pressures *per se* but from political conflict. The origins of these conflicts are frequently complex but contributory factors include inherited colonial boundaries that either cut across traditional ethnic and political divides or enclose disparate and sometimes antagonistic communities. The imposition at independence of political systems inappropriate to countries embarking on the development process has also triggered a multiplicity of internal conflicts, with *coup* followed by counter-*coup*. It is hardly surprising therefore that almost without exception, especially in Africa, countries prone to desertification have experienced recent armed conflict. The inevitable consequence is destruction of infrastructures for protecting the environment, abandonment of large tracts of land, mass migration of populations which places added stress on already sensitive regions and diversion of more and more resources into buying arms.

TIPPING THE BALANCE: DROUGHT AS A TRIGGER MECHANISM

Desertification as a threshold phenomenon

The previous section showed that the world's dryland areas are environmentally marginal for human occupation, and that politics and economics have conspired to add additional layers of stress on top of this. The gradual accumulation of these stresses may eventually reach a point where environments can no longer absorb them and rapid degradation ensues. More probably, however, as stress vs. resistance thresholds are neared, a sudden, perhaps localised increase in stress triggers rapid environmental decline which soon develops its own momentum. Sudden increases in stress take many forms, for example, armed conflict, sudden in-migrations of displaced populations or changes in national economic circumstances. Such triggers are, however, generally confined within national boundaries and for the simultaneous, region-wide degradation that characterised desertification in the 1970s and 1980s an alternative trigger mechanism must be sought. The obvious, and generally accepted as most important, stimulus for desertification is a prolonged period of water shortage or drought. This acts both as a trigger for environmental degradation and also as a catalyst once change is set in motion (Campbell, 1986).

Case study of drought in the Sahel zone of West Africa

It was drought, and its attendant ecological catastrophe in the Sahel zone of West Africa (Figure 7.2) between 1968 and 1973 that first drew global attention to the problem of desertification. The Sahel zone (its name means literally 'edge of the desert') is a classic example of a dryland area with an extremely seasonal climate. During winter months weather is dominated by dry, dusty winds blowing out of the Sahara – the Harmattan. These winds blow across West Africa until they encounter moist south-westerly winds blowing off the Atlantic to form a zone of disturbance known as the Inter Tropical Convergence Zone (ITCZ). In the winter the ITCZ lies along the southern coast of West Africa, but as the year

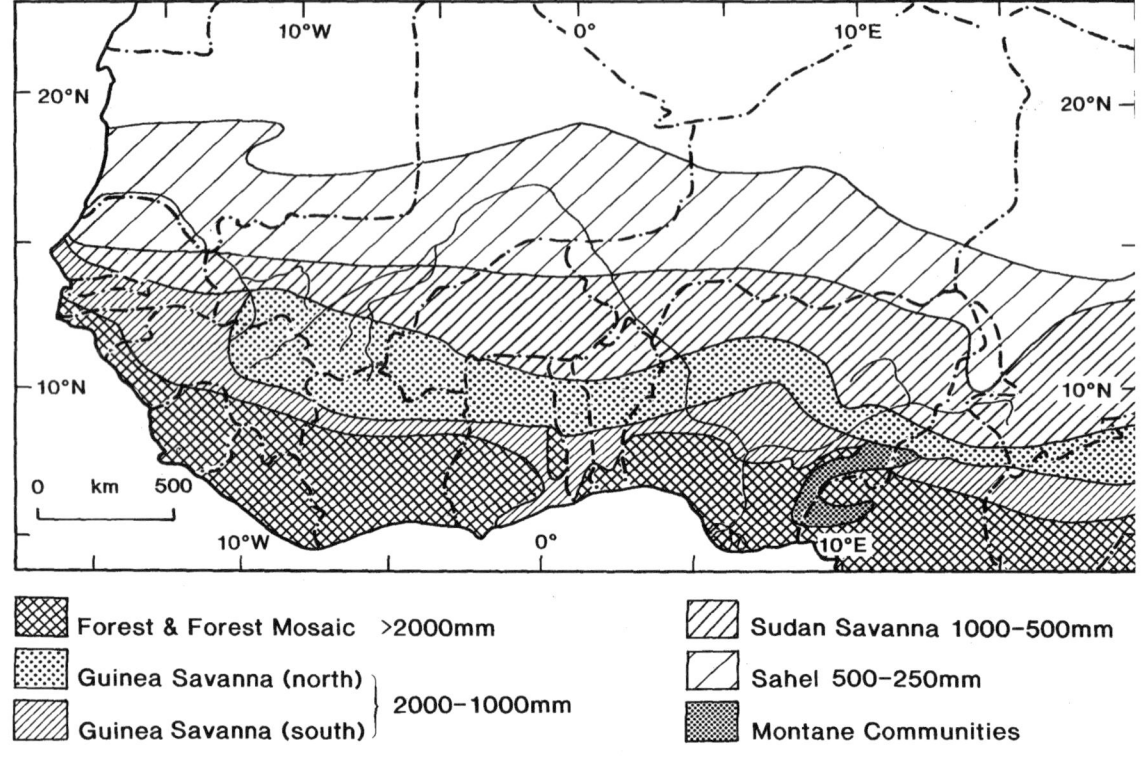

Figure 7.2 Vegetation zones of West Africa

progresses and the thermal equator moves northwards so too does the ITCZ. This replaces Harmattan conditions first with a period of intense storm rainfall and then with less intense rains from the Atlantic air mass. Success for the farmers in the Sahel zone depends on when the ITCZ reaches them and how long it stays to the north, which determines the timing and length of the rainy season. What is clear from numerous climatological studies (e.g. Bunting *et al.*, 1976; Olsson, 1983; Dennett *et al.*, 1985) is that the timing and annual amount of rainfall in this zone is highly variable. The variability is extremely complex and Littman (1991) has, for example, claimed to have identified periodicities of 2–4 years, 6 years, 11–12 years and 25 years; while studies of Quaternary climatic change have suggested that the Sahel drought of 1968–1973 is part of a natural alternation of drier and wetter phases of differing magnitudes and durations (Williams, 1979). During the period 1968–1973 rainfall across the Sahel was respectively 72 per

cent, 98 per cent, 97 per cent, 74 per cent, 69 per cent and 65 per cent of the long-term mean annual rainfall. There have, however, been several other extended dry periods this century between 1907 and 1918, 1940 and 1944, and 1947 and 1949. Evidence from sediments in the river Niger suggest that conditions were most severe between 1907 and 1918 (Grainger, 1982). Between these dry periods there were intervening periods when rains were consistently above the long-term mean (1929–1931, 1952–1955 and 1957–1962). Although it is unlikely that these relatively assured rains drew farmers northwards towards the desert, it is certain that they allowed other pressures – population increase and the spread of commercial agriculture – to push people into areas with a high meteorological drought risk.

Although the very dry years of the early 1970s were succeeded by several years of near-normal rainfall in the mid-1970s (Figure 7.3), these represented only a brief respite from what has turned out to be

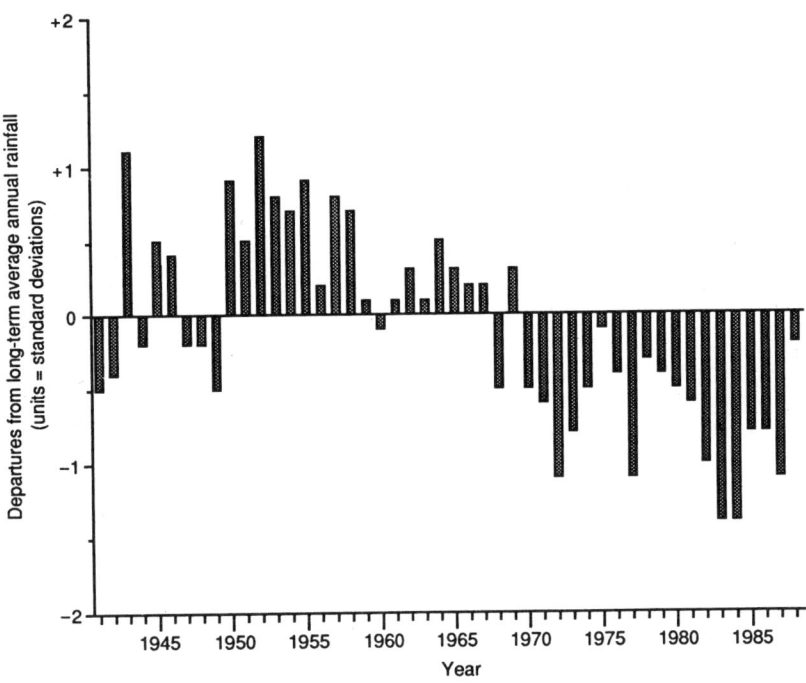

Figure 7.3 Rainfall trends in the West African Sahel showing departures from the long-term mean
Source: Lamb, 1987; Grainger, 1990

a prolonged period of drought extending through the 1980s and into the 1990s (Lamb, 1986; Agnew and O'Connor, 1991; Pearce, 1991). This has naturally raised questions as to whether what we are seeing now is a reflection of natural variability or whether it is the precursor of a long-term shift in climate towards dryer conditions. This debate has been with us since the 1970s, with an early consensus in favour of an eventual return to wetter conditions (e.g. Miles and Folland, 1974; Bunting *et al.*, 1976; Ojo, 1983). Within the last ten years or so, however, there has been increasing evidence of global changes in climate due to anthropogenic emission of greenhouse gases. While there is no consensus yet as to the precise effects of global warming, it seems reasonable to assume that any consistent change in global climate will accentuate recent regional climatic variations in areas such as the Sahel.

Overall global warming is only one possible component in prolonged Sahelian drought and Grainger (1990: 37–64) reviews a number of suggestions as to which factors influence movements of the ITCZ. These include proposals that drought years may be linked to variations in sea surface temperatures (Lamb, 1986) and of feedback mechanisms centred round 'biogeophysical feedback' and increased dust storm activity. Biogeophysical feedback is based on the premise that drought causes a decline in vegetation cover, which increases earth surface albedo. The increased reflection of solar energy means that the bare surface is cooler and this produces less convective uplift which inhibits rain generation. With increased dust storm activity (Middleton, 1985), dust particles at altitude absorb insulation, warm the surrounding air and thus reduce upward airflow and again inhibit rainfall. In both circumstances, however, such feedbacks are the immediate products of desertification and while they may explain the persistence of drought they are unlikely to have triggered drought in the first instance. For a review of the concept that 'drought follows the plough' see Glantz (1994).

The environmental and human impact of drought is very dependent upon the nature and scale of rainfall change. In a study of rainfall changes in Central Sudan, Trilsbach and Hulme (1984) identified a number of variable parameters, each with different impacts (Box 7.1). However, in whatever its manifestation, drought in fragile dryland environments already under stress from concentrated human activity does trigger environmental changes related to runoff, soil moisture, evapotranspiration and groundwater resources. Drought is not, however, a new phenomenon and dryland ecosystems and their inhabitants have previously managed to survive periods of meteorological drought. In the next section traditional strategies for survival in dryland areas are outlined together with those human activities that have led directly to desertification.

REAPING THE HARVEST: CAUSES OF DESERTIFICATION

Background: traditional adaptations

Despite the harshness of dryland environments, there are also advantages that have encouraged people to live there and to return after periods of drought. The light soils are easy to cultivate, and despite being deficient in some nutrients, there are others, especially inorganic elements, that are readily available, and there is abundant sunshine. In other words, provided that water is readily available there is the potential for rapid growth of crops, high yields and abundant pasture. Farmers and pastoralists are aware of the hazards, however, and have practised a number of strategies designed to minimise risks from drought, disease and over-exploitation of the landscape (Grainger, 1982). Included in these is the concentration of rainfed agriculture in areas where rain is most likely to fall in reasonable quantities and with reasonable regularity. Cultivation is also concentrated in areas of better soils and where soil moisture contents are highest. Prominent amongst these areas are the broad floodplains and alluvial sinks which are common on savannah landscapes ('Fadamas' in West Africa). They are areas of deep alluvial soils washed in from surrounding interfluves and receive inputs of soil and nutrients during annual floods. Flood rice and rainfed agriculture in the wet season, followed

---

BOX 7.1   RAINFALL CHANGES IN CENTRAL SUDAN AND THEIR EFFECTS

(a) *Changes in annual rainfall.* Effects included increased out-migration and movements to cities, and declines in yield and productivity of large areas of rainfed agriculture and pastoralism. Wells have been closed by local government because of reduced yields, increased salinity and higher maintenance costs. Within affected areas, villagers were encouraged to move to settlements with reliable water supplies which often resulted in social unrest. All these effects are 'major components of desertification' (p. 295).

(b) *Changes in daily rainfall magnitude-frequencies.* These need not necessarily change with a change in total annual rainfall, but when they do they add another layer of environmental impact. For example, in many of the driest areas, crops depend upon regular 'light' falls of rain to maintain soil moisture levels. Disappearance of these 'light' falls thus has serious implications for the successful growth of crops such as millet and sorghum. Trilsbach and Hulme also noted a decline in the frequency of very intense storms, which may have some benefits through reducing erosion hazard by surface runoff, but also reduces the incidence of major flows in wadis, which in turn inhibits flood-irrigated agriculture and groundwater recharge.

(c) *Rainfall localisation.* Records from Central Sudan showed that even during years when regional rainfall totals were below average, some areas experienced above average rainfall, falling as intense storms. This intense rainfall destroyed crops and caused flooding and waterlogging of soils. Localisation also meant that in some areas the severity of the drought is much greater than regional statistics would suggest.

(d) *Rainfall timing.* Trilsbach and Hulme stressed that the 'temporal distribution of rainfall through the wet season is crucial for ecological and human activities' (p. 297). Rainfed agriculture needs regular rainfall in sufficient quantities to satisfy evapotranspiration requirements. Prolonged rainfall, on the other hand, can lead to waterlogging and attention has been drawn elsewhere in this chapter to the problems of 'false starts' to the rainy season and 'premature' end, which can prevent crops from ripening.

Source: Trilsbach and Hulme, 1984

by dry season cultivation using residual soil moisture has made these areas extremely important to the traditional economies of areas such as northern Nigeria. Their reliance upon flood irrigation does, however, make them vulnerable to development projects which seek to dam and regulate river flows. The adverse effects of such schemes are illustrated by what has happened in the Hadejia-Jama'are valley in northern Nigeria (Kimmage and Adams, 1992). Since the construction of the Tiga Dam in the headwaters of the system there has been a considerable reduction in the areas flooded during the wet season compared to the 1960s and 1970s. This may be partly due to continuing drought in the region, but is also a function of the dam which, although it continues to release water into the system, effectively removes the flood peak that is essential for flood irrigation.

Within cultivated areas, farmers would traditionally choose drought resistant crops and would grow a variety of crops to reduce the risk of complete crop failure. The primary aim of the farmer would be to ensure basic food requirements, but with a limited area devoted to cash crops. Long fallow periods would be used to allow soil regeneration with fallow land used for grazing under a protective vegetation cover to reduce erosion risk. Other erosion control methods could include the use of tied-ridging to encourage infiltration and the use of crop residues to construct windbreaks which reduce both erosion risk and evaporation rates from dry season crops. Farmers would also co-operate with pastoralists, and barter produce and grazing of fallow land for dung left by the animals. Pastoralists would in turn adapt to rainfall variability by controlled grazing of areas where rainfall is most assured and by moving herds to areas where rain has recently fallen.

While never generating great individual profits or large agricultural surpluses, the strategies described above nevertheless ensured a degree of environmental protection and supported a relatively stable population. It is the advent of additional stresses brought about by population increase, social, economic and political change and conflict that have set in motion modifications to traditional agriculture that have led directly to desertification.

Direct causes of desertification

Overcultivation

Overcultivation can occur in a variety of ways; it may result from an intensification of existing practices involving, for example, abandonment of fallow periods or failure to replace nutrients. Alternatively, it may involve bringing new land into cultivation which is marginal because of steep slopes, erodible soils, nutrient deficiency or less reliable rainfall. The effects of a change from natural vegetation to different types of cultivation are illustrated in Table 7.3 which indicates up to a hundredfold increase in erosion rates following a switch from natural vegetation to maize and millet cultivation in Senegal. Finally, there may be a switch in crop types and/or cultivation methods, especially a move to mechanised farming. In all of these instances erosion hazard is increased either by an increase in the erodibility of the soil, often through a breakdown of soil structure, or an increase in environmental erosivity. This frequently involves soil being left bare and exposed to intense rain and high winds. A likely sequence of events that follows overcultivation has been described by Grainger (1982). This begins with a decline in soil fertility and a tendency for surface crusting of exposed topsoil. Crusting in turn increases surface runoff and leads to soil erosion by rills and gullies. In the dry season the topsoil exposed by water erosion is blown away and may lead to the formation or reactivation of dunes which encroach upon arable land and destroy crops by sand-laden winds. Alternatively, Hocking and Thomson (1979) see

Table 7.3 Rates of accelerated soil erosion under natural vegetation, fallow and different crop types in Sefa, Senegal

Vegetation cover	Annual rate of soil erosion (t/ha)
Natural vegetation (burnt)	0.2
Natural vegetation (unburnt)	0.1
Fallow with sparse vegetation	4.9
Groundnuts	6.9
Cotton	7.8
Sorghum	8.4
Maize	10.3
Millet	10.3

Source: data from Roose, 1967

overcultivation as a spiral of decline, where cultivation (and grazing) without erosion control leads to reduced vegetation in fallow periods, which exposes the soil to more erosion which removes more topsoil. Loss of topsoil reduces infiltration and encourages surface runoff and increased erosion. Increased erosion in turn slows the regrowth of fallow vegetation which reduces soil productivity and encourages more intensive cultivation, invariably without erosion control measures. Whatever the sequence of events it is unquestionable that overcultivation leads to soil degradation and subsequently to desertification.

Overgrazing

Grainger in his 1982 overview of desertification identified three principal pastoral systems: nomadic (transhumant), sedentary (village based) and ranching. Ranching is a relatively new departure and pastoral nomadism has been the traditionally predominant form of grazing in dryland areas. Mixed herds of cattle, sheep and goats are kept as insurance against drought and herds are continually moved, either to follow the rains or over historical migratory routes to exploit seasonal pasture. Two main reasons for overgrazing have been identified. First, the size and number of herds increases and, second, the area of land available for grazing is reduced leading to an increase in livestock densities in remaining areas.

The underlying causes of overgrazing are numerous and frequently reinforce one another. They include overpopulation, improved veterinary care, cultivation of former pastures, new boreholes which create local problems, settlement of nomads, the liberation of vassals and slaves of nomads who in turn acquire their own herds, designation of 'grazing reserves' and the spread of commercial and irrigated agriculture which excludes pastoralists. An indication of the magnitude of the problem is given in Table 7.4

Table 7.4 Changes in livestock density in Niger

Livestock	Period	Increase in numbers
Cattle	1938–61	× 45
Sheep and goats	1938–70	× 3
Sheep and goats	1961–70	+ 30%
Camels	1938–60	× 7

Source: Grainger, 1982

where increases in livestock numbers in Niger prior to the 1960s are shown. The environmental consequences of such increases can be dramatic. They may begin with a decline in the annual production of pasture, especially of palatable perennial grasses, which are efficient at binding the soil and ephemeral plants which only grow after rain will eventually dominate. These plants provide less protection to the soil and make the pasture less durable. As areas of pasture decrease, competition for remaining areas increases, soils are trampled and compacted, encouraging overland flow and fluvial erosion. In other areas, reduction of the vegetation cover increases the risk of wind erosion and can lead to the formation or reactivation of dunes. Reduction in pasture eventually leads to a decline in health of the stock and milk and meat yields decline. Cattle and sheep numbers are the first to decline leaving herds dominated by goats which close-crop vegetation in a very destructive way and may eventually be sustained only by herders cutting down the last remaining trees in the area. The end result is often a desolate area stripped of cultivable topsoil with the only indication of the former cover being the exposed roots of isolated tree stumps.

Deforestation

'Clearance of open woodland in the arid tropics is taking place at the rate of four million hectares per year according to FAO estimates' (Grainger, 1982: 28). The effects are particularly bad in Africa which in the early 1980s was losing 2.7 million hectares per year. Clearance can be for cultivation, to provide grazing, for fuelwood or, in response to increased urbanisation, for making charcoal. Grainger (1982) estimated that by the year 2000 the Sahel would be able to supply only 20 per cent of its fuelwood requirements. He drew upon the example of Burkina Faso in the West African Sahel to illustrate this crisis. The capital, Ouagadougou, consumes 95 per cent of national forest production and almost all trees in a 40 km radius of the city have been destroyed. Women, because they do most of this work, may have to walk four to six hours each day to collect wood and urban families may spend as much as a quarter of their income buying fuelwood and charcoal. Despite government controls introduced in the 1980s which have forced some villagers to burn crop residues and sometimes dung, fuelwood still accounts for 90 per cent of energy needs in the average house and an estimated 4 million logs are burnt each day (Sharp, 1990).

Forests protect the environment in many ways. The thick vegetation cover intercepts precipitation, which reduces the amount reaching the soil and also its erosivity. The root mat gives the soil an open structure which encourages rainfall to infiltrate and also helps to bind it together, reducing its erodibility. Erodibility is further reduced by organic matter from litter which improves structure and aggregate stability. Trees also act as effective windbreaks and inhibit wind erosion both beneath the trees and in downwind areas. In addition, soil temperatures are reduced, relative humidities are increased and the environment is generally more hospitable. If managed properly, dryland trees can provide a renewable source of fuelwood as well as numerous tree crops. Hence, even in intensively cultivated areas of the West African Savannah it is normal to find numerous trees that have economic significance. An example would be the mango trees which flower and fruit in anticipation of the rains and provide an extremely important source of nutrition when grain stored from the previous harvest may be very scarce.

Reduction in tree cover removes the protection that is afforded to soils, which either crust under raindrop impact, encouraging overland flow and erosion, or are disturbed by cultivation and/or grazing and eroded by the wind. Their utility as a food and fuel source is lost and increased surface runoff leads to increased frequency and intensity of flooding – often by water heavily charged with sediment which can shorten the life of downstream reservoirs and dams. Greater surface runoff means that less water infiltrates, groundwater recharge is reduced, streamflow between storms and after the rains finish is reduced and wells and boreholes may experience decreased flow or may dry up completely. Less infiltration also results in lower soil moisture contents during the dry season which increases susceptibility to wind erosion. Both wind and water erosion preferentially removes fine material from soils together with many nutrients and much of the nutrient and moisture-holding capacity of the soil. In this way deforestation can initiate a downward, self-reinforcing spiral of soil erosion and degradation as well as major hydrological changes.

Failed irrigation

In 1982 Grainger (p. 24) neatly summarised a major difficulty over irrigated agriculture when he noted that, 'for many years, irrigation has been the developed world's single minded and often simplistic recipe for growing more food in the arid tropics'. It is true that irrigation can dramatically increase crop yields by, for example, allowing continuous cropping and by reducing water stress. As a result individual yields are higher, risks of crop failure are much less, a greater variety of crops can be grown and an all-year-round supply of food as well as cash crops is available. As has already been noted, however, irrigation projects, especially those associated with big dams, have an unfortunate history of problems.

Irrigation schemes normally fail through 'inappropriateness' in one or more of a number of areas. The most obvious of these is *technical* inappropriateness, whereby schemes are ill-suited both for the environment in which they are placed and the purpose for which they are designed. The South Chad basin scheme in Nigeria was designed in the 1970s to irrigate up to 100,000 hectares by exploiting groundwater and lake water in the Lake Chad basin. It was not foreseen, however, that the Sahelian drought would continue and that Lake Chad would have shrunk by the late 1980s to a fraction of its former size with little prospect of it supplying irrigation water in the near future. Elsewhere in Nigeria, it was noted earlier how construction of the Tiga Dam as part of the Kano River Project has adversely affected traditional flood irrigation downstream. Also, areas to be irrigated are invariably already intensively cultivated, often using traditional irrigation techniques. The benefit of new production must therefore be set against the loss of previous agricultural output from the area as well as any farmland that may have been flooded by new reservoirs. Technical appropriateness also includes *health* appropriateness where projects should not increase the risk of waterborne, water-related or water-washed diseases.

Irrigation may also be *socially* inappropriate and cause social conflict because traditional social structures and lifestyles are disrupted as land tenure is reorganised, old responsibilities and traditions are made redundant and new priorities are imposed. *Educational* inappropriateness includes problems such as the maintenance of new equipment and even the simplest machinery will quickly fail if people are not educated in its use and upkeep and structures are not created to prevent its misuse. A failure of *economic* appropriateness may lead to precisely the opposite of what was intended. Schemes intended to benefit small farmers may lead to polarisation and exploitation if speculators are allowed to buy up large areas of land. Governments can also introduce a level of *political* inappropriateness through, for example, constraints upon the crops to be grown and excessive taxation. Failure in any of these areas will lead to bad management, under-performance and ultimately can bring outright failure.

The most common direct cause of failure in irrigation projects is an inability to maintain a balance between the amounts of water applied and the amounts lost by evapotranspiration and drainage. In all successful irrigation schemes more water is supplied than is required for the consumptive use by crops, and excess irrigation water is removed by drainage schemes which ensure that soluble salts are leached from the soil and that the water table does not approach the surface. Most major irrigation schemes are, however, located in flat, low-lying floodplains. These can be used for flood irrigation where water is dispersed at little cost and with low technology systems using gravity. Unfortunately, drainage of these low-lying areas is extremely difficult and in many areas over-irrigation has caused water tables to rise. This results in waterlogged soils and salinisation of the topsoil as salts brought to the surface either in groundwater or by capillary rise are precipitated by evaporation. As an indication of this problem Grainger (1982) cites the example of the Indus basin in Pakistan. By 1976, water tables that were originally at a depth of 24–27 m were at the surface, with 50 per cent of soils showing some evidence of salinisation and 20 per cent highly saline. In addition to salinity problems, the canals built to service these lowland schemes have gentle gradients, reservoirs are frequently very shallow and both are therefore prone to siltation. The Mangla Dam in Pakistan was, for example, constructed in 1967 at a

cost of US$ 600 million but now has an expected reservoir life of only 57 to 75 years. The Tarbela Dam, which was finished in 1975, was constructed with a life expectancy of only 50 years (see Eckholm (1976, Ch. 7) and Rhoades (1990) for reviews).

MANAGEMENT OF DESERTIFICATION

Requirements

Following its 1977 meeting the UN Committee on Desertification made a number of recommendations for action necessary to halt and reverse the spread of desertification. It is perhaps a sad indictment of the international community that most of the recommendations remain relevant. The specific requirements fall into seven broad categories:

Evaluation of desertification and improvement of land management

1 Countries should assess the magnitude and extent of desertification, its causes and effects, by monitoring dryland dynamics, including the human condition. Techniques such as satellite imagery should be used to produce maps of desertification and to monitor change.
2 Land-use planning and management based on ecologically sound methods should be introduced into areas affected or likely to be affected by desertification.
3 Public participation should be integral to any plans to prevent and combat desertification. Account should be taken of the needs, wisdom and aspirations of the local population.

Industrial–urban–agricultural interactions

It was recommended that UN organisations should study and publicise the role of industrialisation and urbanisation in intensifying, preventing or eliminating desertification in arid areas.

Corrective anti-desertification measures

1 Environmentally sound management and development of water resources should be introduced.

Developments should use technologies appropriate to local environments and communities and use local materials. Regional databanks should be created together with flood control and river pollution control schemes. Revegetation of watersheds and other measures to reduce erosion were recommended as well as desalinisation schemes.

2 Degraded rangelands should be improved, with better range management, wildlife management and improved welfare for pastoral communities. Techniques could include deferred or rotational grazing, new breeds, forage reserves, better livestock marketing, price stabilisation, alternative uses for rangeland plants, and investigation of systems for combined forestry, agriculture and livestock raising.

3 Rainfed agriculture should be improved so as not to cause desertification. Strategies indicated included cover crops, rational use of organic and artificial fertilisers, terracing, strip cropping, shelter belts, sand stabilisation, diversification, improved land tenure, vegetation and protection from grazing of sensitive areas with watersheds.

4 Each nation should determine its own priorities for the development of irrigated agriculture. Waterlogging, salinisation and alkalinisation should be combated whenever possible with, for example, the development of salt tolerant crops.

5. Vegetative covers should be restored, maintained and protected so as to stabilise and protect soils, especially on watershed and mountain slopes, dunes and where villages, roads and farms are threatened. Techniques mentioned included fenced reserves around settlements, dune stabilisation using matting, mulches and bitumen coatings, restrictions on firewood gathering, fuelwood plantations, and establishment of green belts – 'mosaics of revegetated areas'.

6 All necessary steps should be taken by governments to conserve flora and fauna in areas subject to or threatened by desertification, including controls on trade in endangered species.

7 National or intra-regional systems for monitoring climatic, hydrological and pedalogical and ecological conditions of land, water, plants and animals should be established or strengthened in areas affected by or prone to desertification.

Socio-economic recommendations

1 Social, economic and political factors that bear on desertification should be analysed and evaluated, particularly inequitable relationships at national, regional and international levels. This will involve educational programmes on the ecological aspects of development.

2 Countries that so wish should adopt demographic and economic policies that will ameliorate problems arising from population growth, especially rural-to-urban migration that leads to labour shortages and poorer land use in rural areas.

3 Peoples affected by desertification should be provided with health services, including family planning, comparable to the rest of the population.

4 Human settlement in drylands should not conflict with land productivity. It should take full account of local climate, building materials, architecture and social habits. They should minimise energy demand, use solar and wind energy where possible and share traditional expertise in design and materials.

5 National systems for monitoring the human condition should be established in countries subject to desertification.

Drought insurance

There should be insurance schemes to cope with drought-related disaster. Schemes should take note of indigenous strategies for coping with drought and should include crop and livestock insurance, fodder, fuel and seed reserves, permanent breeding stocks of animals, refugee pasture areas and stockpiles of tools for relief employment. Regional movements of people across frontiers should be encouraged during severe drought.

Science and technology

1 Indigenous scientific and technical capabilities should be strengthened and imported technologies should be adapted to suit local conditions.

2 Alternative and unconventional energy sources should be encouraged. Wood lots and forest reserves with new woody species should be

planted, charcoal manufacture could be improved, solar power, wind energy, biogas and charcoal, coal and oil from organic wastes should be utilised where possible.

3 Training and education should be undertaken – especially via mass media.

4 National bodies should be established to evaluate and disseminate information on desertification, prepare national action plans, arrange financing, monitor progress and liaise with regional and international bodies.

Integrated development plans

Programmes to combat desertification should whenever possible be formulated in accordance with the guidelines of comprehensive development plans at the national level.

Other recommendations

In addition to those listed above, the conference made further recommendations concerning international co-operation and action, a series of specific recommendations for immediate actions at international, regional and national levels and for implementation of action plans. These recommendations can be found in the appendix to Alan Grainger's excellent book *The Threatening Desert* (1990).

The need for an integrated approach to desertification

Faced with the need to implement recommendations of the type outlined above, governments, international bodies, aid agencies and, to an extent, individuals are faced with a number of choices. Perhaps the most obvious one is between emergency relief versus development assistance which prevents desertification occurring. This choice is not, however, straightforward in that desertification exists and requires remedial action and it is not always possible to legislate for the factors, especially political and economic, that can trigger it. Whichever strategy or combination of strategies is adopted, these will invariably require a technological input. In the case of irrigation it was noted how some high technology

schemes can founder because of technical or environmental inappropriateness. Similarly, there is no guarantee that low or intermediate technology solutions will flourish, unless steps are taken to ensure that they are appropriate for the particular environment concerned. Environment in this context goes beyond the physical surroundings to encompass social, cultural, educational, economic, political and health appropriateness, with an emphasis upon schemes that are labour rather than capital intensive. Development in this context must be sequential with each innovation built upon earlier successes and the surplus capital they generate. A format for such development plans was given by Harrison (1987) in his book *The Greening of Africa*. While it is not specific to desertification, the steps he describes (Table 7.5) are none the less relevant to this problem.

The need for an integrated approach to desertification based upon the implementation of appropriate solutions was emphasised at the United Nations Conference on Environment and Development at Rio de Janeiro (UNCED) in 1992. At this meeting it was generally agreed that in the 15 years since the first Plan of Action to Stop Desertification, the strategies deployed by UNCOD 'were a complete failure'. Despite successes in formulating pilot projects, training professionals and technicians, instigating information campaigns and constructing databases, the plan failed because it championed technical solutions to largely socio-economic and socio-political problems. In addition, remedial measures did not always involve local populations or integrate with other development projects. To address these deficiencies, Agenda 21 supported a new, integrated approach to desertification which emphasises actions that promote sustainable development at the community level. The Earth Summit also called upon the UN to establish an intergovernmental negotiating committee to prepare, by 1994, a Convention to Combat Desertification. Against a background of compromise between western and African countries and major inputs from non-governmental organisations, the final negotiations were held in Paris in June 1994. As a result of this, the 'UN Convention to Combat Desertification in Countries Experiencing Serious Drought and/or Desertification, Particularly Africa' was 'opened for signature' in October,

Table 7.5 A blueprint of environmental protection and improved food production

Stage One: the foundations
1 Improvement of producer prices for food and cash crops plus devaluation or currency deregulation.
2 Security of tenure and use of land, trees and pasture including full title for women.
3 Functional integration of agriculture, livestock, forestry and conservation.
4 Dissemination of no-cash cost and minimal-cost techniques with high returns to labour, capital and land. Areas to be covered include soil, water and nutrient conservation, improvement of soil fertility and structure, forestry management, livestock management, small-scale irrigation, improved timing and spacing of planting and improved labour availability and productivity.

Stage Two
Low-cost techniques that require some nation-wide distribution service, minimal imports and small cash investments. These could include improved crop varieties, phosphate fertiliser, introduced tree species, cross-breeding of livestock, pest control, cheap wells, hand-operated pumps, scooped ponds and improved devices to reduce women's burdens.

Stage Three
Moderate cost techniques involving some imports, higher cash investment by farmers and competent nation-wide supply and maintenance systems. These could include hybrid seeds, moderate levels of adopted nitrogenous and potash fertiliser, selective mechanisation, washbores, motorised pumps and alternative energy sources, for example, biogas.

Source: Harrison, 1987

when an initial 87 governments signed. The Convention envisages direct action at local levels to counter desertification, but also aims to prevent damage by, for example, addressing broader questions of mass migration, loss of plant and animal life and climatic change. Any actions will be underpinned by guiding principles that include the right of countries to exploit their own resources, the participation of local communities in the design and implementation of projects and attention to the particular requirements of developing countries. This will be set against a 'bottom-up' approach to planning, greater co-ordination amongst donor agencies and increased horizontal activity on socio-economic questions such as land tenure.

The main goal in implementing the Plan of Action to Combat Desertification (PACD) remains the same as it was in 1977, namely, to prevent and to arrest the advance of desertification and, where possible, to reclaim desertified land for productive use. However, the policy guidelines and courses of action for implementing the PACD are now built around a series of practical targets to be achieved by 2020. These are aimed at preventing desertification in susceptible areas and remedying its consequences where it has already occurred. Strategies should be carried out as an integral part of the socio-economic development of land resources in drylands and should focus simultaneously on improving land productivity while ensuring 'the rehabilitation, conservation and sustainable management of land and water resources, leading to improved living conditions, in particular at the community level' (UN Convention to Combat Desertification, 1994, Article 2).

The special consideration given to Africa in the Convention was a recognition of the severe problems faced by desertification-prone countries on this continent. These include:

(a) a high proportion of arid, semi-arid and sub-humid areas;
(b) a high proportion of countries and the population adversely affected by desertification and subject to frequent, severe drought;
(c) the large number of landlocked states;
(d) widespread poverty and the general need for external assistance if development objectives are to be pursued;
(e) difficult socio-economic conditions, exacerbated by deteriorating and fluctuating conditions of trade, external indebtedness and political instability;
(f) a heavy reliance of populations upon natural resources for subsistence, compounded by demographic trends, a weak technological base and unsustainable production practices;
(g) a weak infrastructural base and generally insufficient institutional, legal, scientific and educational frameworks.

Whilst Africa exhibits a unique combination of these factors their affects can be found to a greater or lesser degree in most other continents. This is reflected in additional annexes to the Treaty which identify regional implementation strategies for Asia, Latin America and the Caribbean and the northern Mediterranean as areas in which desertification is a particular problem.

INCREASING DOUBT AND FUTURE DIRECTIONS

Enough has been said at desertification conferences and enough written in detailed reports and academic papers to suggest that there are few management strategies that have not been proposed, evaluated and recommended. The problem is not therefore a lack of possible solutions, but the failure to implement them. Conflict in desertification-prone countries appears unending. Moreover, the spread of illnesses such as AIDS has made careful husbandry of the land and generation of a cash surplus increasingly difficult, and infrastructural collapse has resulted in the reappearance of diseases such as rinderpest which had been virtually eliminated by the 1970s.

If improvements are to be made in desertification control, many workers feel that changes will have to come from and be implemented at the grass roots level. It has, for example, been pointed out by Raynaut (1977, summarised in Van Apeldoorn, 1981) that three fundamental changes must be made in the rural economies of affected countries. First, the agricultural economy should be given the means to support its own development through the potential to accumulate internal surpluses. Second, rural societies should be allowed to re-establish control over the functioning of the agricultural system. Finally, the central aim of the agricultural production system should become the most direct possible satisfaction of the needs of rural communities. Rural communities would thus have returned to them far greater control over their response to drought and incentives to maximise the long-term output of their land. The essential resilience of dryland communities and their history of successful land management has been further illustrated by Michael Mortimore (1989) in his studies of West Africa. Although many of these communities have shown a history of south-wards migration to the wetter savannas in times of drought they have also shown a persistence in

occupying or re-establishing occupancy of the arid zone. Especially resilient are combined enterprises of farming and livestock which survive through intensified exploitation of alternative opportunities and spatial mobility. Any management strategy that seeks to take advantage of this resilience must therefore be based upon keeping options open, an emphasis upon heterogeneity, a regional rather than a national view of the problem (Kassas, 1984) and a willingness to exploit the adaptive capacity of the local population (Mortimore, 1988). Nowhere is this adaptive capacity better illustrated than in the Machakos District of southern Kenya (Box 7.2).

An additional factor that makes the assessment of desertification problematic is that, although it is some twenty years since the problem first achieved prominence, there is still very little primary data available and 'the number of studies referring to the Sahelian environment are few and scanty' (Helldén, 1991: 382). What data are available suffer from the difficulty of distinguishing, from short-term observations, differences between permanent degradation caused by human activities and natural fluctuations in vegetation that reflect inherent climatic instability (Thomas, 1993). There is also the problem that much of our ongoing ignorance of the extent of desertification is a reflection of 'the poor quality of research . . . since the UN Environmental Programme (UNEP) claimed a global mandate to address desertification' (Pearce, 1992: 38). Indeed, the view has been expressed that, once UNEP was charged with overseeing anti-desertification measures, it was in their interest to retain desertification's prominence on the political environmental agenda. To do this, UNEP has 'utilised various data at its disposal to publicise the issue, often in alarmist terms' (Thomas, 1993: 329). In support of this contention there is a growing movement that questions the degree and extent of desertification claimed by the UN. This has included papers such as 'Is desertification a myth?' (Binns, 1990) and books such as *Desertification: Exploding the Myth* (Thomas and Middleton, 1994). The view is supplemented by data from satellite imagery which, for example, suggest that previous estimates of desertification may have overestimated the extent of the problem by a factor of three (Thomas, 1993). Studies on the ground, such as that

BOX 7.2   RESILIENCE AND ENVIRONMENTAL RECOVERY IN MACHAKOS

The Machakos district of south-east Kenya covers some 14,500 km^2 of arid and semi-arid land between 1,000 m and 1,800 m above sea level (Thomas et al., 1980). 'Long rains' (March–May) and 'short rains' (October–December) give two short-duration growing periods with a high degree of unreliability and susceptibility to crop failure. None the less, observations over the last 70 years have documented both a significant population growth and in-migration from surrounding uplands. This increased population pressure, combined with a double exposure of cleared land to intense rainfall makes the area prone to erosion and has produced locally severe soil loss (Moore et al., 1979). However, the fact that population and land use have been monitored over 70 years makes Machakos an ideal location to test the commonly held assumptions concerning desertification, frequently based upon short-term observations of problem areas. These assumptions include the beliefs that commercial production harms food output, that development depends overwhelmingly upon government initiatives and aid, that there has been little increase in agricultural production in Africa and that population growth means fewer trees and a degraded environment (Tiffen et al., 1994).

In their detailed study of the area Tiffen et al. (1994) confirm the highly degraded condition of much of the land-scape in the 1930s, but with a series of comparative photographs they elegantly demonstrate widespread environmental recovery by the 1980s and 1990s together with development of a well-managed agricultural economy. In doing so they demonstrate that population increase can be compatible with environmental recovery provided that market conditions make farming profitable. Thus, between 1930 and 1990 there was 'an approximately threefold increase in the value of output per capita, and a tenfold increase in the value of output per hectare. Population growth with new market opportunities has stimulated investment and innovation' (Tiffen et al., 1994: 13).

Development and environmental recovery has been associated with positive actions to reduce erosion through, in particular, the widespread use of terracing (Thomas et al., 1980). Although terracing is expensive its use was facilitated by the 'profitability for high value crops and its demonstrated superiority for maize yields in dry years and dry areas' (Tiffen et al., 1994: 200). Terrace construction was further encouraged by food-for-work and tools-for-work schemes that assisted poorer farmers and groups of farmers. Moreover, because terracing and other soil conservation methods are labour intensive, their successful implementation may actually require high population densities (Pearce, 1994). Ironically, therefore, the experience of the people of Machakos has shown that, when left to their own devices population growth has impelled society to change and has encouraged interactions that allow positive responses to challenges posed by environmental pressures. 'Direct dictation of what farmers should do to limit their families or to develop their land is not the way to assist them to retain the necessary flexibility to face the future' (Tiffen et al., 1994: 285). Instead, governments should facilitate development by aiding information flows, raising prices paid to farmers, encouraging dialogue between government experts and local people and removing impediments to innovation and change.

by Helldén (1988) in the Kordofan area of Sudan, have also cast doubt on the severity of the problem. He found, for example, no systematic evidence of an encroachment of the desert boundary, no expansion of desertified villages, no growth of sand dunes and no northwards migration of cultivation over the last 100 years or variations in crop yields indicative of an expansion of cultivation into marginal lands. He concluded that, despite evidence of severe, short-term drought impacts there was 'no creation of long-lasting desert-like conditions during the 1962–79 period' (p. 11). Thomas and Middleton (1994) summarised the doubts over desertification under four categories. First, that data are 'at best inaccurate and at the worst centred on nothing better than guesswork' (p. 160). Second, changes in dryland

ecosystems in response to natural fluctuations in climate are often reversible, because such areas appear to be 'well-adapted to cope with and respond to disturbance, demonstrating good recovery characteristics' (p. 160). Third, there are many other causes of human suffering and misery in drylands and it is difficult to distinguish the social outcomes of drought and desertification. Lastly, although the UN has been central to efforts to conceptualise desertification, the successes of its anti-desertification measures 'have yet to be reliably demonstrated and, in many cases, appear to have had little relevance to affected peoples' (p. 161).

Finally, as more research is carried out into global climatic trends, especially ocean–atmosphere interactions, a clearer picture is emerging of the factors that control drought (Oguntoyinbo, 1986). Pearce (1991) has, for example, reported work at the Hadley Centre for Climate Prediction and Research that has identified linkages between rainfall in the Sahel and global sea surface temperatures. According to their analysis of temperature records since 1900, low rainfall in the Sahel correlates with periods when the southern oceans are warmer than the long-term average. It seems that under these conditions the warm southern oceans weaken the ITCZ which reduces the uplift generated along it and thus reduces associated rainfall. Other evidence to support linkages between drought and ocean–atmosphere interactions comes from studies of the El Niño phenomenon. This is a warm ocean current that periodically replaces the cold Humbolt current along the western shore of South America. In doing so the south-east trade winds are disrupted and the pattern of atmospheric circulation over the southern hemisphere is displaced. Again, studies of long-term climatic records (Lockwood, 1986) have shown that during 22 El Niño years between 1871 and 1978 monsoonal rainfall over most of India was below normal, and correlations between El Niño events and drought conditions in India are statistically significant. However, whatever the possibilities of early warning of drought, this information is of little use if local contingency planning does not exist, infrastructures for the distribution of aid are not in place and whole regions remain inaccessible because of armed conflict.

REFERENCES

Adams, W.M. (1986) 'Traditional agriculture and water use, Sokoto Valley, Nigeria', *The Geographical Journal* 152: 30–44.

Adams, W.M. (1991) 'Large scale irrigation in northern Nigeria: performance and ideology', *Institute of British Geographers, Transactions* N.S. 16: 287–300.

Adams, W.M. and Grove, A.T. (eds) (1984) *Irrigation in Tropical Agriculture: Problems and Problem Solving*, Cambridge African Monographs No. 3, Cambridge, UK: University of Cambridge African Studies Centre.

Agnew, A. and O'Connor, T. (1991) 'The meteorological scapegoat', *The Geographical Magazine* (Geographical Analysis Supplement) 63: 1–4.

Armstrong, C.L., Mitchell, J.K. and Walker, P.N. (1980) 'Soil loss estimation research in Africa – a review', in M. DeBoodt and D. Gabriels (eds) *Assessment of Erosion*, Chichester, UK: Wiley, pp. 285–294.

Aubréville, A. (1949) *Climats, Forets et Désertification de l'Afrique Tropicale*, Paris, France: Société d'Edition Geographiques, Maritimes et Coloniales (in French).

Binns, T. (1990) 'Is desertification a myth?', *Geography* 75: 106–113.

Bunting, A.H., Dennett, M.D., Elston, J. and Milford, J.R. (1976) 'Rainfall trends in the West African Sahel', *Quarterly Journal of the Royal Meteorological Society* 102: 59–64.

Campbell, D.J. (1986) 'The prospect for desertification in Kajiado District, Kenya', *The Geographical Journal* 152: 44–55.

Dennett, M.D., Elston, J. and Rogers, J.A. (1985) 'A reappraisal of rainfall trends in the Sahel', *Journal of Climatology* 5: 353–361.

Dregne, H.E. (1977) 'Degradation by degrees', *The Geographical Magazine* 49: 708–710.

Dregne, H.E. (1983) *Desertification of Arid Lands*, Advances in Desert and Arid Land Technology and Development, Volume 4, New York, USA: Harwood Academic Publishers.

Eckholm, E.P. (1976) *Losing Ground*, Oxford, UK: Pergamon Press.

Farmer, G. and Wigley, T.M.L. (1985) *Climatic Trends for Tropical Africa*, Research Report for the Overseas Development Administration, Norwich, UK: Climatic Research Unit, University of East Anglia.

Glantz, M.H. (ed.) (1994) *Drought Follows the Plough*, Cambridge, UK: Cambridge University Press.

Grainger, A. (1982) *Desertification*, London, UK: Earthscan Publications.

Grainger, A. (1990) *The Threatening Desert*, London, UK: Earthscan Publications.

Harrison, P. (1987) *The Greening of Africa*, London, UK: Paladin/Collins.

Helldén, U. (1988) 'Desertification monitoring: is the desert encroaching?', *Desertification Control Bulletin* 17: 8–12.

Helldén, U. (1991) 'Desertification – time for an assessment', *Ambio* 20: 372–383.

Hocking, J.A. and Thompson, N.R. (1979) *Land and Water Resources of West Africa*, Edinburgh, UK: Moray House College of Education.

Kassas, M. (1984) 'Seven years after the 1977 UN Conference on Desertification: lessons learned or to be learned', *Environmental Conservation* 11: 195–197.

Kimmage, K. and Adams, W.M. (1992) 'Wetland agricultural production and river basin development in the Hadejia-Jama'are valley, Nigeria', *The Geographical Journal* 158: 1–12.

Kowal, J.M. and Kassam, A.H. (1978) *Agricultural Ecology of Savanna*, Oxford, UK: Oxford University Press.

Lal, R., Hall, G.F. and Miller, F.P. (1989) 'Soil degradation. 1. Basic processes', *Land Degradation and Rehabilitation* 1: 51–70.

Lamb, P.J. (1986) 'Waiting for rain', *The Sciences* May/June: 31–35.

Lamb, P.J. (1987) 'On the development of regional climatic scenarios for policy orientated climatic-impact assessment', *Bulletin of the American Meteorological Society* 68: 1116–1123.

Le Houérou, H.N. (1959) *Récherche Ecologiques et Floristiques sur la Vegetation de la Tunisie Méridionale*, Alger, Algeria: Institut de Récherches Sahariennes, L'Université d'Alger (in French).

Lewis, L.A. and Berry, L. (1988) *African Environments and Resources*, Boston, Mass., USA: Unwin Hyman.

Littman, T. (1991) 'Rainfall, temperature and dust storm anomalies in the African Sahel', *The Geographical Journal* 157: 136–160.

Lockwood, J.G. (1986) 'The causes of drought with particular reference to the Sahel', *Progress in Physical Geography* 10: 111–119.

Mabbutt, J.A. (1984) A new global assessment of the status and trends of desertification. *Environmental Conservation* 11: 100–113.

Meigs, P. (1953) 'World distribution of arid and semi-arid homoclimates', *Arid Zone Hydrology*, UNESCO Arid Zone Research Series 1: 203–209.

Middleton, N.J. (1985) 'Effect of drought on dust production in the Sahel', *Nature* 316: 431–434.

Miles, M.K. and Folland, C.K. (1974) 'Changes in the latitude of the climatic zones of the Northern Hemisphere', *Nature* 252: 616.

Moore, T.R., Thomas, D.B. and Barber, R.G. (1979) 'The influences of grass cover on runoff and soil erosion from soils in the Machakos area, Kenya', *Tropical Agriculture* 56: 339–344.

Mortimore, M.J. (1987) 'Shifting sands and human sorrow: social response to drought and desertification', *Desertification Control Bulletin* 14: 1–14.

Mortimore, M.J. (1988) 'Desertification and resilience in semi-arid West Africa', *Geography* 73: 61–64.

Mortimore, M.J. (1989) *Adapting to Drought: Farmers, Famine and Desertification in West Africa*, Cambridge, UK: Cambridge University Press.

Oguntoyinbo, J.S. (1986) 'Drought prediction', *Climatic Change* 9: 79–90.

Ojo, O. (1983) 'Recent trends in aspects of hydroclimatic characteristics in West Africa', *International Association for Hydrological Sciences Publication* 140: 97–104.

Olsson, L. (1983) *Desertification or Climate*, Lund Studies in Geography, Series A Physical Geography No. 60, Lund, Sweden: University of Lund.

Pacey, A. (ed.) (1977) *Water for the Thousand Millions*, Oxford, UK: Pergamon Press.

Pearce, F. (1991) 'A sea change in the Sahel', *New Scientist* 2 February: 31–32.

Pearce, F. (1992) 'Mirage of the shifting sands', *New Scientist* 12 December: 38–39.

Pearce, F. (1994) 'Deserting dogma', *Geographical Magazine* 66: 24–28.

Rapp, A. (1986) 'Introduction to soil degradation processes in drylands', *Climatic Change* 9: 19–31.

Rapp, A., Le Houerou, H.N. and Lundholm, B. (1976) 'Can desert encroachment be stopped? A study with emphasis on Africa', *Ecological Bulletin* 24, Ch. 12.

Raynaut, C. (1977) 'Lessons of a crisis', in D. Dalby, R.J. Harrison-Church and F. Bezzaz (eds) *Drought in Africa 2*, African Environment Studies Special Report, London, UK: International African Institute.

Rhoades, J.D. (1990) 'Soil salinity – cause and controls', in A.S. Goudie (ed.) *Techniques for Desert Reclamation*, Chichester, UK: Wiley, pp. 109–134.

Roose, E.J. (1967) 'Dix années de mesure de l'érosion et du ruissellement au Sénégal', *L'Agronomie Tropicale* 22: 123–152 (in French).

Sharp, R. (1990) *Burkina Faso: New Life for the Sahel*, Oxford, UK: Oxfam.

Smith, B.J. and Bailie, C.S. (1985) 'Erosion in the Savannas', *The Geographical Magazine* 57: 137–141.

Thomas, D.B., Barber, R.G. and Moore, T.R. (1980) 'Terracing of cropland in low rainfall areas of Machakos District, Kenya', *Journal of Agricultural Engineering Research* 25: 57–63.

Thomas, D.S.G. (1993) 'Sandstorm in a teacup: understanding desertification', *The Geographical Journal* 159: 318–331.

Thomas, D.S.G. and Middleton, N.J. (1994) *Desertification: Exploding the Myth*, Chichester, UK: Wiley.

Tiffen, M., Mortimore, M.J. and Gichugi, F. (1994) *More People, Less Erosion*, Chichester, UK: Wiley.

Trilsbach, A. and Hulme, M. (1984) 'Recent rainfall changes in central Sudan and their physical and human implications', *Institute of British Geographers, Transactions* N.S. 9: 280–298.

UNCOD (1977) *Desertification, its Causes and Consequences*, Oxford, UK: Pergamon Press.

UNEP (1978) *United Nations Conference on Desertification: Corrigendum*, Document Na 80–5770, Nairobi, Kenya: UNEP.

Van Apeldoorn, G.J. (1981) *Perspectives on Drought and Famine in Nigeria*, London, UK: Allen and Unwin.

Verstraete, M.M. (1986) 'Defining desertification: a review', *Climatic Change* 9: 5–18.

Walsh, R.P.D., Hulme, M. and Campbell, M.D. (1988) 'Recent rainfall changes and their impact on hydrology and water supply in the semi-arid zone of the Sudan', *The Geographical Journal* 154: 181–198.

Williams, M.A.J. (1979) 'Droughts and long-term climatic change: recent French research in arid North Africa', *Geography Bulletin* 11: 82–96.

SUGGESTED READING

Binns, T. (ed.) (1995) *People and Environment in Africa*, Chichester, UK: Wiley.

Foster, L.J. (ed.) (1986) *Agricultural Development in Drought Prone Africa*, London, UK: Overseas Development Institute.

Goudie, A.S. (ed.) (1990) *Techniques for Desert Reclamation*, Chichester, UK: Wiley.

Independent Commission on International Humanitarian Issues (1985) *Famine: A Man-Made Disaster?*, London, UK: Pan Books.

Independent Commission on International Humanitarian Issues (1986) *The Encroaching Desert*, London, UK: Zed Books.

Middleton, N.J. and Thomas, D.S.G. (1992) *World Atlas of Desertification*, London, UK: Edward Arnold.

Williams, M.A.J. and Balling, R.C. (1995) *Interactions of Desertification and Climate*, Chichester, UK: Wiley.

SELF-ASSESSMENT QUESTIONS

1 Four direct causes of desertification are generally recognised, overcultivation, overgrazing, failed irrigation and one other:
   (a) Urbanisation.
   (b) Overpopulation.
   (c) Deforestation.
   (d) Climatic change.

2 Desertification first achieved international prominence at a United Nations conference; when and where was this held?
   (a) Geneva, 1968.
   (b) New York, 1977.
   (c) Stockholm, 1972.
   (d) Nairobi, 1977.

3 At the conference mentioned in Q2, numerous recommendations were made for stopping and pushing back the spread of desertification. Select which you think are the three most important steps that need to be taken and briefly note your reasons for choosing them.

4 El Niño is?
   (a) A dry, dusty wind that blows across West Africa.
   (b) A periodic shift in the oceanic circulation pattern off the cost of South America.
   (c) The first storm of the monsoon rains.
   (d) A 22-year climatic cycle associated with sunspot activity.

5 The dry, dusty wind which blows across West Africa during winter months is:
   (a) The Sirocco.
   (b) The Harmattan.
   (c) The Khamsin.
   (d) The Gibli.

6 Desertification is:
   (a) The progressive, outwards spread of desert margins.
   (b) A prolonged period when rainfall is considerably below the long-term average.
   (c) The inability of a community to feed itself.
   (d) A decrease in the biological potential of an area producing desert-like conditions.

8

TROPICAL FOREST ECOSYSTEMS

Fernando Dias de Avila-Pires

'In ecological terms, evergreen forest is the climax vegetation of the equatorial climate.'

P.W. Richards

SUMMARY

This chapter examines basic notions concerning the characterisation of tropical forests in ecological terms, and discusses the main implications of the use and abuse of forests.

Tropical forests are fragile, mature systems maintained by heavy rainfall, where most nutrients are concentrated in the living organisms, and not in the soil. They are found in a belt roughly limited by parallels 23° 27′ North and South of the Equator. Unlike in temperate forests, there is a large diversity in plant and animal species, but a small frequency, meaning that co-specific individuals are found far apart. This type of distribution poses a serious economic problem for the exploitation of forest products, and for the implementation of reforestation programmes.

The main questions involved in forest management are related to the growing demand for tropical timber and forest products, and the increasing population pressure towards new frontiers. Migration also carries the threat of epidemic diseases.

Both deforestation and reforestation influence the health of human populations. Methods for evaluating more precisely the risks involved in human interventions in natural ecosystems are being developed.

During the 1960s and 1970s growing concern over the indiscriminate use of forests and forest products resulted in the establishment of international agreements, and the inception of national and international policies.

Scientists agree that the amount and quality of knowledge already available is adequate for the delineation of management guidelines to assure the sustainable development of tropical forest areas.

The future of the tropical forests depends on political decisions, based on social awareness of their importance for the biosphere and for humanity itself. They are the natural habitat of thousands of species, which will disappear when these forests vanish.

ACADEMIC OBJECTIVES

In this chapter the most serious problems concerning the exploitation, conservation and management of tropical forests are presented and discussed. Introductory remarks define the subject and highlight the main relevant questions. A general definition is given, and the main characteristics of tropical forests described. In addition, forest products are listed, and the need for rational exploitation and conservation is stressed. The aim is to familiarise the reader with the main issues that must be considered for the establishment of national and international policies.

After reading this chapter, the reader should be able to identify, understand and discuss the following topics:

1 What are the distinctive characteristics of the tropical forests?
2 Where do they grow?
3 What are their principal productions?
4 What are the chief problems relating to their exploitation?
5 What are the main issues concerning the management of tropical forests?

ECOLOGY OF TROPICAL RAIN FORESTS

Tropical rain forests are found between latitudes 23° 27' North and South of the Equator. Vertical distribution varies with latitude, showing a telescoping effect as we proceed from the Equator to the Poles. Near the Equator there is a progressive reduction in the number of tree species above 1,000 metres of altitude, until the tropical forest reaches its limits around 4,200 m.

There are several different types of tropical forests, according to soil, climate and orography: lowland, dry land, permanently flooded (igapó), montane, cloud or mist forests, evergreen, deciduous or semi-deciduous.

The actual geographical limits of distribution of tropical forests are shown in Figure 8.1. Actual limits are difficult to trace on account of the rapid rate of deforestation and reforestation in the last 50 years, often with the introduction of non-native species. Our main concern in this chapter will be with the tropical belt of South and Central America to the Gulf of Mexico, Africa, western India and Thailand, and the Malayan region, ranging northwards into the Himalayas, north-east to Indo-China and the Philippines, and south and east through Indonesia, New Guinea, Fiji, with extensions into adjacent archipelagos and eastern Australia, below the Tropic of Capricorn (Schultz, 1995; Richards, 1952, 1996).

Factors that permit and prevent the establishment of forest vegetation are orography (or relief), altitude, climate and proximity of the ocean. South America illustrates well the importance of these factors. The warm waters of the Brazilian current evaporate, and humid air is swept over the continent by the trade winds. A mountain chain near the coastline acts as a barrier. The warm air soars and is cooled above 1,000 m, where humidity condenses in clouds. As a result, the eastern slopes are covered in dense mountain tropical rain forest. In the Pacific on the western coast of South America, the cold oceanic Humboldt current runs parallel to the coastline, from Tierra del Fuego up to the Gulf of Guayaquil, in Ecuador, where it meets with the El Niño current from the north, and both divert to the Galapagos archipelago. Over the sea, the air loses its humidity, blowing dry over the coast, absorbing what little humidity there is on land. As a result, a desert belt runs from Chile to Ecuador, between the Andean cordillera and the sea. The Andean mountain forests grow on the eastern slopes, and merge with the Amazonian forest (Meigs, 1966).

There is a mean annual insolation on the ground of 100–160 joules/cm^2 against over 220 in tropical arid lands. Mean annual temperature is around 27°C and with a relative humidity of 80 per cent. Rainfall averages 2,000 mm per year, but there is a short 'dry' season, lasting no more than two or three months. Montane forests in higher altitudes are not 'tropical' in terms of climate, in spite of being located inside the tropical belt.

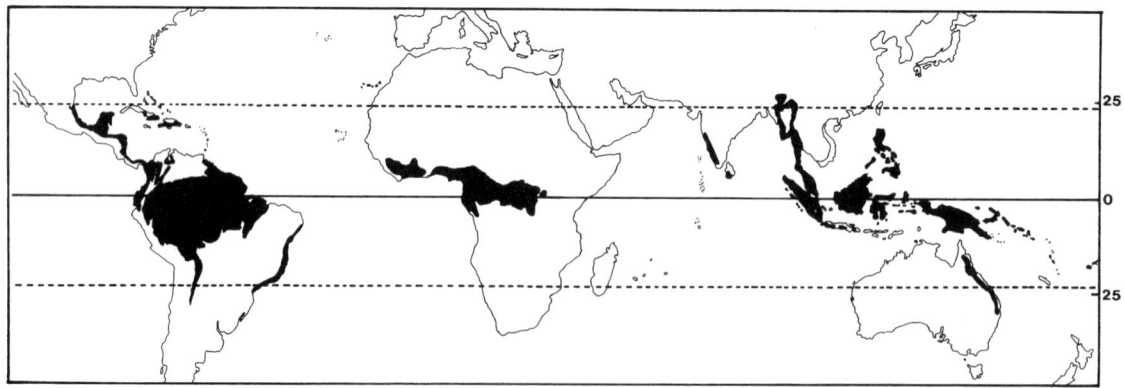

Figure 8.1 World distribution of tropical rain forests
Source: adapted from Richards, 1952

As a climax, for example, the final or mature stage in a plant community successional development, with heat and humidity supplying the main elements for organic growth during the whole year, edaphic conditions are in general not restrictive or limiting. A tropical forest is a climatic climax. This means that the layer of humus or topsoil may be shallow, as most nutrients are concentrated in the living trees, not in the soil. Heavy rainfall precipitations pre-empt competition for water. Swampy or flooded areas provide an opportunity for single-species dominance in limited areas: the fruits of certain palms are protected against rotting, and may germinate submerged in water. The notion that tropical forests are uniform in plant and soil composition and structure is misleading. Soil studies performed so far are not adequate to be representative or to permit generalisations. Maps on scales of 1 : 100,000–1 : 500,000, which are appropriate for geological, biogeographical or taxonomical analysis, have been used by researchers and by economists and politicians for their own interpretations. These macro-level generalisations have often been mistaken and in many cases have had disastrous consequences (Gentry, 1993).

Biomass is not correlated with soil fertility: nutrients are stored in plant tissues, trunks, leaves, fruits and roots. Production and consumption of organic matter along the ecological chain are equivalent. The tree roots are shallow, even where the soil is deep. To support tall trunks and wide canopies, tabular, lateral roots are common. Light is the main factor in competition, and in succession. At river banks, the effect of light is easily detected, as the riparian vegetation differs in floristic composition and luxuriance from the formations of the hinterland. Tree architecture – or *habitus* – depends on the amount of light that reaches the leaves. When growing outside a forest, the trees change their *habitus*, usually branching lower than when in forest stands. Light is filtered through the layers of vegetation, and rainwater does not fall directly upon the forest floor.

For the same reason, forest trees, shrubs and lianas that grow on riverbanks receive more sunlight. The *vegetational* landscape the river traveller sees on the margins is richer, denser and is not representative of that of the hinterland. Its species composition or *flora* is also distinct.

Tree tops in some forests form a continuous canopy, offering shelter and food for a host of plant and animal species many of which never descend to the forest floor (Moffett, 1994). The description and analysis of such communities restricted to the topmost strata had to wait until the recent development of special methods of study. Earlier attempts involved the use of a small smoking petrol engine, raised to the tree tops by means of a pulley (fogging). A sheet spread under the tree caught the animals that fell as the smoke reached the canopy. Helicopters have been used, but the downdraft disturbs a wide area and scatters leaves and small animals. Both methods also have the disadvantage of not permitting identification of the microdistribution of animals along vertical and horizontal layers. Ladders and catwalks built at different heights are preferable (Lowman and Nadkarni, 1995). Usually, three storeys, layers or strata are recognised by botanists, but they are usually difficult to distinguish by the casual observer, thanks to the presence of species at intermediate height, and individuals in different stages of growth. The second stratum is commonly the highest layer of tree crowns that form a continuous canopy, 15–30 m above the ground. The tallest trees, 40–50 m in height, grow widely spaced, and soar above what, from an aeroplane, looks like the top of a green cauliflower. The analysis of the vertical distribution of forest animals is an important issue in the management programmes, as a method of screening and identifying pollinators. Flowers of forest trees may be pollinated by insects, birds or bats. The study of vertical distribution will also lead to the elucidation of the epidemiology of certain diseases transmitted by vectors, in particular those carried by mosquitoes that feed on birds and mammals (Deane *et al.*, 1969). Through long-term, careful observations during the 1960s, Deane was able to record the strata frequented by different species of mosquitoes. Eighty-eight per cent of all individuals belonging to the species *Anopheles darlingi*, responsible for malaria transmission in humans, were captured at ground level (up to 5 m) and only 12 per cent in traps set in platforms built 15 m from the ground. Other mosquito species that live near the canopy are responsible for the transmission of other kinds of *Plasmodium* parasites to primates, rodents and birds. The use of 'sentinel' laboratory animals, kept in cages

Figure 8.2 Schematic representation of a forest structure, floristic distribution and vertical stratification. A simple method of notation

at different heights, allows the isolation of viruses carried by animal hosts living in distinct vertical strata and transmitted by vectors. A simple method for the representation of forest structure, floristic composition and stratification is shown in Figure 8.2 (adapted from Richards, 1952).

Analysis of treetop fauna resulted in the re-evaluation of our previous estimates of total number of animals on the globe, from around 1.5 million species to something between a conservative 10 million and a possible 50 million.

Contrary to what occurs in temperate regions, there is a large diversity of species per unit of area, and a small number of individuals of each species that grow widely spaced, for example, with a low *frequency*. This patchy distribution has an important effect upon conservation, exploitation and pollination as well as disease transmission as it has a direct relationship with local diversity. One hectare of lowland Amazonian forest may contain more species diversity than the whole of the temperate forests of the USA and Canada together. Estimates of timber potential are more difficult to make, and economic exploitation involves the destruction of unwanted trees. Collection or harvesting of forest products also involves more work and longer trails to reach each tree of a particular species. The same is true for pollinators.

BOX 8.1   THE PLIGHT OF FOREST ANIMALS

In 1937 Landsteiner and Wiener discovered the agglutinogen factor Rh (Rhesus factor) while trying to immunise guinea pigs and rabbits with the blood from an Asian forest rhesus monkey. Landsteiner had described the ABO system early in 1902 and was awarded the Nobel Prize in 1930 for his contribution to medicine and physiology. He was also the first scientist to use monkeys as experimental animals in polio research. Non-human primates had been the object of general curiosity since early times, companions of pirates and court ladies since the Renaissance and the object of basic and applied research in psychology since the pioneer works of Klüver in 1930.

During the 1960s forest primates became the prime choice of medical laboratory researchers. They were used to test new vaccines and drugs, for vaccine production, for organ transplants to man and for other kinds of experiments. Early in 1960 India banned the export of monkeys, as they were considered sacred animals. Chimpanzees from Africa and five species of primates from South America became the target of traders, who supplied laboratories and pet lovers, before conservation legislation was adopted by both exporting and importing countries. It was estimated that 1,800,000 specimens of primates entered the USA during the period 1952–64. It was also calculated that the death of animals resulting from the primitive methods of capture and poor conditions of housing and transportation amounted to up to 90 per cent of that number (Middleton *et al.*, 1972; Roth, 1965).

The introduction of the Marburg virus in Germany, which resulted in the death of several laboratory workers, caused the country severely to restrict the import and utilisation of monkeys as experimental animals. It was also a warning of the dangers of keeping them as pets.

The next decade saw the adoption of national and international legal measures to control the animal traffic and trade.

In 1974 a leading zoologist M.J. Dourojeanni demonstrated that 'in the Peruvian Amazonia wild life production value is, under actual [present] circumstances, more important than wood production value'. The market value of skins, hides, meat, trade of live animals for laboratory use, pets and zoos, plus tourism and sport hunting was higher than the price of lumber or timber. The same is true for the other forests in the world, a fact that has not been taken into serious consideration in impact assessment studies (Robinson and Redford, 1991).

Lianas are abundant where the forest has been disturbed, along forest fringes, and where large trees fall to the ground leaving a large gap open to the skies, where some pioneer species will grow before being replaced by the regrowth of forest trees.

Epiphytes grow attached to tree trunks and branches, without interacting with their hosts, and with high requirements of sunlight. They rely on rainfall for water supply, and on the scarce nutrients that float on the winds or percolate from the foliage. The leaves of some species form a water reservoir, where whole communities of small organisms are found, from microscopic forms to tree-frogs and snakes. The plant may feed on residual organic matter that collects in these small water tanks.

Conservation issues

Waterways are the main routes for human penetration, daily displacements and travel, and an important source of food for local populations and for commerce. Although abundant, river fish populations fluctuate with the seasonal floods. In general, they support moderate exploitation, but not industrial harvesting. Their preservation depends upon the conservation of the whole forest ecosystem and, in particular, that of the riparian, marginal or gallery forests.

There is a growing concern over the possible implication of the reduction of forest cover in the global warming trend of the Earth's atmosphere (Peters and Lovejoy, 1994). In the last 150 years there has been an increase of 25 per cent in the concentration of CO_2. Water vapour also contributes to that effect, but its concentration varies over space and time. Depending on air temperature, its concentration may be up to 7 per cent of the volume of the atmosphere in humid regions, as over the Amazonian forest, while it accounts for only 1 per cent over deserts. Although a large number of scientists involved in climatic research claim that human activities are leading to a definite increase in global temperature, there are some differing opinions from those who claim that the available models are too crude and simple to be trusted.

As to the impact of deforestation upon the regional climatic patterns, there is clear evidence of its effect. It leads to the reduction of soil permeability and, in consequence, to its storage in underground layers. The Amazon basin alone contributes with 20 per cent of the fresh water that reaches the oceans. Furthermore, forest cover accounts for some 50 per cent of the rains, through the process of evapotranspiration. Solar radiation upon the topsoil will increase, and the cycles of carbon and nitrogen will be affected. A longer dry season will result in profound changes in the floral and faunal composition.

The loss of the remaining tropical rain forests will result not only in changes that will be felt in all continents around the world, but also in the extinction of thousands of animal species that have not even been described, let alone studied. It will be an indictment of the incapacity of mankind to manage its own heritage.

Historical hallmarks

Forest preservation has been a strategic issue since ancient times. Game animals were reserved for the king, and forests were preserved to supply timber for the building of ships. Noble woods have been the object of legal definition. In 1923 the First International Congress for Nature Protection was held in Paris. It was an important step towards the conservation of natural areas. By then, the need for conservation had become a preoccupation of scientists and earlier, utilitarian considerations had given way to ecological arguments. But it was during the 1960s and 1970s that important discussions concerning the future of tropical forests gained momentum. The main events brought together intergovernmental, non-governmental and professional agencies and associations. The International Union for the Conservation of Nature and Natural Resources, based in Morges, Switzerland, became the leading international agency in the field of conservation. Its technical committees convened individual natural scientists from many countries. IUCN was responsible for the clarification of theoretical concepts related to all aspects of nature conservancy, for the establishment of the criteria to define the status of threatened animal and plant species, and for the preparation of lists of endangered species in the world.

In March 1968 the Latin American Regional Conference on the Conservation of Renewable Natural Resources met in San Carlos de Bariloche, Argentina, to discuss the implementation of official national policies concerning the protection of natural habitats and the control of plant and animal traffic. An official list of the endangered species of each country was adopted, and an inter-American convention was drawn up and eventually signed by all governments.

In September 1968 UNESCO headquarters in Paris hosted the Conférence Intergouvernamentale d'Experts sur les Bases Scientifiques de l'Utilization Rationelle et de la Conservation des Ressources de la Biosphère (Intergovernmental Conference of Experts on the Scientific Basis for the Conservation of Resources of the Biosphere). It was sponsored by the United Nations and its agencies UNESCO, FAO and WHO, with the co-operation of the International Biological Program of the International Union for the Conservation of Nature (IUCN). The proceedings were published in 1970 under the title *Utilization et Conservation de la Biosphère* (UNESCO, Paris). The experts showed particular concern for the need to adopt conservation policies to protect tropical forest ecosystems.

In December of the same year, experts in ecological problems relating to economic development met in Virginia, USA, for a conference on the Ecological Aspects of International Development. A book resulted from the discussions: *The Careless Technology: Ecology and International Development* (Farvar and Milton, 1972).

In September 1970, thanks to the initiative of E.M. Nicholson of the International Biological Program, a conference was organised by the FAO, World Bank, UNDP, United States Development Agency, Canadian Development Agency, Conservation Foundation, IUCN and IBP. An important book was published, incorporating the conclusions: *Ecological Principles for Economic Development* (Dasmann *et al.*, 1973).

These initiatives contributed to the establishment of the foundations of what would become the field of political ecology, as defined during the Stockholm Conference of the Biosphere, held in 1972.

In February 1974 a meeting was held in Caracas, Venezuela under the auspices of the IUCN, UNEP, FAO, UNDP, United Nations Economic Commission for Latin America and the Organization of the American States (OEA). A document was published under the title *The Use of Ecological Guidelines for Development in the American Humid Tropic* (IUCN, 1975). This document defined the ecological principles for the development of the humid tropics in the Americas, with emphasis on the rational exploitation of the tropical forests.

In May 1974 a second meeting took place in Bandung, Indonesia to discuss the situation of tropical forests in South-East Asia.

By that time, international treaties for the conservation of natural resources were in operation, and public opinion had been alerted to the importance of preserving what became known as our 'vanishing heritage'. But scientific opinions were divided as to the need for more research to subsidise official policies. According to some authors, it was necessary to proceed with a complete inventory of all resources before any course of action could be adopted. Others supported the view that there would be no time for it, as the rate of deforestation and land reclamation was increasing drastically. The leading experts at the time, though, were in favour of swift action.

Budowski (IUCN, 1975) held that

> Scientists have been guilty of keeping too much knowledge to themselves. . . . This led nowhere and has actually increased the gap between scientists and the practical decision-makers, who are faced with other kinds of pressures besides scientific ones. . . . Nevertheless, all of us here would probably agree that enough scientific facts . . . are already available to provide a series of most useful aids to decision-makers.

Dasmann *et al.* (1973) agreed that

> It is customary in books concerned with ecology to point out the need for further research. This need is apparent, but should not be overemphasised. The greater immediate need is for effective application of ecological principles that are already known. Enough ecological knowledge is now available to permit a far better job to be made of development than ever in the past.

We may summarise our relationships with nature – and natural resources – by recognising a change, during the 1960s, from a preoccupation with the preservation of threatened species against extinction to the more realistic preservation of whole ecosystems. Great efforts were made to show that economic development was not incompatible with the idea of conservation of natural resources: it actually depended upon a rational management of nature.

In the following decade, some important changes in our aims, objectives and perspectives took place. Satellite technology then permitted monitoring of human interference in natural habitats and precise mapping of forested areas in all continents. Human ecology became a new and important field of research, multidisciplinary in nature and transdisciplinary in its ultimate goal.

Social and cultural anthropologists of the 1960s had tried to incorporate ecological principles, concepts and models into their methods of community analysis. They were faced with the need to devise a new model incorporating the basic contradictions between entrepreneurial production, globalisation, individuality, the progressive energy requirements of technological societies, and the balance of cyclic nutrient circulation and energy transfer that characterises the natural ecosystem, where the energy budget (production/consumption) approaches the equilibrium. The ever-increasing needs and costs of production must come from outside the local system, in the form of fossil fuels, agricultural and forest products. It is up to the human ecologists to try and develop a new systems theory, based upon a new environmental ethic and a progressively globalised economy, which will guide humanity towards sustained development.

During the 1980s, a considerable joint effort was made to define the new concept of sustainable development. In 1980 the *World Conservation Strategy* (IUCN/UNEP/WWF, 1980) formulated a new world policy for integrated economic/ecological development based on three general principles: the maintenance of essential ecological processes, the preservation of genetic diversity and the sustainable utilisation of species and ecosystems. In 1987 the World Commission on Environment and Development (WCED, 1987) published a report emphasising the global relationships between economy and environment conservation, or rational utilisation of natural resources. In the same year, the United Nations Environment Programme adopted a tentative agenda for the year 2000 (UNEP, 1988).

The United Nations Conference on Environment and Development held in Rio de Janeiro in June 1992, 20 years after the Conference of the Biosphere,

in Stockholm, was preceded by a number of important meetings.

In 1991 a follow-up to the *World Conservation Strategy* of 1980 was published and presented to the governments of 60 countries (IUCN/UNEP/WWF, 1991). This work resulted from the contributions of hundreds of experts and institutions around the world, concentrating on the more practical aspects of sustainable development.

For many years medical ecologists and social anthropologists had been building a body of knowledge concerning the social and environmental factors that affect human health. Apart from this, the interactions of human health and environment had never been analysed in depth, at an international level, in the context of global development. The Rio Conference of June 1992 presented a major opportunity for a detailed and informed summary on what was to be understood as environmental health, as dependent on the continued availability of natural resources and integrity of the environment. The report of the World Health Organization Commission on Health and Environment (WHO, 1992) was presented as a sequel to the WCED report in 1987 on *Our Common Future*. It addressed issues such as health and development, demographic problems, food and agriculture, water, energy, industry, human settlements and urbanisation, international problems of biodiversity and environmental pollution. Strategies were devised and recommendations made.

At the Rio Conference the issue of species and ecosystems preservation was discussed under a new perspective. The preservation of biodiversity – including species diversity and individual genetic variance – became the focal point of renewed efforts to prevent the deterioration and loss of our living heritage. The Biodiversity Convention, the Report of the Commission of the European Community (EC, 1992), and the Declaration on the Forests (United Nations, 1993) epitomise the world's preoccupation with the future of nature's richest environment (de Klemn, 1993; Glowka *et al.*, 1994).

MANAGING TROPICAL FORESTS

Forest management is the rational exploitation of natural and man-made forests, in such a way as to maintain productivity. Here we will be concerned with the management of natural tropical forests. Exploitation includes logging for the use of timber, sap and latex extraction (rubber *Haevea*, balata *Couma*), fibres (*Sabal*), fruits, leaves and nuts (Para nuts, betel, cacao, palmetto), oils (*Elaeis, Orbignya, Aniba*), resins (*Myocarpus*), medicinal plants, paper pulp, poles, fuel and tannin (*Schnopsis*). Harvesting forest products at certain times of the year, such as Brazil nuts, and hunting and fishing for local food consumption does not cause major disturbance to the ecosystem. Palmetto collection involves the felling of the palm trees, which need replanting.

Management may involve selective deforestation, afforestation or introduction of species, including tree crops, fruit trees, cocoa, coffee and tea (Anderson, 1992). Not all practices listed in manuals guarantee the integrity of the forest ecosystem. We must be aware that the diversity of flora and fauna in tropical forests, as compared with temperate forests, makes it difficult to intervene without changing the whole environment.

The bewildering complexity of tropical forest formations, rich in species composition and variable in species frequency, makes their rational exploitation dependent on careful planning. The small number of individuals of commercially interesting species per hectare has led to the common malpractice of destroying wide tracts of forest to get at the choice trees.

Good management involves a correct knowledge of forest ecology and of the requirements of commercial species in terms of soil, light, temperature, water, spacing and pollinators. The main goal will be to increase the frequency of valuable trees through one of the following methods:

1 Improving the chances of survival and normal growth of choice species in their heterogeneous environment through shading, elimination of competitors or parasites and other sylvicultural practices.
2 Increasing the number of choice trees per hectare through selective thinning of the natural vegetation and technically oriented natural regeneration of commercial species, which is preferable to the transplantation of young shoots from a nursery.

Attempts were made at substituting whole tracts of native mixed forest for uniform stands of choice species of market value. This practice implies the loss of protection against the dissemination of parasites afforded by the mixed composition of the natural formations. Trees of distinct species act as screens for parasites, while homogeneous stands offer an easy ground for the increase in numbers and spread of micro-organisms and insect pests.

The disaster that jeopardised the Ford venture in the Amazonia, early in the twentieth century, must be kept in mind. In Fordlandia, the rubber trees that grow naturally scattered amongst other forest trees became easy prey for parasites when cultivated in single stands. It took many years of experiments with genetic selection and grafting to produce parasite-resistant plants. A more recent disaster was the cultivation of *Gmelina* trees in homogeneous stands in Amapá, northern Brazil.

The harvesting of some natural products such as rubber, sap, nuts and fruits is compatible with forest preservation, or even depends on it. Commercial ventures must take into serious consideration the advisability of substituting mixed forests for artificial stands of single species, whether of imported or native origin.

Conveyance of logs to processing plants merits special consideration and care. Floating trees downriver may involve losses of up to 15 per cent. The use of primitive saw-mills is responsible for a further loss of over 50 per cent of the raw material that reaches the plant.

Population expansion and the demand for new lands place great pressure on forests. In general, migrant populations do not have the necessary knowledge, the appropriate tools and the necessary health requirements to cope with the characteristics of the new environment. Cutaneous leishmaniasis, malaria, African trypanosomiasis, intestinal parasites, leprosy, arbovirus and hepatitis are some of the main health problems. Native populations have long been selected, and show special types of adaptation, as in the case of sickle-cell anaemia in Africa, which is a favourable factor in areas where a certain type of deadly malaria parasite is prevalent.

It has been fully established that both deforestation and reforestation influence the health status of local populations though the changes in the population dynamics of vectors, hosts and reservoirs of parasites – including viruses and bacteria. The management of such systems must be based on the qualitative and quantitative analysis of the risks involved, both in economic development and in the conservation and restoration of forest ecosystems.

One of the main pressures comes from the growing demand for tropical timber or other forest products. Selective logging affects the forest through the need to open space for logging for forest exploitation. The cost of operation is responsible for a great loss of biomass. Tropical forests are complex, and reconstitution of deforested land is difficult. Sowing by air has been used with success, but the final objective is to recompose *vegetation*, not *flora*. The final paisagistic aspect will be of a forest, but species composition and frequency will probably be different. Commercial enterprises often consider as *reforestation* the substitution of native forests for imported species of commercial value. Little is known about the rates of growth and management of tropical trees,

---

BOX 8.2   THE EFFECTS OF LOGGING AND FIRE

Early in 1953 Ducke and Black, working in the Amazonian forests, remarked that cutting and burning had distinct effects upon the flora, in terms of regeneration and succession. A recent field study in a forested region of northern Brazil (Vieira et al., 1996) proved they were right. Data from a ten-year analysis of the impacts of exploitation, deforestation and agriculture upon the flora, vegetation and fauna in a tropical forest at Braganta, Par, Brazil, shows that although natural succession take place, the end result is an impoverished ecosystem, even after 40 years of vegetational succession.

Modern agriculture began in Braganta by the end of the nineteenth century, introduced by former rubber workers. Thirty years later the region was densely occupied, with 17 families per square kilometre. By 1960 a quarter of all cereal production in Par came from the Braganta region, so that by 1991 only 15 per cent of the original forest cover remained standing. The red/yellow latosol depleted by agricultural practices recuperates its fertility after a period of 8–10 years, which is the length of each cultivation cycle.

The analysis of tree diversity in primary forests and in capoeiras or secondary forests indicates a marked reduction in the number of species after deforestation, as the following data show.

During this study 268 tree species were identified in remnants of primary forests found in the Braganta region. In the secondary forests, growing in abandoned fields formerly used for agriculture, the number of tree species increased regularly with the passing of time, from 62 after four to five years to 112 after about forty years without cultivation. Of these, from 50–60 per cent were trees found in the primitive forests. Of the original 268 species, there are 173 (64.5 per cent) that do not occur in secondary growths and are thus threatened with extinction. Tens of species are threatened by selective logging.

Ninety-five species (35.4 per cent) of native tree thrive on the modifications in the environment brought about by agriculture. They reproduce quickly, producing small light seeds which germinate after fire. Ninety-eight species are found in regenerated forests, but not in primary forests.

The change in forest composition and biodiversity has a marked effect upon the fauna. In primary forests, 84 per cent of the trees produce seeds which are eaten by birds and mammals, while only 79 per cent of trees in secondary forests are apt to feed these animals.

A study of the bird fauna in Braganta shows how the changes in the forest cover affect them. In primary forests, 56 species of birds are found in the undergrowth, 71 per cent of these being insectivorous. In 10-year-old secondary forests, only 40 species were observed, 23 of those being found in primary forests, of which 65 per cent were insectivorous. In 20-year-old secondary forests 38 bird species were found, 31 being found in primary forests, 65 per cent of those being insectivorous. This means that the dissemination of the seeds of native trees, and especially the heavier ones, is jeopardised by the reduction in populations of birds that are capable of transporting them.

The utilisation of secondary forests may be a valuable option to reduce the economic pressure upon primary forests. A combination of methods used in agriculture and forestry, combining cultivation with secondary forests, may be a good alternative for the local population. The secondary growth may be improved with the introduction of trees of rapid growth and economic importance.

The data show that even after five cycles of cutting and burning, secondary forests accumulate biomass at a rate of 4 tons per year of dry organic matter, absorbing over 2 tons of carbon from the atmosphere per year, which helps to improve general climatic conditions.

strengthening the case for the option of the intro-duced species.

Economic estimates of forests seldom take into consideration the value of the associated fauna, which may be higher than that of the wood. All managing actions must consider the whole forest ecosystem as an integrated unity: flora, fauna, micro-organisms, soil, underground mineral resources water and the role of vegetation in the maintenance of mesoclimatic conditions. They should also take into account the presence and impact of human native populations. Up to the present, partial evaluations and interven-tions have been the rule.

The construction of hydroelectric plants involves the destruction of large tracts of native flora and animals. Conservation practices, enforced by inter-national financing agencies such as the World Bank, demand that a study of impact be made, and com-pensatory measures be taken. The rescue of forest animals presents problems which have not been solved. The usual practice has been to capture animals as the dam is filled and transport them to the margins. Despite the public appeal of these practices, results are disappointing. Theoretical principles, corroborated by field studies, show that they result in population imbalances where animals are introduced, reducing the chance of survival and jeopardising natural equi-librium in existing populations. Very often, rivers act as zoogeographical barriers, isolating distinct sub-species. Animals found swimming in the middle of the current are taken randomly to any margin, irrespec-tive of their provenance, disturbing the natural pat-terns of distribution. Measures to compensate for the inevitable death of animals must be adopted instead of pseudo-rescue procedures.

The opening of roads into forested areas gives access to human penetration and hence the intro-duction of domestic animals and plants, non-planned colonisation, illegal exploitation and destruction of natural resources. The roads are built across animal territories, disturbing their natural patterns of distri-bution.

Deforestation for the raising of cattle or for crop production shows weak results. The thin layer of nutrients does not support the demand for agricul-tural yield, and the heavy rains wash the surface soil of their mineral nutrients.

In 1813 H.v. Carlowitz stated that a judicious application of art and science is demanded for the orderly production and conservation of forest prod-ucts, in order to guarantee a continuous, permanent and stable yield. The concept of sustainable use of forest resources is being rediscussed, expanded and applied tentatively for the establishment of national and international conservation policies. It tries to circumvent the radical views – on the one hand, to preserve everything and, on the other, to use up and deplete all resources. The answer lies with a proper plan of ecological zonation and planning stipulating the delimitation of exploiting activities according to the characteristics of each region. It also involves a new ethic whereby consumers must feel responsible for the preservation at their source of the products they use.

The question of the minimum areas necessary for preserving genetic diversity has been raised, with the extension of the theory of island biogeography to isolated patches of land, as national parks and other preserved habitats. Size and distance from the main-land determine the composition, dynamics and survival of island animal and plant populations. The derived mathematical equations also apply to isolated patches of forests, and to the isolated biota found on mountain tops.

CONCLUSIONS

This chapter presents basic facts concerning the geo-graphy, natural history, conservation and economy of the world's tropical forests.

The proper management of tropical forests must take into careful consideration planning for the rational use of natural resources, such as trees, animals and micro-organisms. It should include a study of the impact of harvesting, removal or substi-tution of species and formations upon the forest ecosystem. It must consider local human populations as part of the system. Above all, it must be planned from a systemic or integrated point of view.

Tropical forests are sensitive systems, maintained by heavy rainfall and sunlight. Tropical vegetation is made up of a rich and varied flora, and differs from temperate forests in species composition, types of association and frequency patterns.

BOX 8.3   OIL EXPLOITATION AND ENVIRONMENTAL CONSERVATION IN THE AMAZON

The year 1958 marked the beginning of oil exploitation in the Brazilian Amazon, with a successful well producing a modest output in the town of Labrea (see figure below). Twenty years later, large deposits of gas were discovered in the valley of the river Juruá, a tributary of the Amazon river (see figure), in Western Amazonas. In 1986 oil was found between layers of gas and water. By December 1993 over 2 million cubic metres of oil had been produced. By 1995 134 wells had been perforated.

(a) river Amazonas, (b) river Juruá, and (c) town of Labrea, in Brazil

Exploring, exploiting and transporting oil in the Brazilian Amazonian forest poses a number of problems: the sedimentary basin covers 1 million km²; seismic prospection involves independent crews of 350 people each and covers 150–200 km per month. They are supported and supplied by helicopter. Base camps are established on riverbanks (the river Juruá shows a difference in level of 18 metres from the 'dry' to the 'rainy' season). Wells are located in clearings averaging 1–4 hectares, and their crews work in shifts of 15 days' duration. All equipment, personnel and housing installations must be transported by helicopter. To support all the field work, a centre of operations with heliport and airport must be established, staffed and serviced.

In 1988 a group of ten scientists visited the base camp on the Juruá river and toured field installations, including oil wells, riverbank camps and prospection sites. The purpose was to evaluate and advise on actual and potential environmental impacts, and to offer suggestions concerning the safe transport of oil and gas over thousands of kilometres of jungle and water. Contrary to what was expected, the jungle sites had a minimal impact upon forest vegetation and fauna. Left to themselves, the clearings would disappear through rapid floristic succession. Paths cut through the forest were barely visible from the air, and closed up as soon as they were abandoned. The actual problems were of a distinct nature. Base camps were actually small towns of thousands of people, brought in from other regions, which had a strong impact upon the existing small towns and settlements nearby. They acted as a powerful source of attraction for vagrant individuals and wandering traders over a wide area. Within two years, unemployed former workers of contractors, gatherers of forest products, river traders and prostitutes all settled across the river in shanty towns, in improvised camps with no basic conveniences, assistance, sanitation or law and order. These free settlements would reach in two years a population of 3–4 thousand people.

Other than the question of impact upon the environment, the human problem was also both real and urgent. In 1989 the basic document produced by the scientists (*Carta de Manáus*) became the general directive of a programme

→

for environmental management adopted by PETROBRAS, a governmental enterprise which had the monopoly of oil exploration in Brazil. It became a general policy of PETROBRAS that environmental management should involve the obligation to promote sustainable development in the areas where it operated. Base camps were moved to isolated places; the evaluation of impacts became standard procedure; contingency plans were drawn up to cope with accidents and disasters; health, education, scientific research and the transfer of experience were ensured. It was agreed that PETROBRAS could not be charged with the support of vagrant populations, but that it should take the necessary steps to promote their assistance by federal, state and municipal agencies. In 1995 the group of scientists revisited the sites and were most happy to see that their recommendations had been adopted with success.

Overexploitation, selective cutting, forest removal and mass human immigration are some of the problems that threaten the survival of these forests and their inhabitants.

Scientists agree that there is enough knowledge for action, but hardly enough time to act. The extensive knowledge gained from the study of temperate forests and sylvicultural methods is inadequate for the management of tropical forests, but some basic general principles may apply. Furthermore, there is abundant literature available concerning tropical forests.

Finally, everyone, wherever they live, must feel responsible for the conservation or judicious use of forest lands and products. The repercussions of ecological problems are far-reaching, both in distance and in time.

REFERENCES

Anderson, A.B. (1992) *Alternatives to Deforestation*, New York, USA: Columbia University Press.

Crosby, A.W. (1986) *Ecological Imperialism. The Biological Expansion of Europe, 900–1900*, Cambridge, UK: Cambridge University Press.

Crosby, A.W. (1994) *Germs, Seeds & Animals. Studies in Ecological History*, Armonk, New York, USA: M.E. Sharpe.

Dansereau, P. (1951) 'Description and recording of vegetation upon a structural basis', *Ecology* 32, 2: 172–229.

Dasmann, R.F., Milton, J.P. and Freeman, P.H. (1973) *Ecological Principles for Economic Development*, London, UK: John Wiley.

Deane, L., Ferreira-Neto, J., Okumura, M. and Ferreira, M.O. (1969) 'Malaria parasites of Brazilian monkeys', *Revta Inst. Med. Trop. S. Paulo* 11, 2: 71–86.

de Klemn, C. (1993) *Biological Diversity, Conservation, and the Law*, IUCN Environmental Policy and Law Papers, 29, Morges, Switzerland: IUCN.

Dourojeanni, M.J. (1974) 'Impacto de la producción de la fauna silvestre en la economía de la Amazonía peruana', *Revista Forestal del Peru* 5, 1–2: 15–27 (in Spanish).

EC (1992) *Rapport de la Commission de Communautés Européennes à la Conférence des Nations Unies sur l'environnement et le Développement, Rio de Janeiro, juin 1992*, Luxembourg: Office des publications Officielles des Communautés Européenes (in French).

Farvar, M.T. and Milton, J.P. (1972) *The Careless Technology: Ecology and International Development*, New York, USA: Natural History Press/Doubleday.

Gentry, A.H. (ed.) (1993) *Four Neotropical Rain Forests*, New Haven, Conn., USA: Yale University Press.

Glowka, L., Burhenne-Gulmin, F. and Synge, H. (1994) *A Guide to the Convention on Biological Diversity*, IUCN Environmental Policy and Law Papers, 30, Morges, Switzerland: IUCN.

IUCN (ed.) (1975) *The Use of Ecological Guidelines for Development in the American Humid Tropic*, Morges, Switzerland: IUCN.

IUCN/UNEP/WWF (eds) (1980) *World Conservation Strategy: Living Resource Conservation for Sustainable Development*, Gland, Switzerland: International Union for Conservation of Nature and Natural Resources, United Nations Environment Programme and World Wildlife Fund.

IUCN/UNEP/WWF (eds) (1991) *Caring for the Earth*, Gland, Switzerland: International Union for Conservation of Nature and Natural Resources, United Nations Environmental Programme and World Wide Fund for Nature.

Lowman, M.D. and Nadkarni, N.M. (eds) (1995) *Forest Canopies*, New York, USA: Academic Press.

Meigs, P. (1966) *Geography of Coastal Deserts*, Paris, France: UNESCO.

Middleton, C.C., Moreland, A.F. and Cooper, R.W. (1972) 'Problems of New World primate supply', *Lab. Prim. Newsl.* 11, 2: 10–17.

Moffett, M.W. (1994) *The High Frontier: Exploring the Tropical Rainforest Canopy*, Cambridge, UK: Harvard University Press.

Peters, R.L. and Lovejoy, T.E. (eds) (1994) *Global Warming and Biological Diversity*, New Haven, Conn., USA: Yale University Press.

Richards, P.W. (1952) *The Tropical Rain Forest*, London, UK: Cambridge University Press (2nd edition forthcoming, 1996).

Robinson, J.G. and Redford, K.H. (eds) (1991) *Neotropical Wildlife Use and Conservation*, Chicago, Ill., USA: University of Chicago Press.

Roth, T.W. (1965) 'Progressive primate procurement', *Lab. Anim. Care* 14: 524–526.

Schultz, J. (1995) *The Ecozones of the World*, Heidelberg, Germany and New York, USA: Springer-Verlag.

UNEP (1988) *Environmental Perspective to the year 2000 and Beyond*, Nairobi, Kenya: United Nations Environmental Programme.

UNESCO (ed.) (1970) *Utilization et Conservation de la Biosphère*, Paris, France: UNESCO (in French).

United Nations (1993) *Agenda 21: Rio Declaration and Forest Principles*, Geneva, Switzerland: United Nations.

Vieira, I.C.G. *et al.* (1996) 'O renascimento da floresta no rasto da agricultura (Forest regeneration after agriculture)', *Ciência Hoje* 20 (119): 38–44 (in Portuguese).

WCED (1987) *Our Common Future*, Oxford, UK: Oxford University Press.

WHO (1992) *Our Planet, Our Health: Report of the WHO Commission on Health and Environment*, Geneva, Switzerland: World Health Organization.

SUGGESTED READING

Brookfield, H.C. and Byron, Y. (eds) (1994) *South-East Asia's Environmental Future*, Malaysia: Penerbit Fajar Bakti.

Constanza, R. (ed.) (1991) *Ecological Economics: The Science and Management of Sustainability*, New York, USA: Columbia University Press.

Davis, S.D., Heywood, V.H. and Herrera-MacBride, O. (eds) (1994) *Centres of Plant Diversity. A Guide and Strategy for their Conservation*, 2 vols, Morges, Switzerland: IUCN.

Ehrlich, P.R. and Wilson, E.O. (1991) 'Biodiversity studies: science and policy', *Science* 253: 758–762.

Fearnside, P.M. (1993) 'Deforestation in Brazilian Amazonia', *Ambio* 22, 8: 537–545.

INPE (1992) *Deforestation in Brazilian Amazonia*, São José dos Campos, Brazil: Brazilian Institute for Space Research.

Johnson, N. and Labarle, B. (1993) *Surviving the Cut: Natural Management in the Humid Tropics*, Washington DC, USA: World Resources Institute.

McDade, L.A., Bawa, K.S., Hespenheide, H.A. and Hartshorn, G.S. (eds) (1994) *La Selva: Ecology and Natural History of a Neotropical Rainforest*, Chicago, Ill., USA: University of Chicago Press.

Mickler, R.A. and Fox, S. (eds) (1988) 'The productivity and sustainability of southern forest ecosystems in a changing environment', *Ecological Studies* 28, New York, USA: Springer-Verlag.

Nepstad, D. and Schwartzman, S. (eds) (1992) *Non-timber Products from Tropical Forests: Evaluation of a Conservation and Development Strategy*, New York, USA: New York Botanical Garden.

Panayotou, T. and Ashton, P.S. (1992) *Not by Timber Alone. Economics and Ecology for Sustaining Tropical Forests*, Washington DC, USA: Island Press.

Poncel, L. (ed.) (1993) *Tropical Forestry Handbook*, Berlin, Germany: Springer-Verlag.

SAREC (1995) *Tropical Diseases, Society, and the Environment*, Stockholm, Sweden: SAREC.

Terborgh, J. (1992) *Diversity and the Tropical Rain Forests*, New York, USA: Scientific American Library.

Whitmore, T.C. and Sayer, J.A. (eds) (1992) *Tropical Deforestation and Species Extinction*, London, UK: Chapman and Hall.

SELF-ASSESSMENT QUESTIONS

1 Tropical forests are found:
   (a) at sea level, around the equator;
   (b) from sea level to 1,000 m, in the intertropical belt;
   (c) in the lowlands of Africa and Asia;
   (d) in all continents, except Australia and Antarctica.
2 Tropical forest trees:
   (a) grow to form a continuous canopy 50 m above ground;
   (b) are stratified in layers, easily recognised from the ground;
   (c) grow to different heights, usually forming three layers, with the highest trees conspicuous above a continuous canopy 20 m below;
   (d) are interspersed with dense, thorny, close vegetation at ground level, as can be seen from riverboats.
3 Tropical forest soils:
   (a) have a shallow topsoil layer, most nutrients being concentrated in the living matter;
   (b) have a deep upper layer of organic matter resulting from the decomposition of dead trees;
   (c) are rich in nutrients to support their luxuriant growth;
   (d) are permanently soaked from the direct impact of heavy tropical rains.
4 The sustained economic exploitation of tropical forests:
   (a) may profit directly from the accumulated scientific data and methodology used in temperate developed countries;

(b) causes no immediate problems, due to the rapid regrowth of forests in the tropical climate;

(c) is a problem to be considered in the future;

(d) does not need to wait for the production of further scientific knowledge to be viable.

5 The main threats to tropical forest ecosystems come from:

(a) population expansion, the wide use of forest products and the construction of roads and hydroelectric plants;

(b) natural disasters;

(c) destruction by native forest inhabitants;

(d) the lack of basic scientific knowledge concerning floral composition and vegetational systems.

6 The conservation of tropical forests is:

(a) a local problem, the responsibility of regional authorities;

(b) not an important or urgent issue;

(c) to be considered as secondary to its transformation to agricultural land;

(d) an international concern, and the responsibility of us all.

9

WILDERNESS MANAGEMENT

Bruno Kawasange

SUMMARY

The discussion in this chapter will explore the importance and management complexity of wilderness areas and other related protected areas such as national parks and reserves. Apart from giving the underlying principles that guide the categorical classification of an area either as a wilderness or as a national park, the chapter outlines the various ways that protected areas contribute to scientific, social and economic welfare. It is shown that protected areas contribute to life support systems, preservation of biological diversity (i.e. the wealth of life forms on earth, the millions of plants, animals and micro-organisms, the genes they contain and the intricate ecosystems they help to build into the living environment), sustainable utilisation, conservation of natural heritage and recreation and tourism.

Given the fact that maintenance of essential ecological processes and life support systems is vital to human beings, protected areas are discussed in the framework of ecosystems, emphasising their dynamic nature, their strong internal linkages and the ways various human activities affect their equilibrium. Pressures and hazards emanating from local populations, either within or adjacent to protected areas, will be discussed. It is quite apparent that due to human encroachment, the majority of protected areas in the world have suffered and continue to suffer from habitat conversion and reduction of potential habitats for some animal species. The illegal harvesting of resources in protected areas (e.g. poaching), particularly in tropical protected areas, is identified as the major source of reductions in wildlife populations, especially those of large mammals such as elephants and rhinoceroses. This situation has raised alarm and calls for a programme to protect endangered species and preserve natural areas.

The dangers inherent in promoting protected areas as tourist attractions will also be discussed, indicating the potential effects of recreation by tourists on the environment and various recreational activities. The need for comprehensive planning in the management of protected areas will be highlighted as one of the means of ensuring resource sustainability. Also, in reconciling the dilemma between use and preservation of protected area resources, emphasis is given to the linkage between the management of protected areas and other disciplines in order to create professional integration and interaction. The environmental education component is discussed in relation to protected areas. It is recommended that environmental education must be combined with conservation in order to achieve management objectives.

Organised in this way, it is to be hoped that readers, whether students of environmental science and engineering, ecologists, resource managers, engineers or lawyers will find areas that are of interest to them. Hopefully, the knowledge acquired will help to enhance environmental conservation endeavours.

ACADEMIC OBJECTIVES

This chapter aims to assist readers in developing a better ecological perception of wilderness areas and further understanding of the environmental effects of human activities on such areas. The intention is to acquaint readers with some of the definitions of wilderness and protected areas, and the general principles of ecology that need to be considered and integrated into the formulation of plans for the development, conservation and protection of wilderness areas. Because wilderness areas are considered part of the total human environment, an interdisciplinary approach and the need to involve local communities is stressed.

INTRODUCTION

The last few decades have been marked by a tremendous interaction between man and the environment. This has led to the increasingly large-scale exploitation of nature, the outcome of which has been detrimental to the ecological balance. Population growth, coupled with the ever-increasing use of natural resources in economic activity, has led to the depletion and degradation of natural resources. Large numbers of animal and plant species have been, and continue to be, exterminated with a corresponding reduction in critical habitats and other essential requirements.

Among the key problems facing mankind today is the task of adjustment to a deteriorating environment. Many problems are apparent, but a key ecological question is the conservation of natural protected areas and their biological resources. There have been various approaches to these ecological issues, including the infusion of ecological awareness into various academic disciplines. For example, the present professional training of engineers, lawyers, architects and planners continues to be enriched by ecological inputs concerning issues of land use, natural resources and human welfare, all covering broad aspects of the environment. The aim is to create a conducive environment in which problem-solving by various professionals is balanced by a holistic point of view. Environmental science and engineering may be regarded as an alternative way of bridging the gap between the need to train specialists who are technically competent and the need to understand the importance of contextual knowledge.

Despite a variety of environmental problems, most nations accept the desirability of protecting outstanding examples of their natural heritage and acknowledge their contribution to the world-wide effort of protecting living resources and conserving biological diversity. Wilderness areas are the most common form of protection, but they have been complemented by many other categories of protected areas such as national parks and reserves. Many countries recognise several different types of protected area within one overall national system, each with different conservation and management objectives including some that are heavily utilised.

DEFINITIONS OF SOME CATEGORIES OF PROTECTED AREAS IN THE WORLD

Wilderness areas

Although it is not the objective of this chapter to trace the origins of the word 'wilderness' and its cultural evolution, nor to review the extensive literature on the many values and philosophies of wilderness, it is important to present a general definition of the term. For the purposes of this chapter the following general definition after Hendee *et al.* (1978) has been adopted in which they define a wilderness as 'an area where the influence of modern humans is absent (or at least minimised as much as possible)'. But how minimum is minimum is an aspect that requires careful consideration since it is subject to criticism and debate, especially when the term 'modern human' is included in the definition. Maybe much more significant to the definition is the question whether there are still any areas remaining where human influence is absent or minimal?

In fact, what are commonly observed in many protected areas are so-called 'wildlife zones', where limited visitor access is permitted but management is primarily aimed at maintaining an undisturbed nature, or a desired balance, or the natural status quo. In such areas recreation management is limited to the provision of rough trails and sometimes primitive camp sites. Given the difficulties in delimiting areas that still retain features consistent with an ideal wilderness area, the above definition tends to sound more like a sociological definition based on what people expect from a wilderness area.

At the other extreme, wilderness has been defined in legal terms, in various countries. For example, in the USA, the Wilderness Act of 1964 (Sec. 2c) defines a 'wilderness area in contrast with those areas where man and his own works dominate the landscape; it therefore recognises it as an area where the Earth and its community of life are untrammelled by man, where man himself is a visitor who does not remain'. Here again we have a question whether there are still areas existing on the Earth where people are only visitors, since humans either live in or on the boundaries of the majority of protected areas (if not all) as administrators or managers.

Whatever definition of wilderness area we adopt as the most suitable, it is clear that deriving a universally acceptable definition is not easy because perceptions of what constitutes wilderness vary widely. However, by combining sociological and legal definitions with information contained in the following discussion about wilderness management, the reader should be able to determine the main features of a wilderness area or other category of protected area.

It is now widely accepted by conservationists and nature lovers that there are no wilderness areas (even liberally defined) left to protect in many countries. In East Africa, for example, it is argued that there is 'wildness' without 'wilderness'. This simply means that, although there is still a wide diversity of wild animals and plants, humans, none the less, have interfered with their 'naturalness' or, in this case, their wildness. Despite these difficulties, many countries have set aside areas as national parks or equivalent reserves, where the definition is again either sociological or legal, depending on the objective of the country concerned.

The most important objective is the protection and provision of the habitat to wildlife and sometimes recreation and/or provision of grazing land for livestock, the latter leading to problems which have manifested themselves in some African countries. For example, in Ngorongoro Conservation Area in Tanzania, grazing has become part of wilderness management where Masai and their livestock are allowed to live inside the protected area even though there are still some few serious conflicts arising from time to time. There are cases where big game have been reported to have damaged trees and, due to overstocking, destroyed vegetation; other cases include wildlife killing or spreading diseases to livestock.

In many cases, foresters do not know what is happening with the wilderness and its surroundings, and livestock officers do not know much about the cattle kept and grazed in these wilderness areas and probably leave the whole affair to protected area officers who in turn are not experts in livestock management. Such ignorance has adversely affected the decision-making for many of these areas following poor understanding of the situation.

Exclusive use of some rangelands by either wildlife or livestock exists in the form of national parks, forest and game reserves. In a few of the protected areas, such as game controlled areas, communal grazing by sharing resources between people and their livestock is also practised. Traditionally, wild animals have been hunted to provide a variety of products such as ivory, hides, skins and trophies. They have also been an important source of meat for pastoralists and farmers. However, conflicts have always arisen between the two: livestock and wildlife. While livestock support the landholders' livelihood, game is regarded by the landholder as either a threat or a competitor for the same available resources. This conflict has been intensified by two opposing and seemingly mutually exclusive groups, i.e. the pro-game and the pro-livestock supporters. The conflict need not have arisen, or at least not as strongly as it has. For both physiological and ecological reasons, biologists have tended to advocate a more widespread habitation of the rangelands by wildlife, thus gaining strong support from the tourism industry. At the same time for financial and economical reasons, livestock owners have campaigned vigorously for the exclusion of non-paying wildlife on their property. This has raised general problems of multiple use in resource management (Clawson, 1973). Most of the land in Africa has traditionally supported both livestock and wildlife. Each has played a prominent role in the lives of the people using the land. The use of land by animals with different feeding strategies has benefited the ecology of wilderness areas in many ways including an increase in production per given area since no one single species can fully utilise all the resources within such an area. However, conflict between livestock and wildlife production exists. A number of other benefits can be realised through the proper use of wilderness areas such as hunting, general tourism and wildlife photography.

National parks

A national park is an area designated for the protection of natural and scenic areas of national or international significance for spiritual, scientific, educational, recreational and tourism purposes. These are relatively large natural areas not materially altered by human activity where extractive resource use is not allowed. The area should perpetuate, in a natural state, representative samples of physiographic regions, biotic communities, genetic resources and species and provide ecological stability and diversity. Examples include Serengeti National Park in Tanzania, The Royal Chitwan National Park in Nepal, and The Iguazu National Parks in Argentina and Brazil. While there are a confusing number of different names describing protected areas in different countries, there are, in fact, relatively few basic objectives for which areas have been established and are managed.

The World Conservation Union (IUCN) has set international benchmark standards for protected area management and, more recently, the IUCN Commission on National Parks and Protected Areas (1991) listed five categories ranging from Wilderness Area (Category I) to Protected Landscape (Category V). In this system, national parks are designated Category II and their management objectives are defined as the protection of 'natural and scenic areas' with the intention of preserving 'biotic communities, genetic resources, and species'.

Human activities in wilderness areas (Category I) are restricted to research and, in national parks, to research, education and recreation. Protected areas (Categories III through V) allow progressively more human use and more manipulation of ecosystems to meet species management objectives and human needs. Human habitation and commercial resource use are consistent with Category V (Protected landscape). Exclusion of most human uses makes integration of conservation and development a special challenge in Category I and II areas. The following is a summary of the system of categories and their management objectives according to Lee (1992).

I Strict nature reserve/wilderness area

To maintain essential ecological processes and to preserve biological diversity in an undisturbed state, in order to have representative examples of natural environment available for scientific study, environmental monitoring and education and for the maintenance of genetic resources in a dynamic and evolutionary state. Research activities need to be planned and undertaken carefully to minimise disturbances.

II National park

Refer to the definition on p. 167.

III Natural monuments

To protect and preserve outstanding natural features because of their special interest, unique or representative characteristics and, to an extent consistent with preservation, provide opportunities for interpretation, education, research and public appreciation.

IV Habitat and wildlife management areas

To ensure the natural conditions necessary to protect significant species, groups of species, biotic communities or physical features of the environment where these require specific human manipulation for their perpetuation. Scientific research, environmental monitoring and education are the primary activities associated with management of this category.

Following the resolutions made during the Earth Summit in Rio de Janeiro, especially Chapter 26 of Agenda 21 (Stanley, 1993), many countries in the southern part of Africa decided to involve local communities in the management of their natural resources. This, in a way, forced these countries to make a few modifications in the definition to suit the new areas and thus defined these areas as communal land that contains wildlife and other natural resources to which local people have user rights as well as a mandate to protect them. From these resources they may also retain a significant proportion of revenue for the purpose of development activities within their community.

This is a new effort to find optimum conditions for resource management through giving the resources a focus of value as well as giving the communities living with or adjacent to these resources higher benefits than are available to communities not bearing the costs they may also sometimes impose.

V Protected landscapes/seascapes

To maintain significant areas which are characteristic of the harmonious interaction of nature and culture, sites providing opportunities for public enjoyment through recreation and tourism, and to support the normal lifestyle and economic activity of these areas.

Ideally in all protected areas, objectives and activities should be related to environmental protection and to economic and social development. All categories are of value, each with a different role, and only several taken together can adequately cover national and global resource management needs.

In the subsequent discussion the term 'protected area(s)' instead of wilderness area(s), national park(s) or equivalent reserves will frequently be used as representative of all natural areas which have been set aside by various countries for the conservation of their natural attributes. Conservation in this case is in accordance with the definition provided by the World Conservation Strategy which refers to 'the management of human use of the biosphere so that it may yield the greatest sustainable benefit to the present generation while maintaining its potential to meet the needs and aspirations of future generations'. Therefore, where attention is directed towards a particular protected area, clear reference will be made to such a category.

CHARACTERISTICS OF WILDERNESS AND OTHER PROTECTED AREAS

Wilderness areas, national parks and reserves as ecosystems

When addressing wilderness areas and equivalent reserves at the level of an ecosystem, let us follow Odum's (1971) definition of an ecosystem, namely,

> a unit that includes all of the organisms (i.e., the community) in a given area interacting with the physical environment so that a flow of energy leads to a clearly-defined trophic structure, biotic diversity, and material cycle (i.e., the exchange of materials between living and non-living parts) within the system.

Wilderness areas, national parks or any equivalent reserves are therefore broad ecosystems and their main functions in ecological terms are to emphasise interdependence and causal relationships, that is the coupling of components to form functional units. The emphasis is on the dynamic nature of ecosystems, the strong internal linkages and the ways various human activities have affected and continue to affect the 'naturalness' of 'wilderness' ecosystems. Only when ecosystem dynamics, including interrelationships with humans, are fully understood can an assessment of the consequences of alternative management strategies be made.

A wilderness ecosystem or a national park has a series of attributes that are fundamentally the same as those in any other ecosystem. At any given time, there is a particular *composition* and *structure*; that is there is an array of plant and animal species in various proportions. Within the ecosystem various processes such as photosynthesis, transpiration, consumption and decomposition are occurring. The lineages or flows between the different parts of an ecosystem are characteristics of special importance to resource managers. Energy, water, nutrients and other substances do not remain indefinitely in any one place or state but flow through the ecosystem along a series of pathways (Odum, 1971). These attributes are extremely important as humans exert their most profound influences on ecosystems when they alter the paths and rates of these flows. As an example, imagine what would happen to a forest or grassland ecosystem if somehow decomposition rates (the breakdown of dead organic matter with the release of energy and nutrients) were significantly reduced? Answers to this question are quite diverse but maybe the simplest one would be 'paralysis of the entire system'. What follows then is a state of disorder in the entire system.

Due to the pervasive effects of modern humans on ecosystems, we all recognise that there are no completely unaltered ecosystems left on this planet. It is very clear that even in areas perceived as 'pristine wilderness', human activities have affected several key attributes of such ecosystems. First, they have affected their *functional ability*; that is their capacity to perform key functions such as fixing and cycling energy, conserving and cycling nutrients and providing suitable habitats for an array of inhabiting species. Second, the *structure* or spatial arrangements within such ecosystems is affected – whether it is savannah, meadow or some other type. Third, the *composition* and *population structure* is altered, that is the number of species and their relative abundance, as well as the densities of individual species and their age and size–class distributions. Finally, human activities looked at from the ecosystem level have changed the basic *successional patterns* of many protected areas.

Human influence on protected ecosystems has threatened the world's biological diversity. Let us examine this aspect by first defining the term 'biological diversity'. Put simply, this is 'the wealth of life forms on Earth, the millions of plants, animals, and micro-organisms, the genes they contain, and the intricate ecosystems they help build into the living environment'. The Earth's survival depends on maintaining this diversity and protected areas play an important role in this. One measurable unit of biological diversity is the species and from this one indication of the extent of biological diversity in an area is the number of species present. However, life on Earth contains much greater variety than can be measured by number of species alone. Each species contains its own variety, such as different subspecies, varieties, races or breeds as well as differences between populations and individuals within populations. Populations join to form communities and these combined with the abiotic environment form ecosystems. Many species survive in only one type of ecosystem (Granville, 1984) so discussion involving biological diversity usually recognises the concept at three distinct levels: genetic diversity, species diversity and ecosystem diversity as explained in the next sections.

Genetic diversity

Genes are the biochemical packages passed on by parents that determine the physical and biochemical characteristics of their offspring. Although most genes are passed on unaltered (which is why offspring resemble their parents) some changes and variations do occur. Such changes may be easy to detect such as the size or colour of a body part or they may be invisible, for example, increased susceptibility to disease. Such genetic diversity has made it possible to produce new breeds of crop plants and domestic animals and, in the wild, has allowed species to adapt to changing conditions. Genetic variation within a species can currently be detected by physical characteristics or biochemical tests.

Species diversity

A group of organisms genetically so similar that they can interbreed and produce fertile offspring is called a species. Horses and zebras are different species. They are genetically similar and can interbreed but all their offspring are infertile. Species are usually recognisably different in appearance, allowing an observer to distinguish one from another, but sometimes the differences are extremely subtle. Species diversity is usually measured in terms of the total number of species within discrete geographical boundaries. The degree of species diversity is a little better known than the extent of genetic diversity, but current estimates of the number of species only serve to emphasise our degree of ignorance.

Ecosystem diversity

The functional relationships within and between communities and their environment are frequently complex, but they are the mechanisms of major ecological processes such as the water cycle, soil formation, nutrient cycling and energy flow. These processes provide the sustenance required by living communities and bring about their critical dependence on their biotic and abiotic environment.

Two different phenomena are frequently referred to under the heading of ecosystem diversity:

(a) The variety of species within different ecosystems: the more diverse ecosystems contain more species.
(b) The variety of ecosystems found within a certain biogeographical or political boundary.

The World Conservation Strategy states that the preservation of biological diversity is a matter both of insurance and investment necessary

(a) to sustain and improve agriculture, forestry and fisheries production;
(b) to keep future options open
   ● as a buffer against harmful environmental change;
   ● as the raw material for much scientific and industrial innovation; and
   ● as a matter of moral principle.

Economic, scientific, social and cultural dimensions of wilderness areas and other protected areas

Economic value

The economic value of protected areas is quite evident in many countries especially where these areas have been developed for tourism. Tourism is an industry that, with proper planning and investment, can show spectacular growth, and protected areas can contribute to this. Tourism development in and around protected areas can also be one of the best ways of bringing economic benefits to remote areas by providing local employment, stimulation of local markets and improvement of transportation and communication infrastructure. At the national level, tourism brings valuable foreign exchange into the country and at the local level stimulates profitable domestic industries such as hotels, restaurants, transport systems, souvenirs, handicrafts and guide services. In some categories of protected area, hunting and other forms of legal direct harvesting of wild resources are of appropriate and economic importance and, where harvesting is conducted with proper control and the application of sensible quotas, can generate substantial revenue. However, there are dangers inherent in promoting the idea of protected areas as tourist attractions and these will be discussed later.

Scientific value

Human influence on the planet is already so great that it is hard to imagine that any wilderness area is totally 'natural' or stable. Protecting the things of nature, whether they be a single important species or a whole representative ecosystem, requires intervention in the form of management to ensure that the appropriate environment is maintained. Protected areas in many countries have been the focus for ecological studies because they usually include the most intact examples of natural habitats. Recent research seems to be concentrating on ecology and the distribution of the more endangered species, including captive breeding requirements as well as biological inventories and habitat assessments for management purposes. In many protected areas, however, basic fauna surveys and plant inventories are still lacking and basic applied research on this is needed for management purposes. More detailed ecological investigations of particular species or habitats are also required to ensure their best conservation management.

Social value

Protected areas may also provide expanded services to local people who benefit from their proximity to a protected area which, in turn, helps to reduce their dependence on the adjacent protected areas for harvestable products. Directly or indirectly, a protected area can enhance employment opportunities in the region and local people may be employed directly by the management authority or in catering to visitors or providing ancillary services. Whenever possible, local people should be employed as reserve staff or as concessionaires in preference to outsiders. Out of the revenue realised from the protected areas, government can provide many forms of special assistance to rural people including:

(a) agriculture and grazing improvement schemes;
(b) road improvement schemes;
(c) water, sewage and electricity services;
(d) schools, clinics and dispensaries;
(e) direct grants for land improvements;
(f) loans or credit facilities for individual farmers.

Cultural value

Protection of natural reserves can result in the preservation of locally important cultural sites and traditional practices which would otherwise be destroyed. Some protected areas, such as national parks and equivalent reserves, protect cultural sites of great significance to the local people surrounding the protected area and, at the same time, serve as potential attractive features. As a result, nature conservation often benefits from the protection of such important features. Examples include the Tikal National Park of Guatemala, Tassili N'Ajjer National Park in Algeria and Angkor Wat National Monument in Cambodia. Some sites within a protected area may be spectacular and well publicised, while others may be less conspicuous and kept secret by the people for whom they have a special significance. The former often attract visitors and it is usually a relatively simple matter for the park manager to provide for appropriate access. This is more difficult in the case of secret sites as the manager may not know of their existence or of their importance and may be unwilling to prevent their use or may permit their desecration.

A sensitive manager will try to establish the location of those sites (or other cultural features such as traditional pilgrimage routes) and provide adequate protection without inquiring too intrusively about their cultural significance. In any case, the proper use of cultural sites calls for close liaison with a range of interested parties from professional scientists to informed and influential members of local communities.

PRESSURES AND HAZARDS TO WILDERNESS AND OTHER PROTECTED AREAS

Threats to effective management of protected areas in the form of poaching, habitat destruction and general encroachment are reported from almost every protected area in the world. This is largely due to rapid population growth and the conflict between traditional hunting practices and modern legislation which has been imposed upon people with different cultural values and outlook, especially in tropical protected areas. Combined with other natural factors, the natural equilibria of many protected areas have been altered. Even for those that have remained fairly intact, isolation due to habitat alteration in surrounding areas threatens the viability of migrant species requiring large ranges. Severe reduction in herbivores' habitats has led to overgrazing and habitat destruction inside protected areas in countries where population densities are high.

Local population

Local people may be considered as those directly affected by the establishment of the protected area, and often include many who are not permanent inhabitants of the area or its vicinity. The success of the management of protected areas depends on the degree of support and respect awarded to the protected area by neighbouring communities. Where protected areas are seen as a burden, local people can make protection impossible, but when seen as a positive benefit, the local people tend to ally themselves with management in protecting the area from threatening developments. The establishment and maintenance of protected wilderness areas therefore needs to pay adequate attention to the interactions

between local people, the management authority and the natural environment under protection. To provide for long-term natural interactions, the guidelines outlined in the following sections are potentially useful.

Use of local knowledge

People who have a long history of use or occupancy of areas considered for protection or already under protection have a familiarity with the species, communities and ecological processes which cannot readily be gained through surveys, inventories or baseline studies by external experts. In particular, long-term trends or fluctuations in the abundance and distribution of wild species, past influences and changes, values and usefulness for human purposes can be determined most easily from local knowledge. Taking the example of Africa, living in balance with the environment was an integral component of African cultures. For example, the concentration of plains herbivores, which are now conserved, existed because of a tolerance for wildlife in some African cultures in which individuals were taught to co-exist with the natural world around them and to see themselves as part and parcel of the system. Many traditional African religions referred specifically to the preservation of natural things and made it a taboo to kill more than what was needed for survival. According to Lusigi (1984), wildlife that supported life and gave spiritual satisfaction was hunted for food and clothing but was also used in tribal ceremonies and rituals. It is therefore clear that consultation with the people is very important for gaining essential knowledge for both conservation and the avoidance of conflict.

Local participation in planning for protected areas

Planning for protected areas should involve people who are most likely to be directly affected, positively or negatively, by implementation of protected area status. Every effort should be made to achieve the desired conservation objectives with minimum disruption to traditional ways of life and maximum benefits to local people. A simple conservation rule that has local adherence and support may accomplish more than a protected area that has none. The decade through which the state has been the exclusive owner and manager of wildlife areas now seems to have passed due to the fact that the current process of policy reform provides a legal framework which paves the way to community participation in managing most of the nation's wildlife and possibly other natural resources. But for these community programmes to be more effective, there are certain issues that need to be fully addressed, like the issue of land ownership (land tenure), an issue that concerns the community directly because, for a community to manage its resources in a sustainable manner, benefits arising from a protected area that is adjacent to it need to be appreciated. According to Achim and Elisabeth (1995) in their unpublished paper 'The commons without the tragedy?', in which they reviewed strategies for community-based natural resource management in southern Africa,

> effective management of natural resources is best achieved by giving the resources a focused value; that is, for differential inputs there should be differential benefits; there must be a positive correlation between the quality of management and the magnitude of derived benefits. The unit of who decides should be the same as the unit of production, management and benefit; finally, the unity should be as small as practicable, within ecological and socio-political constraints.

The five principles should then be drawn based on the cultural and socio-economic requirements of the people.

Local involvement in management and conservation

Looking at Agenda 21 (Stanley, 1993) on the issue of participation of local communities in the management of wilderness areas, many countries have so far tried to involve local people living adjacent to these protected areas in planning, management and conservation practices. Taking the example of Africa, especially the southern zone, there are several good programmes that can be mentioned. These programmes include Administrative Management Design for Game Management Areas (ADMADE), Zambia; Natural Resources Management Programme (NRMP), Botswana; Communal Areas Management Programme

For Indigenous Resources (CAMPFIRE), Zimbabwe; Community Based Natural Resources Management (CBNRM), Namibia; Selous Conservation Programme (SCP), Serengeti Regional Conservation Programme (SRCP), Tanzania. Most of these programmes have, to certain extent, tried to adopt some of the principles discussed by Murphree (1993) and Achim and Elisabeth (1995). Some of the countries have incorporated these principles in their respective policy and legal frameworks so as to create better conditions that will provide proper and easy establishment of these community-based management institutions which will bring sustainable development to their societies.

Economic benefits

The economic benefits derived from tourism or other forms of use of a protected area must be shared by agreement with local people. Negotiations with local people are important in giving them a greater role in maintaining the protected status of the area.

Overlooking the above consideration has occasionally led to conflicts with local communities living adjacent to protected areas, especially over the aspect of land use. With the increase in human population density all over the world, there has been an increase in the demand for land and for resources associated with land. Consequently, many protected areas suffer human encroachment and habitat conversion which has left the majority of protected areas in isolation and is a serious problem for the long-term survival of mammalian species that inhabit small protected areas.

According to island biogeography theory, the number of species in any isolated area, whether it is an oceanic or habitat island, is determined by an interaction between species colonisation and extinction. Species colonisation is influenced primarily by the distance of the isolated area from a potential gene pool source while species extinction is influenced primarily by the population size of the species in question which, in turn, is influenced by the size of the area. It is therefore apparent that as protected areas become increasingly isolated, the distance of a protected area from a potential gene pool source increases while the effective size of the conservation

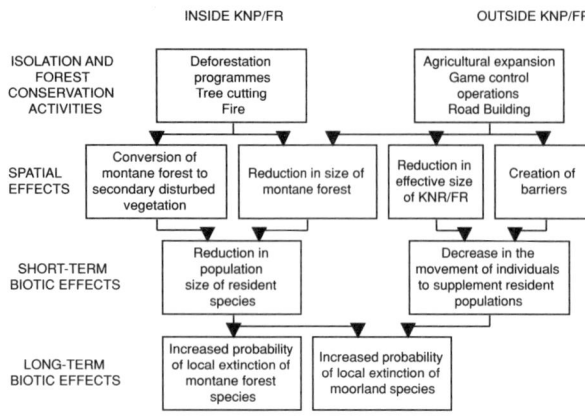

Figure 9.1 The potential short-term and long-term biotic effects of isolation and forest conversion activities inside and outside Kilimanjaro National Park and Forest Reserve (KNP/FR) on resident large mammal populations
Source: based on Newmark *et al.*, 1991

area decreases. The predicted effects of a reduction in colonisation and decrease in the effective size of a protected area are an increase in the number of local extinctions of species in a protected area. Evidence in support of this model comes not only from studies of fauna communities in a variety of habitat islands (see Diamond, 1984) but also from large mammal communities in protected areas in North America and East Africa (Miller and Harris, 1977; Soule *et al.*, 1979; Western and Ssemakula, 1981; East, 1981, 1983).

For example, the results of the study we conducted between 1988 and 1989 to study the local extinction of large mammal species in Mount Kilimanjaro National Park and the adjacent Forest Reserve in northern Tanzania were consistent with the findings of other scientists. By comparing the historical status of a species as documented in the literature with its current status, we determined the effects of habitat conversion and the increasing isolation of the protected area using a model (Figure 9.1) to describe the process. However, the model should not to be viewed as a perfect model for studying the implications of increasing isolation and habitat conversion upon large mammal fauna in protected areas, but as a means of assisting managers in monitoring future changes in their parks' fauna. It should be viewed as a series of hypotheses (Newmark, 1984).

Poaching or illegal harvesting of a protected area's products

The designation of wilderness areas, nature reserves or national parks as places where people can no longer live, and can only use as visitors who come to look but not to interfere, has been difficult for people to accept who have always lived in wild country and consider themselves part of it. Logging, poaching and trapping of wild animals for trade, meat, skins, feathers, horns, antlers or ivory are threats to the conservation and survival of many species as well as to the ecosystems and the integrity of some protected areas. The pressure arises from the needs and demands of local populations around protected areas.

For example, over the last two decades elephant populations in Africa have been drastically reduced, mainly due to poaching for ivory. For both white and black rhinoceroses, the current populations represent no more than a fraction of what they were in the last century. Poaching is primarily responsible for this dramatic reduction. In many countries, certain categories of protected area include provision for the continuance of traditional ways of life (e.g. the Maasai people in Ngorongoro Conservation Area, Tanzania and Amboseli National Park, Kenya, and Bedouins in Jebel Elba National Park, Sudan). In addition, certain activities such as controlled fishing may also be carried out legally (e.g. in the Kenyan Marine Reserve and in the Hadejia-Nguru wetlands, Nigeria). In Nepal's Royal Chitwan National Park, for instance, local people are allowed into the park during a specific two-week period each year to harvest thatch grasses, the only traditional roofing material available in the region. However, population expansion within subsistence communities has increased the level of conflict between traditional peoples and area management. Dependence on wildlife for food and other products in protected areas is likely to increase in coming years and it is important that these wild resources be used wisely (Mishra, 1984).

Tourism – the negative impacts on protected areas

The reader should already have an idea as to how tourism can bring significant economic benefits to the country, and tourism is discussed more fully in

Volume 2, Chapter 14, 'Sustainable tourism'. Notwithstanding that, it is important to emphasise and indicate briefly the dangers inherent in promoting the idea of protected areas as tourist attractions, especially where decision-makers are led to believe that national parks and equivalent reserves exist primarily for economic gain. Quite often, when these expectations are not fulfilled they may begin to look for more profitable alternative uses for the land. According to Thorsell (1984), development of large hotels, highways or golf courses designed to attract more visitors can diminish a park's natural value and eventually turn it into an area for which the main objective is mass tourism rather than conservation. Table 9.1 lists certain potential negative effects of tourism reflecting experience in East Africa's protected areas and shows that unless carefully controlled, the volume of visitors may have a deleterious impact on protected areas and may eventually destroy the very resource on which tourism depends.

Table 9.1 Potential environmental effects of tourism in protected areas in East Africa: the types of negative visitor impact that must be controlled

Factor involved	Impact on natural quality	Comment	Examples
Overcrowding	Environmental stress, animals show changes in behaviour	Irritation, reduction in quality, need for carrying capacity limits or better regulation	Amboseli
Overdevelopment	Development of rural slums, excessive human-made structures	Unsightly urban-like development	Mweya, Seronera Keekorok Ol Tukai
Recreation			
Powerboats	Disturbance of wildlife and quiet	Vulnerability during nesting seasons, noise pollution	Murchson Falls
Fishing	None	Competition with natural predators	Ruaha, Nile
Foot safaris	Disturbance of wildlife	Overuse and trail erosion	Mt. Kenya Kilimanjaro
Pollution			
Noise (radios etc.)	Disturbance of natural sounds	Irritation to wildlife and other visitors	Many areas
Litter	Impairment of natural scene	Aesthetic and health hazard	Many areas
Vandalism	Mutilation and facility destruction	Removal of natural features, facility damage	Sibiloi
Feeding of animals	Behavioural changes of animals	Removal of habituated animals, danger to tourists	Masai-Mara, Ruaha
Vehicles			
Speeding	Wildlife mortality, dust	Ecological changes	Amboseli, Mikumi
Off-road driving, night driving	Soil and vegetation damage	Disturbance to wildlife	Ngorongoro, Amboseli
Miscellaneous			
Souvenir collection	Removal of natural attractions	Shells, corals, horns, trophies, rare plants	All areas
Firewood collection	Small wildlife mortality and habitat destruction	Interference with natural energy flow	All areas
Roads and murram pits	Changes, natural scars if not well sited and constructed	Aesthetic scars, ecotones damaged	All areas
Power lines	Destruction of vegetation	Aesthetic impact	Tsavo, Bale Mts.
Artificial water holes and salt provision	Unnatural wildlife concentrations, vegetation damage	Replacement of soil required	Aberdare
Introduction of exotic plants and animals	Competition with wild species	Public confusion	Many areas, Mt. Kilimanjaro, airstrips

Source: after Thorsell, 1984

BOX 9.1   SCIENCE AND ENVIRONMENTAL EDUCATION AS TOOLS FOR GOOD MANAGEMENT: THE GALAPAGOS ISLANDS, ECUADOR

The Galapagos Archipelago is located 960 kilometres west of continental Ecuador. It is composed of 19 islands and 45 islets which emerged from the sea and were never joined to the continent. The islands were discovered in 1535 by Fray Tomas de Berlanga; in 1832 Ecuador took possession of them.

In 1835 the English naturalist Charles Darwin visited the islands and 24 years later he published his book *The Origin of Species*. Since then Galapagos has been considered a natural evolutionary 'laboratory', unique for the many endemic species (found nowhere else in the world) such as the Galapagos tortoise, marine and terrestrial iguanas, Scalesia trees and Galapagos penguin that inhabit these islands. Probably the most important ecological problem of this park are the introduced species (e.g. rats, goats, guava trees and wasps) that compete for food or habitat with the endemic species.

A national institution and a non-governmental organisation working together to manage the national park

Galapagos constitutes an interesting example of how a governmental organisation, the Galapagos National Park Service (GNPS), and a non-governmental organisation, the Charles Darwin Research Station (CDRS) have worked together with a common objective – in this case the conservation of the islands.

The GNPS is the legal authority in the Galapagos National Park and as such regulates the activities in this area. The CDRS is an international institution which has served for more than thirty years as scientific adviser of the government of Ecuador in matters regarding the conservation of this region. The scientists of the CDRS work on projects that are relevant to the management and protection programmes designed by the GNPS authorities.

Tourism in the Galapagos National Park – a challenge for good management of the region

More than 40,000 tourists visit the islands each year (Jackson, 1995). Tourism has impacts and benefits for the islands. One of the benefits is that it provides employment for local people. The fact that Galapagos has an organised system of nature trails and that the GNPS must approve itineraries of the tourist boats helps to keep to a minimum the negative impacts tourism may cause on the ecosystems.

The increasing number of tourists visiting the islands represents a challenge for the managers of the park, who must balance the need to conserve the islands in their natural state with the need to have a tourist industry to help the economy of the Galapagos Province.

Environmental education as a tool for effective management in Galapagos

The programmes of conservation in Galapagos must have the positive support of the estimated 15,000 people living in the five islands of the Province of Galapagos. The region needs strong environmental education programmes, with a focus on the need for conservation and sustainable development for the archipelago. The GNPS and the CDRS are working towards the strengthening of such programmes. Daily radio programmes, TV spots, workshops with teachers, a conservationist club for children and seminars with members of the local population are some of the activities carried out by these institutions. Effective environmental education programmes constitute an important tool for the effective management of this unique national park.

Source: contributed by P. Ponce (Ponce, 1995)

WILDERNESS MANAGEMENT

From the definitions of wilderness given and based on the previous discussion, it is quite apparent that a wilderness area is supposed to be one where human influence is absent (or at least kept to a minimum). By introducing the concept of 'management', however, we are implying human control over such areas, which causes many people to react negatively because they equate management with bulldozers and environmental manipulation. Wilderness area management, though, is the opposite and involves managing people in the interest of the natural environment. This is stated more fully as:

> Wilderness area management is essentially the management of human use and influence to preserve naturalness and solitude. It includes everything that a person responsible for a wilderness does in administering the area; for example, the formulation of the goals and objectives for individual areas and all policies, standards, and field actions to achieve them.

BOX 9.2   TOURISM AND THE GORILLA

Gorilla is a peculiar species in terms of habitat requirements. It is mainly a high altitude equatorial forest-dweller, a characteristic that makes it unique in its distribution and abundance. This uniqueness has created two major contrasting events which need to be discussed if sustainable development and protection of natural resources is the prime objective.

The first event is the development of tourist industry in countries that possess this peculiar natural wealth, the second is the need for economic development of that country. The two events seem to grow fast and at a similar pace. Gorilla is a major attractor of tourists, and governments are trying hard to tap every single coin this generates. In Rwanda, for example, mountain gorilla attracted so many visitors between 1988–1989 that all quotas were filled months in advance (Myra, 1995).

Other countries that benefit in the same way include Uganda and Zaire, which together attracted more than 72,000 visitors in 1992 (Myra, 1995: 68). This investment was capable of paying for the anti-poaching patrols conducted in relevant areas; it also provided employment for local citizens and thus reduced the continuous clearing of the forest for cattle raising. Furthermore, the investment contributed significantly to the economy of the country. In 1991 the government of Rwanda had a revenue of US $ 1m in entrance fees, ploughing back a very small fraction of this into the park for its general operations (Lindberg, 1991). However, the growth of tourism in Uganda, Zaire and Rwanda has been greatly affected by political turbulence and continuous ethnic wars. It is very difficult to ascertain what the status of the gorilla will be once the situation in these countries stabilises. It is predicted that the effect of these wars on gorilla will be severe, possibly to the extent of its outright extinction. The decline in tourist activities has led to a drop in revenue collection from conservation activities, a factor that has created another catastrophic problem for gorilla, in that people not previously engaged in agricultural activities will now need land for agriculture. Other catastrophic problems include the landmines already deployed, and to be deployed, in the forest, which are likely to destroy both the forest and the gorilla; famine as a result of continuous war has created more poachers than before, due to the emergence of a lucrative black market in wild meat; increasing trade in live animals (especially in gorilla babies sold to zoos); illegal grazing; illegal woodcutting for timber, firewood and building-poles.

These activities have contributed significantly to deforestation, soil erosion, loss of soil nutrients and fertility; they have also impoverished reproduction potential of the land, and consequently disrupted gorilla breeding cycles and their propagation. These factors, coupled with loss of habitat, have led to a decline in their numbers.

Though eco-tourism was possible in these countries before the civil wars, their current political instability threatens the survival of these endangered species, and to a greater extent the tourist industry that could have flourished thanks to the existence of gorilla. It is the duty of the countries concerned and the world at large to decide now how best this peculiar natural resource can be managed for the benefit of all; for survival of the species and for future generations.

Debate over conservation of mountain gorilla has so far shown that efforts are underway to protect the species from extinction. Currently the Dian Fossey Gorilla Fund is conducting surveys aimed at ascertaining the status of the species. According to Schreuder (1995) the impenetrable Bwindi forest, in the cool misty highlands of southwestern Uganda, is estimated to have about 300 mountain gorilla, a number that is believed to be about half of the world's remaining population. The remaining 350 mountain gorilla are thought to be living in the tropical rain forest covering Zaire and Rwanda. With only 650 mountain gorilla alive today, the task of saving them and their rain forest home becomes more urgent each day. Part of the effort to preserve this endangered species in its natural habitat rests in learning how well gorilla have been able to survive the devastating war in Rwanda and the refugee crisis in Zaire (Anon., 1995).

This definition introduces a key person, the 'manager', who is charged with the task of doing what is needed when it is needed in order to meet a given set of objectives for the wilderness. Successful implementation, however, depends on many factors, some of them outside the manager's control.

Whatever the case, management of wilderness areas needs to be guided by formal plans that prescribe the what, where, when, how and why of proposed management actions. It is through management plans for individual wilderness areas that the hierarchy of applicable direction, that is legislation, departmental regulation and agency policy, is translated into action. The above definition, although specific to wilderness areas, can be extrapolated to cover national parks and strict nature reserves which also fall under the category of protected areas.

Wilderness management is a technically demanding task with enormous responsibilities. The manager must understand ecological principles, appreciate the

---

BOX 9.3 THE AFRICAN ELEPHANT

African elephants are among the largest wild animals of the terrestrial ecosystem: males can commonly weigh up to 6 tons while females can range between 4–5 tons. The name 'elephant' comes from a Greek word *elephas*, which means ivory, referring to the animals' tusks. Both males and females possess tusks which are used to perform various tasks in their daily life – breaking twigs, removing tree bark when feeding, digging when searching for water during dry seasons, etc.

The African elephant is well distributed throughout Africa south of the Sahara desert, in the Savannah grassland and forests. Those found in forested areas live in Cameroon, Congo, Ivory Coast and other central and western African nations. Being herbivores or plant eaters, elephants normally feed on grasses, fruits, leaves, bark and twigs. Because of their large size they are bulk feeders, and as such about 60 per cent of what they take as food passes through without being digested. As a result of this elephants spend more than 16 hours a day foraging for nearly 160 kg of food, and drink about 60 litres (18 gallons) of water each day.

Basically, elephants are social animals, living in groups of small herds composed of either female groups (cows and young, led by an older, experienced matriarch), or bachelor groups (pre-adult males and older males, or 'bulls'). Both groups act as a training ground for the younger animals. The herds tend to work together in every aspect of life from feeding to taking care of young; when a member of the group dies, all other members might participate in mourning and covering the carcass with twigs.

Other peculiar characteristics are their lifespan and rate of maturation: elephants have a lifespan of 60 years and puberty is reached at 15. The gestation period is 22 months; a cow usually gives birth to a single calf (although very rarely twins do occur) and can give birth every four years. The reproductive cycles of this species usually correspond to the seasonal food and water availability.

The status of the African elephant is still questionable in that by 1978 they were estimated to number in the region of 1.5 million, but as of now there are estimated to be in the region of 600,000 remaining in the wild. A number of factors have been advanced to account for the decline in numbers, among them habitat destruction and fragmentation caused by the behaviour of the animals themselves, and that caused by the people living next to areas that form the habitat of elephants. The major factor always associated with the decline is poaching for ivory. This was highly viable in the late 1970s, when ivory trading was lucrative.

Several African nations have tried to implement conservation programmes, which have included the setting aside of protected areas where wildlife rangers could patrol and safeguard the few remaining elephants. However, poor facilities and limited resources for running these programmes made the situation even worse, by creating a rift between the local citizens and protected areas management. The political instability of many African countries makes it a big challenge for these countries to implement effective, long-term elephant conservation programmes, apart from the effort made in 1989 by the US Fish and Wildlife Services in petitioning for the upgrading of the species from 'threatened' to 'endangered'. The current CITES notion of allowing other African nations to utilise their stocks heralds a new era in the decline of the elephant population in the southern part of African. In view of the events in West Africa, political instability could be the major factor, coupled with the destruction of the forest for timber and energy. These two factors have contributed significantly to the decline in populations of the forest elephant, as opposed to the southern African species, which dominates the wooded grasslands, areas that are still available to them.

---

ecological processes operating in the protected area and accept the concept that protected area management is a specialised form of land use. Without management plans derived from an orderly planning process, wilderness management can be no more than a series of unco-ordinated reactions to immediate problems. Unplanned management can be recognised by a shifting focus from problem to problem as each becomes pressing, inconsistent and conflicting, and a loss of overall direction towards wilderness preservation goals ensues.

Formal plans protect the consistency and continuity of management direction even if philosophies, perceptions and definitions of wilderness differ widely between individual managers. Formal plans that establish clear objectives for an area, and the policies and actions by which objectives will be pursued, are essential if wilderness management is to be guided towards consistent outcomes. Good planning is central to good protected area management, but it is merely a management tool, not an end in itself. It is an ongoing process involving the formulation, submission and approval of management objectives, how these are to be attained, and standards against which to measure their achievement. Good planning leads to good management; poor planning or lack of planning impedes successful management. The first step in planning must be the formulation of clear, sensible objectives within the policy framework of the protected area management authority (Thorsell *et al.*, 1986).

It is now accepted as a basic principle of protected area management that every protected area should have a management plan. The management plan guides and controls the management of protected area resources, the uses of the area and the development of facilities needed to support that management and use. It facilitates all development activities and all management actions to be implemented in an area. A management plan is a valuable tool for identifying management needs, setting priorities and organising the approach to the future.

To manage protected areas, such as ecosystems, with any degree of efficiency and safety, the manager must first know and understand a great deal more about the way in which the various ecosystems operate. Six basic areas can be identified for which the manager will require accurate, scientifically collected, biological information before drawing up a comprehensive plan for the long-term management of the wilderness area:

(a) Inventory: what plants, animals and other natural resources are present?
(b) Quantification: how many of each species are present and how are they distributed in space and time?
(c) Ecological relationships: who eats whom? Competes with what? What depends on what?
(d) Species needs: as much information as possible should be gathered on the particular habitat requirements (shelter, food, minerals and water needs) of species of special management significance.
(e) Dynamics of change: studies are needed on colonisation of disturbed areas, succession of plant communities, changes in river flow, evolution of swamps, invasion by new species and population trends within species.

(f) Predictive manipulation of ecosystems: where the natural processes of change are contrary to the objectives of management, the manager will want to prevent change or effect its direction.

Linkage of protected area management with other disciplines

Understanding the human-dominated ecosystem sufficiently well to provide definite answers for business people, planners, administrators and legislators is dependent upon the information provided by or available from natural resources managers (of resources such as soil, forests, wildlife, fish, etc.) being integrated correctly. The complexity of ecological and biological problems in protected areas has meant new areas of specialisation demanding a wider spectrum of knowledge than formerly. Of course, ecologists and protected area managers are not expected to be specialists in all areas and need to be assisted by other people who possess different knowledge essential for an integrated approach to the tasks at hand.

It is clear that some aspects of current planning, economics, landscape architecture and architectural and engineering practice interact with those of the protected area manager. Although protected area managers may have scientific tools and theories based on ecological field observations and experimentation, to bring these analyses into a problem-solving framework for improving areas under management, they still have to rely on advice and input from other people. Managers must generally rely heavily on specialists such as archaeologists, anthropologists, oceanographers, limnologists, entomologists and soil scientists.

Dorney (1989) points out that a good environmental manager (this could be the manager of a protected area) focuses on ecological science and then puts it in direct and meaningful interaction with the disciplines of planning, landscape architecture, civil engineering, law and economics. These relate to landuse planning, natural resource policy and allocation or management of land. However, ability to integrate these separate inputs in a decision-making context is crucial for the manager. Thus, in protected area management, much as in total environmental management, it is desirable to have a multidisciplinary team for protected area planning and management because exponents of different disciplines can, and usually do, perceive the same problem in a different light as a result of their training and background.

Wilderness, national parks and reserve areas as sustainable ecosystems

The idea of the sustainable use of protected areas is gaining acceptance and widespread support from aid planners, governments, protected area managers and conservation groups. But what does it mean? The World Commission on Environment and Development (WCED), set up in 1983 by the United Nations, defined sustainable development as 'that [which] meets the needs of the present without compromising the ability of future generations to meet their own needs'. From a conservation point of view, many would argue that sustainable use should be defined in part by its effectiveness in maintaining key ecological attributes including:

(a) ecosystem processes;
(b) the structure of habitats;
(c) biogenetic diversity.

Ecosystem processes include soil formation and movement (wash, creep, etc.), the cycling of water and nutrients, the accumulation of nutrients in vegetation and animal 'biomass' and their movement, the respiratory function of plants (gases produced by microbes, photosynthesis of plants and so on) and physical attributes such as absorption or reflection of heat and light. Habitat structure means the structural complexity and diversity of the trees, spatial distribution of soil types, the associated microclimates, variation in vegetation structure along the sides of streams and so on. Biogenetic diversity and ecosystem complexity include species richness (the number of species present in the area compared to other areas), the degree of co-evolution (avoidance of competition over available resources, predator–prey relationships, lifestyle connections between, for example, bats and birds and certain plants via pollination, etc.), the genetic variety within species and the diversity between species groupings and associations which are found in particular communities.

It is very important to distinguish these ecosystem properties as they are differently affected by

management and this defines the degree to which a certain sort of management is 'acceptable' or 'unacceptable'. Sustainable utilisation is somewhat analogous to spending the interest and keeping the capital. The World Conservation Strategy points out that a society that insists that all utilisation of living resources be sustainable ensures that it will benefit from these resources indefinitely.

Over the years, African communities evolved a form of co-existence with the wildlife around them which permitted both to survive. The neglect of these survival strategies is a tragic loss which should be redressed in the future.

The concept of carrying capacity in protected areas

Any discussion about the sustainability of protected areas brings us to the important concept of carrying capacity. Carrying capacity is the ability of wilderness, national parks and reserves to absorb the impacts of use and retain their natural qualities. The use an area can tolerate without unacceptable impact occurring has a limit because as increasing use or damaging patterns of use develop at specific places or during particular times, those qualities that define the protected area's resources will disappear. As discussed earlier, change from natural ecological processes occurs inevitably in protected areas, be it wilderness areas or national parks. The objectives or policies guiding the management of such areas are not aimed at holding ecosystems in any kind of static condition, but at allowing natural change to occur with an absolute minimum of human-induced change. As described by Hendee *et al.* (1978), the standards of ecological integrity and solitude that are established for an area help define the carrying capacity of an individual protected area. For protected areas, the concept of carrying capacity has two important parameters:

(a) Physical and biological dimensions which describe the size and kind of ecosystem that can be sustained without undue evidence of unnatural impact.
(b) Social or psychological dimensions which refer to the levels and concentrations of human use an area can accommodate before the kind of 'naturalness' that helps define the wilderness,

national park or reserve area is diminished. Unless use is limited to levels an area can tolerate without changing unnaturally, the long-term goals of protected areas will not be achieved.

The importance of wilderness areas to environmental education

The success of wilderness management objectives depends partly on the ability of the reserve manager to communicate the conservation message to potential supporters. Moreover, the attitudes of local people (those most affected by the restrictions imposed by the reserve) mellow as they begin to appreciate the reserve and its benefits. Using the reserve for teaching purposes is of benefit to both the reserve and the local community and, through schools and youth groups, the management can extend the message to the younger generation through environmental education. Environmental education as discussed by sections 12, 13 and 14 of the IUCN 1980 World Conservation Strategy is the process of recognising values and clarifying concepts in order to develop skills and attitudes necessary for an understanding and appreciation of the interrelatedness among people, their cultures and their biophysical surroundings. Environmental education has been found to foster clear awareness of and concern about economic, social, political and ecological interdependence in urban and rural areas. It has also been found to provide every person with opportunities to acquire the knowledge, values, attributes, commitments and skills needed to protect and improve the environment and to create new patterns of behaviour towards the environment in individuals, groups and society as a whole.

Protected area managers should contact local schools directly or through local education offices to offer facilities for field excursions or classroom lessons about a reserve. Reserves may establish special facilities where organised groups of young people and others can attend appropriate education courses under the supervision of park staff.

Local village extension services

Usually the people living nearest to the reserve are the greatest potential threat to its integrity but they can also be the greatest asset for its protection. Winning the support of local people at the grassroots level and the speed with which this occurs will depend very much on whether local communities do indeed benefit from the park and not just lose access to resources that would otherwise be available to them. The manager must take the job of winning friends outside the boundaries of the protected area by carrying this message to surrounding villages. This process of extension work may include posting notices and posters, holding slide shows or film shows in villages, holding discussions with neighbouring land users or farmers' committees, and giving talks in local schools and to other concerned groups.

It is emphasised by Thorsell *et al.* (1986) that environmental education has to be wrapped in a conservation package and should be integrated into other types of extension materials with the aim of achieving multiple objectives such as:

(a) explaining why it is important to establish protected areas;
(b) showing why this particular area has been selected;
(c) indicating what benefits are derived by the local community and local economy;
(d) identifying the alternative sources of land and forest which villagers may exploit (if applicable) or, in some cases, explaining entitlement to compensation payments;
(e) developing a local sense of pride in the richness of local nature;
(f) emphasising the government's determination to make the reserve a success;
(g) pointing out that breaking the law for selfish ends is a community offence, not just an offence against the government.

Based on such guidelines and using protected areas as teaching grounds, environmental education will help to communicate the message about protected areas and people's receptivity to the concepts of conservation will be heightened.

CONCLUSIONS

Generally, the management of wilderness as discussed in this chapter carries much of the African experience especially in the way some of the terms have been defined. Much of the discussion is still quite relevant, however, to areas where natural vegetation forms the climax habitat and thus the wildlife present. At the same time, the messages and basic definitions contained in this chapter can also be extrapolated to suit other types of area that have, for example, secondary vegetation defining the characteristics of the climax vegetation but are still considered as wilderness and, as such, are maintained and managed following the same principles as have been applied to those areas discussed in the text.

REFERENCES

Achim, S. and Elisabeth, R. (1995) 'The commons without the tragedy?: strategies for community based natural resources management in southern Africa', Unpublished Background Paper presented during the SADC, March 1995, Annual Regional Conference on Natural Resources Management Programme, SADC Wildlife Technical.

Anon. (1995) *Save the Mountain Gorillas*, Englewood, Col., USA: The Dian Fossey Gorilla Fund.

Clawson, M. (1973) 'Multiple land use', in R.L. Reid (ed.) *Proceedings of the III World Conference on Animal Production*, Melbourne, Australia, May 1973, Sydney, Australia: Sydney University Press, pp. 1–15.

Diamond, J.M. (1984) '"Normal" extinctions of isolated populations', in M.H. Nitecki (ed.) *Extinctions*, Chicago, Ill., USA: University of Chicago Press, pp. 191–246.

Dorney, R.S. (1989) *The Professional Practice of Environmental Management*, L.C. Dorney (ed.) Berlin, Germany and New York, USA: Springer-Verlag, pp. 45–73.

East, R. (1981) 'Species-area curves and populations of large mammals in African savannah reserves', *Biol. Cons.* 21: 111–126.

East, R. (1983) 'Application of species-area curves to African savannah reserves', *Afr. J. Ecol.* 212: 123–128.

Granville, L.L. (1984) 'The survival of species genetic diversity', in J.A. McNeely and K.R. Miller (eds) *National Parks, Conservation and Development. The Role of Protected Areas in Sustaining Society*, Gland, Switzerland: IUCN, pp. 56–58.

Hendee, J.C., Stankey, H.G. and Lucas, R.C. (1978) *Wilderness Management*, Miscellaneous Publication No. 1365, USA: Forest Service, US Department of Agriculture.

IUCN (1980) *World Conservation Strategy: Living Resource Conservation for Sustainable Development*, Gland, Switzerland: IUCN-UNEP-WWF.

Jackson, M.H. (1995) *Galápagos, A Natural History*, Calgary, Alberta, Canada: University of Calgary Press.

Lee, H. (1992) 'Protected area standards', in *African People, African Parks. An Evaluation of Development Initiatives as a Means of Improving Protected Area Conservation in Africa*, Gland, Switzerland: IUCN, pp. 4–9.

Lindberg, K. (1991) *Policies for Maximising Nature Tourism's Ecological and Economical Benefits*, International Conservation Financing Project working paper, Washington DC, USA: World Resources Institute.

Lusigi, W.J. (1984) 'Future directions for the Afrotropical realm', in J.A. McNeely and K.R. Miller (eds) *National Parks, Conservation and Development. The Role of Protected Areas in Sustaining Society*, Gland, Switzerland: IUCN, pp. 137–146.

Miller, R.I. and Harris, L.D. (1977) 'Isolation and extirpation in wildlife reserves', *Biol. Conserv.* 12: 311–315.

Mishra, H.R. (1984) 'A delicate balance: tigers, rhinoceros, tourists and park management vs. the needs of the local people in Royal Chitwan National Parks', in J.A. McNeely and K.R. Miller (eds) *National Parks Conservation and Development. The Role of Protected Areas in Sustaining Society*, Gland, Switzerland: pp. 147–218.

Murphree, M. (1993) *Communal Land Wildlife Resources and Rural District Council Revenue CASS*, Harare, Zimbabwe: University of Zimbabwe.

Myra, S. (1995) 'Gorilla tourism in Rwanda', *Journal of Sustainable Tourism* 3, 2: 67–68.

Newmark, W.D. (1984) 'A land-bridge island perspective on mammalian parks', *Nature* 325: 430–432.

Newmark, W.D., Rutazaa, A.G. and Grimshaw, J.M. (1991) 'Local extinctions of large mammals within Kilimanjaro National Park and Forest Reserve and implications of increasing isolation and forest conversion', in W.D. Newmark (ed.) *The Conservation of Mount Kilimanjaro*, The IUCN Tropical Forest Programme, Gland, Switzerland: IUCN, pp. 35–46.

Odum, E. (1971) *Fundamentals of Ecology*, 3rd edition, London, UK and Philadelphia, Pa., USA: Saunders College Publishing, pp. 8–36.

Ponce, P. (1995) 'Teaching in Galapagos. A learning experience', *Loyola New Orleans Magazine* 5, 3: 39–40.

Schreuder, C. (1995) 'Debate over conservation of mountain gorilla', *Chicago Tribune*, 13 June, Chicago, Ill., USA.

Soule, M.E., Wilcox, B.A. and Holtby, C. (1979) 'Benign neglect. A model of faunal collapse in the game reserves of East Africa', *Biol. Conserv.* 15: 259–272.

Stanley, J. (ed.) (1993) *The Earth Summit: The United Nations Conference on Environment and Development (UNCED)*, London, UK: Graham & Trotman Limited, pp. 415–417.

Thorsell, J.W. (1984) *Managing Protected Areas in Eastern Africa: A Training Manual*, College of African Wildlife Management, Tanzania: Mweka, pp. 86–88.

Thorsell, J.W., MacKinon, J., MacKinon, K. and Child, G. (eds) (1986) *Managing Protected Areas in the Tropics*, Based on the Workshops on Managing Protected Areas in the Tropics, October 1982, Bali, Indonesia: World Congress on National Parks.

Western, D. and Ssemakula, J. (1981) 'The future of the savannah ecosystems: ecological island or faunal enclaves?', *Afr. J. Ecol.* 19: 7–19.

SUGGESTED READING

IIED/ODA (1994) *Whose Eden? An Overview of Community Approaches to Wildlife Management*, London, UK: International Institute for Environment and Development.

Estación Científica Charles Darwin (1994) *Galápagos Nuestras Islas*, Quito, Ecuador: Imprenta A & B Editores (in Spanish).

Hendee, J.C., Stankey, H.G. and Lucas, R.C. (1978) *Wilderness Management*, Miscellaneous Publication No. 1365, USA: Forest Service, US Department of Agriculture.

MacKinnon, J.K., Child, G. and Thorsell, J. (eds) (1986) *Managing Protected Areas in the Tropics*, Based on the Workshops on Managing Protected Areas in the Tropics, October 1982, World Congress on National Parks, Bali, Indonesia: IUCN.

McNeely, J.A. and Miller, K. (1984) *National Parks, Conservation and Development. The Role of Protected Areas in Sustaining Society*, Gland, Switzerland: IUCN.

Newmark, W.D. (1984) 'A land-bridge island perspective on mammalian parks', *Nature* 325: 430–432.

Odum, P.E. (1971) *Fundamentals of Ecology*, 3rd edition, Philadelphia, Pa., USA: Saunders College Publishing, Chapters 6, 8 and 9.

Steele, P. (1995) 'Ecotourism: an economic analysis', *Journal of Sustainable Tourism* 3, 1: 30–44.

SELF-ASSESSMENT QUESTIONS

1 What is environmental conservation as applied to protected areas?
   (a) The protection of unique scenic areas in their wild state and allowing intensive use for the rest of the area.
   (b) The management of the human use of the biosphere so that it may yield the greatest sustainable benefit to the present generation while maintaining its potential to meet the needs and aspirations of future generations.
   (c) The wise use of resources so that we may solve current economic, social and cultural demands.
2 According to the IUCN system for classification of protected areas, wilderness areas and national parks fall under Categories I and II respectively. What are the limits of human activities in these two areas?
   (a) Apart from the restriction on the contact use of the natural resources in both categories, other activities such as research and recreational and educational activities are allowed in each area.
   (b) In wilderness areas (Category I) human activities are restricted to recreation while in national parks (Category II) there is more human use and more manipulation of ecosystems to meet species management objectives and human needs.
   (c) In wilderness areas human activities are restricted to research, while in national parks they include research, education and recreation.
3 Define 'biological diversity' and list its three components.
   (a) Plants and animals of economic importance distributed in forest, agricultural, coastal and freshwater systems. Three components:
   ● endangered plants and animal species;
   ● plants and animals of educational and scientific value;
   ● animals that attract visitors and generate revenue for protected areas.
(b) Plants and animals that constitute essential ecological processes necessary to sustain human life and vital for its immediate needs such as food production, health and economic development. Three components:
   ● recycling of chemical nutrients by plants;
   ● cycling of oxygen and carbon occurring throughout the biosphere;
   ● breakdown of dead organic matter and regeneration of soils.
(c) The wealth of life forms on Earth, the millions of plants and animals and micro-organisms, the genes they contain and the intricate ecosystems they help build into the living environment. Three components:
   ● species diversity;
   ● genetic diversity;
   ● ecosystem diversity.
4 What is an ecosystem?
   (a) A biological unit consisting of a group of plants and animals interacting with each other.
   (b) A unit that includes all the organisms (i.e. the community) in a given area interacting with the physical

environment so that a flow of energy leads to a clearly defined trophic structure, biotic diversity and material cycles within the ecosystem.

(c) A terrestrial or aquatic environment whose ecological and biological functions are independent of other systems.

5 List three key ecological attributes.

(a) Biotic components, abiotic components and species diversity.

(b) Food chain, biological processes and energy recycling.

(c) Ecosystem process, the structure of habitats and biogenetic diversity.

6 What are the potential environmental effects of overcrowding from tourists in protected areas?

(a) Disturbance of wildlife, noise pollution and the mutilation and destruction of facilities and behavioural changes in animals.

(b) Environmental stress, changes in the behaviour of some animals, reduction in environmental quality.

(c) Irritation to visitors but not to animals, some ecological changes, soil and vegetation damage.

7 It has been argued that wilderness management should not mould nature to suit people. What alternative do you think management should take?

(a) Management of human use and influences so that natural processes are not altered.

(b) A kind of management that ensures that ecological processes are not interfered with and nature is left to take its own course.

(c) Management that takes care of natural attributes and does not destroy the wild aspects and scenery of the protected area.

8 With reference to protected areas, give a definition of sustainable development.

(a) Development that meets the needs of the present without compromising the ability of future generations to meet their own needs.

(b) Development that utilises the available resources to solve social, economic and cultural demands.

(c) Development that aims to preserve breeding stocks, population reservoirs and biological diversity.

9 What are the predicted effects of a reduction in species colonisation and the decrease in effective size of protected areas?

(a) Reduction in the population size of mammals.

(b) An increase in the number of local extinctions of mammals in a protected area.

(c) An increase in emigration of mammals.

10 What is environmental education as applied to protected areas?

(a) Education about the environment.

(b) Education that aims to provide every person with opportunities to acquire the knowledge and skills needed to improve the environment.

(c) The process of recognising values and clarifying concepts in order to develop skills and attitudes necessary to understand and appreciate the interrelatedness between humans, their culture and their biophysical surroundings.

10

RURAL ENVIRONMENTS

Pham Hoang Hai and Nguyen Ngoc Khanh

SUMMARY

Rural areas are important parts of any territory on almost every continent of the planet. They supply most of the cereals and food. Rural areas have their specific characteristics of spatial and territorial distribution of farmers' production, their living activities and of society organisation. These specific characteristics continuously affect the existence and development of the rural environment.

Rural environments are distinguished from urban and industrial environments by their characteristics, methods of study and the measures required to solve specific problems such as agricultural production, daily life, social and cultural standards, problems of soil, water and air pollution, the problems of use of nature and natural resources in rural areas.

The socio-economic development and cultural standards determine in each community, in each government, in each territorial region different impacts on the natural environment. As a consequence, mitigating measures and the ways to solve problems of rural environments are different too.

The problems of rural environments are diverse. Some of them need management at a global scale (starvation, squalid poverty, illiteracy, diseases, wars), while others need to be solved at the national or regional level or may require macro- or micro-activity strategies. However, the management of the rural environment and development needs to be paid attention by governmental and international organisations, and society as a whole.

ACADEMIC OBJECTIVES

After completion of this chapter, students should be able to recognise the main characteristics, to exchange ideas and to discuss problems of rural environments, resulting from human activities, production, economic and social developments. The research objectives of rural environmental studies are:

- to contribute to the general concept of rural environments;
- to reveal basic differences, both in methodological approach and in solving problems of rural environments in developed, developing and less developed countries;
- to study problems emerging from agricultural production, forestry, industry, transport, water supply, tourism and other socio-economic pressures in rural environments;
- to put forward measures and strategies for countries and for the world to conserve rural environments as well as for their sustainable development.

DEFINITION OF RURAL AREAS AND RURAL ENVIRONMENTS

Rural areas are defined as areas where mainly agricultural activities are performed by farming communities. Food and cereals are produced in the rural areas of the world to feed a growing global population which now stands at about 6 billion. These areas are relatively free from the congestion, overcrowding, air pollution, industrialisation, etc. of urban areas. Rural areas are also characterised by low-intensity infrastructure for road traffic, utilities (electricity, water supply, sewerage, etc.) and housing.

Farms and villages are the fundamental micro-economic units of rural areas where local, regional or national tradition and culture are preserved and passed on from generation to generation. Both family ties and community spirit are stronger in rural areas than in urban areas. Moreover, rural people have an innate attachment to land that is very characteristic of farming communities everywhere.

The standard of living in rural areas is generally lower than in urban areas, especially in developing countries where it can be very low in the mountainous rural areas. And this relatively lower standard of living in the rural communities often translates into greater social inequality, illiteracy, poverty, disease and a lower cultural standard.

The rural environment essentially entails an agro-ecosystem and a rural organisation whose characteristic problems emanate from the way in which natural resources are used for agricultural production. Problems can also arise from economic development, lifestyle and social evolution. Natural conditions have a substantial impact on specific aspects of agricultural production and on the formation of different agro-ecosystems. For example, agro-ecosystems in plains are flat, such as rice-fields on alluvial soils, while they are sloped, with different characteristics, in mountainous terrains. These characteristics can affect both crop production and agro-biomes of the ecosystem. The more favourable the ecological conditions (near the equatorial regions, for example), the greater the capacity for crop rotation and intensive farming. In general these examples show that rural environments are more closely connected to natural conditions than urban environments.

The condition and quality of the soil are the prime resources of rural areas. Quality of both air and water has an important effect on agricultural productivity and on the quality of agricultural products.

The main rural environmental problems result from soil structure and soil quality; availability of adequate water sources; favourable climatic conditions; sources of agricultural crop genes; use of fertilisers and insecticides; methods of cultivation; and problems associated with quality of life such as educational and cultural attainments of farmers, availability of health care of acceptable quality, availability of adequate sanitation, etc.

DIFFERENCES BETWEEN RURAL ENVIRONMENTS IN DEVELOPED AND DEVELOPING COUNTRIES

In developing countries agriculture is a significant (or even the most important) segment of the national economy, and the countryside generally accounts for a considerable part of the common social life of the nation. Farms are small in size, often divided into even smaller individual units measuring only 0.2–0.5 hectares (Tran, 1995), while methods of cultivation are mainly traditional, non-mechanised and rely on animal power and manual labour. In developed countries, on the other hand, farms are large or very large, often measuring thousands of hectares. Agriculture is mechanised and industrialised, and, most often, it is not the dominant sector of the national economy.

Of particular concern in the rural areas of developing countries is infrastructure, which is generally very weak. In particular, roads connecting villages are often unpaved and made of earth. They are subject to erosion and are at risk of being washed away, even by regular rain. Most of these roads were built for the exploitation of forests and mineral resources, and not planned to connect villages or settlements. In developed countries rural transport networks are generally built or upgraded according to a planned, overall infrastructure development programme.

While ready access to electric power is taken for granted in the rural communities of industrialised countries, electricity is lacking in most rural communities in developing countries. It is encouraging to note, however, that in Asia and elsewhere many of those communities are now increasingly generating their own electric power using micro-hydro, biogas and other methods (Ngoc Khanh, 1992).

In developed countries, social problems of rural communities such as public health, sanitation, cultural and educational needs, medical care, etc. are addressed by local or national governments. In many cases humanitarian and charity organisations also contribute substantially. But rural communities in developing countries enjoy such benefits to a much lesser extent, mainly because of the limited resources of local and national authorities. Moreover, farmers' income in the poor developing countries amounts to a fraction of that of their counterparts in developed countries. In many African countries, for example, the annual income of a farmer is still less than US$ 500 (*Statistical Yearbook,* 1995).

Environmental problems in developed countries are also different from those in developing countries. In industrialised countries the use of chemicals to increase crop productivity causes pollution of both soil and water, as well as the destruction of organic matter in soil; in developing countries excessive exploitation of soil without adequate scientific know-how leads to the depletion of nutrients. An estimated 6–7 million hectares of agricultural land has been lost in this way due to erosion, leading to desertification, while another 15 million hectares have lost their productivity for other reasons. Thus the world has lost about 20 per cent of its fertile land in the last 40 years (UNEP/IRSK, 1990). The current trend in developed countries is to ban the use of more harmful insecticides while gradually reducing the use of the less harmful ones. But these harmful chemicals are still widely used in the developing countries where they are causing local ecological and health damage.

RURAL ENVIRONMENTAL PROBLEMS

Agricultural production and environmental problems

Problems of hunger, poverty and food shortage are indeed pressing, especially in some of the developing countries, and appropriate solutions are urgently needed (FAO and IIRR, 1995).

However, problems of low food security and starvation occur unequally among different countries and often among different regions of the same country. Such problems are now concentrated in the developing countries around the Sahara Desert in Africa, and in parts of South-East Asia, South America and the Caribbean. The world agricultural growth rate is declining, with a present value of only 2 per cent compared with 3 per cent in the 1960s. Its expected value in 2010 is estimated at 1.8 per cent. Meanwhile, the world population is increasing at a rate of 1.5 per cent per year, and at a much higher rate in some regions and in some developing countries. In sub-Sahelian Africa, for example, it is 3 per cent per year. Given this scenario, a food crisis appears to be inevitable in the near future (FAO and IIRR, 1995)

The large demand for food, and the economic structure of the market have pushed up agricultural production through intensive farming and increased use of fertilisers, insecticides, etc. However, this trend is likely to cause (indeed is already causing) a wide range of pollution problems with serious adverse impacts on the natural environment, especially in the rural areas of developing countries. For example, excessive exploitation of agricultural soils by continuously rotating cultivation without fallow diminishes, and can even destroy, their self-restoring ability. This, combined with the relatively low level of applied scientific know-how in developing countries to deal with such problems, is a serious threat to the future of agricultural productivity in those countries. So far, about 15 per cent of the world soil has been degraded by human activities, while about 66 million hectares (representing about 30 per cent of total agricultural land area of the world) of irrigated land have been salinated by intensive cultivation (WRI, 1988). Every year an estimated 6–7 million hectares of agricultural land loses its productivity because of erosion and degeneration, while about 1.5 million hectares are waterlogged, acidified and salinated (Ngoc Chau, 1996). Soil degeneration, often leading to desertification, is a matter of serious concern in the dry regions of south Asia and around the Sahara Desert where an estimated 5.5 million hectares of agricultural land has been affected, representing a loss of tens of billions of US dollars per year (Ngoc Chau, 1996).

Besides cultivation, intensive animal husbandry practices and the use of agricultural machinery also influence soil degeneration, reduce the self-restoring ability of soils, and make pasture lands unsustainable. The problem is being compounded, especially in the tropical forest regions, by deforestation whereby forests are removed to provide grazing land for farm animals or for expanding agriculture. In particular, the ecological stability of millions of hectares of natural forests has been disturbed by the 'cut-and-burn' practice. This method of agriculture is especially damaging when it is carried out on the sides or tops of hills or mountains, or on the upstream areas of rivers. The result is that soil is washed out by erosion.

Application of science and technology to improve or increase agricultural production can also have adverse impacts on the environment. Excessive irrigation has already exhausted the available water resources in many relatively dry regions of the world. Also, excessive application of insecticides and chemical fertilisers with residual toxicity has polluted both soil and water in many parts of the world and influences the quality of agricultural produce. Perhaps more importantly, these practices have caused, and continue to cause, loss of biodiversity and increased risk of plant diseases. Humans have also been placed at risk of poisoning, diseases (such as encephalitis), lymph abnormalities, etc. by the excessive and improper use of pesticides.

In conclusion, agricultural activities and practices in developing countries can have serious adverse impacts on nature and on the environment. The major impacts are degradation and loss of agricultural soils, exhaustion of water resources, especially in relatively dry areas, loss of biodiversity, pollution of soil and water, and impacts on human health.

Most of these adverse environmental impacts also occur in the developed countries of Europe and North America. However, those countries are generally much better equipped to deal with such impacts, thanks to their advanced scientific and technological know-how and resource base. Moreover, regulations, directives and their enforcement machinery are in place to ensure that environmental impacts are minimised. Major problems arising in those countries are economic in nature, emanating from the overproduction of cereals and foodstuffs which is a major problem in the European Union. Problems also arise from the need to subsidise farmers, for example, under the Common Agricultural Policy (CAP) of the European Union.

Overproduction in developed countries often has important economic implications for developing countries: stringent conditions are sometimes attached to bilateral aid involving the export of foodstuffs, and this can depress the prices of local products to the detriment of farmers in the recipient developing countries.

With the support of international organisations and through bilateral aid programmes, significant progress has been made in developing countries in the last few decades towards increasing agricultural production and, at the same time, reducing the adverse impacts of agriculture on the environment. Farmers are now more skilled in important issues such as land management, conservation of soil productivity, plant protection, maintenance of natural genes, conservation of traditional plant and animal species, and protection of soil and water against pollution. Much still remains to be done to promote sustainable agriculture and sustainable development in the rural areas of practically all countries of the world. Particular attention needs to be given to the following, especially in the developing countries:

- Strategic plans for land use in agriculture.
- Concrete measures to raise the level of technology in cultivation and animal husbandry for controlling the use of pesticides and chemical fertilisers and for the preservation of the traditional natural sources of genes.
- Social and economic encouragement and support to promote agricultural production with due attention to environmental protection.

Forestry and rural environmental problems

Forests are a valuable natural resource. Forest products such as wood command a high value in the construction of dwellings, furniture, ships, etc., and as charcoal. Other forest products such as animals, flowers and fruits, medicinal herbs, etc. are equally important. Every year forests supply millions of cubic metres of wood and tens of millions of units of other products with a total value of several billion dollars.

Perhaps more importantly, forests play an important role in determining the climate and in ensuring natural circulation of moisture in the biosphere. As a rich source of carbon, they also perform the following major functions: limiting the spread of water flows; protecting soil against erosion and preventing the loss of the fertile surface; limiting sedimentation; and preserving earth's temperature regime.

Forests vary a great deal in both form and structure, extending as they do from coastal regions to the mountains, and from the tropics to the boreal regions. They can be evergreen or deciduous, small or broad-leaved, humid or dry, dense or open. The total area of primary forests of the world, together with grasslands with trees and shrubs, is estimated at 53 million km², which amounts to about 40 per cent of the earth's land area (WRI, 1988).

However, excessive (and often mindless) exploitation of forests to date, especially of tropical forests, is a matter of much concern. In spite of this the destruction of forests, and more worryingly the destruction of tropical forests, continues unabated. It is estimated that each year a tropical forest the size of Denmark is destroyed, mainly for timber, mining and ranching. Other forests are being lost or degraded for firewood, cultivation, urbanisation, road construction, etc.

The cumulative impact of deforestation is very serious, especially for the rural communities and the rural environment: deforestation leads to loss of biodiversity, loss of habitat of flora and fauna, loss of water retention capacity of soil, erosion and loss of the fertile topsoil, and changes in microclimate unfavourable to agriculture. Moreover, a reduced or degraded forest means a permanent loss of a natural source of wood, fruits and vegetables, meat, plants, herbs and other forest products.

There is clearly an urgent need for a strategic policy aimed at sustainable forestry management to ensure that forest products of economic value are exploited without damaging the essential integrity of forests, and that their natural biological capacity is preserved. Such a policy for sustainable management needs to be international. Moreover, the policy must have the full support of national authorities that are also responsible for its maintenance.

Soil degradation, pollution and land management in rural areas

While air and water pollution, and general environmental unsustainability caused by human activities are major problems in rural areas, the pollution of soil assumes particular attention in farming communities. The major causes of soil pollution and degradation are as follows:

(a) Inappropriate land use in agriculture, over-exploitation of land to maximise productivity, lack of technical know-how, inappropriate methods of cultivation, and insufficient attention to soil preparation and restoration.

(b) Excessive and inappropriate use of chemical fertilisers and pesticides which cause high levels of toxicity in both water and soil.

(c) Airborne pollution, as well as untreated or improperly treated solid and liquid wastes from industrial plants in and around rural areas, or from neighbouring industrial zones.

(d) Mining activities.

(e) Irrigation and multi-purpose hydroelectric works.

(f) Transportation including road construction.

(g) A wide range of natural and human causes such as drought, floods, inundation, cultural standards and level of socio-economic development.

(h) Degradation and removal of forests through excessive exploitation for short-term financial gain, and absence of effective measures to prevent soil erosion and surface washing, especially on slopes of mountains and hills.

In many places soil pollution is high, especially in some regions of Europe, Asia, and in North, South and Central America where untreated or partially treated industrial wastes, as well as the use of chemical fertilisers, pesticides and other man-made products without adequate technical safeguards, are responsible for the contamination and degradation of agricultural land.

Soil contamination or degradation has serious economic implications for rural areas where agriculture is the main activity. Moving towards sustainable agriculture and sustainable forest management is based upon an effective programme of action which addresses the major causes of soil pollution and degradation listed above.

Water supply and freshwater usage in rural areas

Water is a unique dissolving substance that is instrumental to the nutrient cycle of all living organisms and plays an important role in maintaining and combining ecosystems.

At present, the world's water resources are facing what is little short of a crisis. With growing demand for water and electric power, enormous efforts and large amounts of capital are being invested in the construction of dams and dikes to collect and conserve water, and to transport it to drier areas. There are notable examples of successful projects of this kind which bear testimony to technical achievements of the highest order. Multi-purpose hydroelectric projects in particular bring enormous benefits in terms of power generation, flood control, river flow moderation, water transport and irrigation. The Hoa Binh hydroelectric plant on the Da river in Vietnam, for example, supplies 40–50 per cent of the total power demand of the country; it also controls floods of the Da river, provides water for irrigation and facilitates water transport (Nguyen Thuong Hung *et al.*, 1995).

Agricultural areas of the world, irrigated by dams, dikes and multi-purpose hydroelectric projects, have increased nearly threefold since 1950; today they account for about one-third of the world's agricultural production (Ngoc Chau, 1996). Although constrained by shortage of funds at present, the world's agricultural land under artificial irrigation is nevertheless expected to increase to 23 million hectares by the year 2010.

Such projects bring economic benefit. In many cases, however, serious and unexpected or unpredicted damage to both national economy and the environment also occurs. One economic impact, which can sometimes be serious, is increased national debt; such large projects are usually financed with loans from the World Bank or with bilateral loans that have to be repaid with interest. Environmental damage can be caused by removal of, or disturbance to, the natural habitats of estuarine fishes and marine organisms; erosion of river banks and of coastal lines; reduced deposition of alluvial matter in the downstream areas; removal and resettling of population to make way for the impounding reservoir and the environmental consequences thereof; loss of agricultural land; salinisation of estuaries; and the loss of entire farming communities together with their traditional way of life, customs and practices.

In artificial irrigation only about 40 per cent of the surface water is used effectively, while the remainder is lost due to evaporation and seepage. This evaporation leaves a polluting saline layer on the soil surface which affects about 25 per cent of the irrigated area (Pescod, 1992). The use of low-efficiency irrigation systems without adequate technical know-how has led to salinisation and degradation of once fertile agricultural land in many parts of the world. Moreover, the depletion of water resources and declining availability of water is also a matter of increasing concern. In India, for example, the average per capita water consumption has declined from 5,000 m^3 in 1955 to 2,500 m^3 in 1990, and it is expected to reduce further to 1,500 m^3 by the year 2010 (Pescod, 1992). Furthermore, diversion of rivers in central Asia in the 1960s to increase the production of cotton has led to the halving of the size of the Aral Sea (Ngoc Chau, 1996), and to serious and irreversible contamination of a large area of land there. This is a typical example of the massive damage that can be inflicted on the environment through ignorance, inadequate know-how and the rush for short-term profit.

Intensive farming and intensive use of chemical fertilisers and pesticides to increase crop productivity are the main reasons for water pollution in agriculture. It is interesting to note in this context that in some of the newly industrialised Asian countries, such as Thailand, Taiwan, Indonesia and the Philippines, the level of fertiliser and nitrate pollution of water has already reached that of the industrialised countries.

The quality of the world's water is generally declining through pollution and overuse. This is a matter of particular concern in the low income countries where nutrients from wastewater and manure continue to pollute water sources, sometimes causing eutrophication of water bodies. In Thailand, for example, raw manure from cattle breeding farms is discharged into rivers. This leads to rapid and abundant growth of water hyacinths and different types of algae (Ngoc Chau, 1996). Fish and other aquatic animals are adversely affected, and the diversity of aquatic ecosystems is reduced. The productivity of fisheries on rivers, lakes and ponds, on which many in the low income countries depend, is also reduced. Discharge of untreated or partially treated waste into rivers and lakes by nearby industrial plants also compounds the problem considerably.

As a result of shortage and pollution of surface waters, groundwater is now being depleted through increasing abstraction, and this is a cause for concern. For example, groundwater in Jordan is expected to be exhausted in the next 40 years, while a source beneath cereal fields in the mid-east region of the

USA can only be used for the next 20 years (Ngoc Chau, 1996). Also, over-exploitation of groundwater sources using pumps depresses the groundwater level as wells are dug deeper and deeper. In India for example, this causes serious salinisation. In Tay Nguyen, which is the central plateau of Vietnam, the level of groundwater has been so lowered by over-abstraction with suction pumps that many local farmers have had to abandon their coffee plantations (Hoang Niem, 1995).

Increasing shortage and pollution of fresh water in the rural areas of developing countries is of great concern. This is mainly because traditionally, people in the farming communities collect water from streams, rivers, lakes, wells or ponds; it is then used without any treatment for all purposes including drinking and cooking. Maintenance of an acceptable water quality at the source is very important. But unfortunately the waters of streams, rivers, lakes and ponds are polluted by human activities to an extent that impacts on human health. Moreover, water quality at source varies seasonally too, and often depends on the stretch (upstream or downstream) of the river or stream, extent of vegetative cover of the watershed, physical and chemical characteristics of watersoil, and on the type and intensity of anthropogenic activities on the watershed. Traditional cultivation practices and customs (such as bathing and washing in streams and ponds) can also contribute to water pollution.

Also of much concern is the increasing shortage of fresh water in high mountainous regions, especially in the dry seasons, and salinisation of water in the rural plain and coastal areas. Sometimes water has to be imported from other regions, as in some of the coastal districts near Ho Chi Minh City in Vietnam for example, because much of the drilled water in those districts has been seriously salinated (Hoang Niem, 1995). Furthermore, the relatively large population growth rate in the rural areas of the developing countries of south Asia and Africa makes the problem even more serious.

Problems of urbanisation in rural areas

Urbanisation of rural areas is a current developmental trend found in many countries of the world. This creates problems emanating from the relocation or expansion of production and service industries in the rural areas. These problems are of particular importance in the developing countries of Asia and South America. Although urbanisation tends to be efficient in terms of land use, it nevertheless puts agricultural land under pressure of urban development. In contrast, in many cities of Africa there is a growing trend of 'ruralisation of cities', arising from the fact that many of the poor city dwellers are obliged to grow as much food as possible for their own consumption or for sale.

The world's overall rural population is about 80 per cent of the total, although this share may be more than 90 per cent in some of the low income agricultural countries. The development of new urban centres offers, at least theoretically, the opportunity to concentrate populations in urban areas equipped with necessary social organisation and infrastructure (e.g. roads, housing, utilities). But this kind of urbanisation often reflects the traditional rural structure, that is the village structure: each element of the structure comprises a house for living in, a garden and a pond (or a swimming pool). The size of each element is only about 0.2–1.0 hectares in low income developing countries, while it is much larger in the developed countries.

Such planned and new urban centres require capital investments. This is often a problem in low income developing countries, especially for rural communities in the mountainous regions, and can only be solved by improving the prosperity of the entire national community. Furthermore, due attention must be given to the proper management of the characteristic urban environmental problems of wastewater, solid waste, public health, etc. in the planning of cities.

High population growth rates, which can reach 3 per cent in the rural areas of low income developing countries and more than 4 per cent in the mountainous regions of those countries (Cleaver and Schreiber, 1992), are a serious problem. In particular, high population growth rate is putting agricultural land in both mountainous and delta regions under increasing pressure because of growing demand for cereals and food. There is also growing demand for water, education, health care, and cultural and social facilities. The creation of new, well-planned urban centres might address these problems. Wider access to education and family planning, which these centres would make possible, would help curb the current high population growth rate.

Industrialisation of rural areas

Industrialisation of the rural areas of low income developing countries is a relatively recent phenomenon. The need to increase both agricultural productivity and farm efficiency provides the impetus for rural industrialisation. Application of technology for hoeing, manuring, sowing, transplanting and harvesting has led to the mechanisation of agriculture, even in remote areas. Agricultural mechanisation has created a whole range of service industries, such as for cattle food processing and distribution, fertiliser processing and distribution, preparation of special soils for high-productivity cultivation, micro-organic manure production and distribution, etc.

Food processing industries constitute the other major element of industrialisation of rural areas. These diverse industries, which process meat, fish, milk, fruits and vegetables, produce finished or semi-finished products from local raw materials. In addition to adding value to local produce in this way, these industries make full use of both local labour and farmers' free time. The rural community as a whole benefits as a result.

The processing of semi-finished goods, development and production of handicrafts, and the production of traditional local commodities continue to contribute significantly to raising farmers' incomes in many of the rural communities in low income developing countries. Both out-of-school children and retired people continue to benefit from such activities, and all these activities contribute to decrease poverty and starvation in many of the poor regions of low income developing countries.

Nevertheless, there is an urgent need for the developed countries to support the poor in the developing countries in this process of rural industrialisation by supplying necessary capital and technical know-how. Aimed primarily at increasing the food fund, such support would alleviate poverty and hunger.

Rural transport and communication

Together with population growth, industrialisation and modernisation, the provision of adequate transportation and communication facilities is now becoming increasingly important to meet the growing demands of travel, transportation of goods, recreation and tourism, and for servicing the economy in general. Rural transportation and communication facilities are developing rapidly. The result is that the provision of infrastructure for the development of transportation and communication facilities, such as construction of new roads and extension or upgrading of old roads, bridges, airports, etc. and increasing volumes of traffic, are creating a whole range of environmental problems that were either absent or of little concern hitherto in many of the rural areas, even in the mountainous rural areas.

Apart from the loss of agricultural land to road-building and road upgrading which improved transportation entails, exhaust from motor cars, trucks and motorcycles containing harmful gases such as CO_2, NO_x, hydrocarbons, airborne lead (from petrol) and volatile organic substances pollute the air. Pollution resulting from this sector contaminates soil, water and air, thus degrading the overall quality of the environment.

However, the impact of transport on the environment depends on the level of economic development in the country in question. While these impacts are limited in poor, developing countries, they are much more marked in the industrialised countries of Europe and North America. And even in developing countries, such impacts are becoming increasingly serious, due to the rapid development of transport and communication infrastructure. World-wide, an estimated half a million hectares of agricultural land is being lost each year to the provision of transportation and communication infrastructures. Data on environmental impact are, however, scarce and often unreliable in many developing countries. Systematic research and monitoring is needed, to establish how and to what extent environmental impacts of transportation and communication development are impinging on the quality of life, so that appropriate remedial measures can be undertaken.

Tourism and rural environmental problems

Tourism is an important sector of the economy which is growing fast in many countries of the world. As yet unexplored areas and regions of outstanding scenic beauty, picturesque lakes and landscapes, rich forests, natural mineral springs, beaches, etc. in many of the developing countries offer much potential for development as holiday resorts and tourist sites.

However, like other productive industries, expansion of tourism also has adverse impacts on the environment. Construction of holiday resorts and tourist facilities has resulted in the loss of forests and agricultural lands. Tourism, which is characterised by high seasonal concentrations of tourists, often degrades the surrounding environment through the cutting or removal of forests, by making further demands on agricultural lands, and through pollution caused by solid and liquid waste generated by tourists. Moreover, tourist sites and objects, which attract tourists in the first place, are in danger of being damaged or even destroyed by the sheer number of tourists visiting them. This is an increasingly serious problem in many countries such as Greece, Turkey and India.

When new tourist sites and facilities are developed this should be compatible with the requirements of sustainable tourism. Only in this way can national heritage be preserved for posterity, thus securing the long-term economic benefits of tourism. But this can only be done with determination and a clear focus on the long term. It also needs public education on matters of environmental protection, together with strong arguments to convince the operators of tourist sites that it is in their long-term interest to aim for sustainable tourism development, even though it might not appear attractive to them on a short-term basis.

Living conditions, education and culture in rural areas

At present there are some 200 million ethnic people, accounting for about 4 per cent of the world population, living in communities with their own culture and their own historic versions of land possession rights (Pacey, 1980). In general these communities are small and remote from major population centres. People living here have their own world-view, often live in harmony with nature, and derive their livelihood from hunting, fishing, gathering and grazing.

The income of people in rural communities in developing countries, especially ethnic communities, tends to be low as compared with their urban counterparts. As a result, the caloric value of their daily diet, consisting mainly of vegetables, is about 8,500 kJ (*Statistical Yearbook*, 1995) which compares with more than 14,500 kJ per day per person in developed countries, where people in the rural communities enjoy far greater incomes with a much higher standard of living. The quality of life of rural people in developing countries is also low because of problems of nutrition, water supply, health care, housing, sanitation, employment, education, etc., and these problems are becoming increasingly more serious because of the high population growth rates. Their difficult situation is often aggravated by foreign debt and civil wars which sometimes erupt from local, regional or national rivalries, as in parts of south Asia, and in Rwanda, Sierra Leone and Liberia among others.

The impact of these deprivations is obvious in the rural communities of many developing countries. Expenditure of much manual labour to earn less than a living wage, coupled with low-energy diet, translates into low life expectancy – less than 45 years in many instances, as in Guinea-Bissau, Mozambique or Guinea. Educational deprivation is reflected by high illiteracy rates, as in Sierra Leone (79 per cent), the Central African Republic (77 per cent), Burkina Faso (82 per cent), Guinea and Somalia (76 per cent), Zambia (73 per cent) and in parts of Asia and South America.

Provision of adequate community health care presents a problem that is difficult to solve, as will be seen from typical data presented in Table 10.1 (Pacey, 1980). The cause of the problem is that qualified medical doctors prefer to work in urban areas, where earnings are more lucrative, than in poor rural areas. A workable solution to the problem entails targeted community programmes with emphasis on primary and preventive care, together with family planning advice. Traditional remedies using herbs and medicinal plants also have an important role to play.

Table 10.1 Number of patients per medical doctor in some developing countries

Country	Number of patients per doctor
Afghanistan	8,935
Bangladesh	5,762
Bhutan	9,100
Hong Kong	919
Nepal	19,895
Thailand	4,479

Less than 20 per cent of the rural population in many parts of Africa, Asia and South America has access to drinking water, which is a major problem in the rural communities of developing countries. In some of the more developed countries this proportion rises to about 50 per cent, which compares with 75–80 per cent in the developed countries of Europe and North America (Pacey, 1980).

The problem of rural electrification in developing countries is gradually being solved. Whereas rural communities in developed countries are connected to the national electricity grid, their counterparts in developing countries enjoy access to electricity only if they are located in the low regions or in the plains. Supply of electricity to rural communities in the mountainous regions is still a problem. Increasingly these communities are using mini-generators and sometimes micro-hydro and other methods to produce their own electricity.

Traditional culture and religions are preserved and practised in the rural communities of developing countries, and passed on from generation to generation. The debit side of this is that superstitions and unhygienic habits and practices still persist, especially on matters relating to health. As a result of this and of scarcity of modern medicine and medical advice, large numbers die each year from malaria, typhoid and tuberculosis, mainly in rural communities in the mountainous regions. Life is especially difficult for women, and for children who also suffer from malnutrition. The quality of life of these communities is further diminished by natural calamities such as earthquakes, volcanic eruptions, typhoons, avalanches, floods, droughts, etc.

MANAGEMENT AND DEVELOPMENT OF RURAL ENVIRONMENTS

The formulation and implementation of an appropriate agricultural policy aimed at increasing productivity, with minimal or preferably no damage to the environment, is central to the overall development of rural communities in developing countries. Only in this way can living standards be raised and quality of life enhanced, in accordance with the requirements of sustainable development. It is proposed that the aforementioned policy should contain the following major elements:

1 Agricultural production and increased productivity should be given priority, and cultivation should be restricted to best available agricultural land.
2 Crops for cultivation should be selected carefully for their perceived demand, and with attention to the extent to which they could be harmonised with local ecological conditions. Crop cultivation should be combined with cattle-breeding where possible, to promote biological productivity.
3 Forests should be planted on degraded and deforested areas to prevent or reduce soil erosion.
4 Agroforestry and agro-industries should be encouraged with due attention to environmental considerations.
5 Biological pest control methods should be adopted, and crop species resistant to pests should be selected for cultivation.
6 Application of chemical fertilisers and pesticides should be strictly controlled and minimised. They should be applied strictly according to manufacturers' instructions.
7 Participation in international programmes aimed at preserving genetic materials of both crops and farm animals should be encouraged.
8 Farmers' knowledge and understanding of water resources and their management for ensuring sustainable supplies of acceptable quality should be upgraded. Collaboration for water use and management with relevant organisations in countries sharing waters of the same river(s) should be strengthened.
9 Rural communities should be encouraged to adopt modern agriculture without deforestation

and should be educated about the ecological benefits of protected forests. Steps should also be taken to inform them about the need for and importance of environmental protection and about the fundamentals of sustainable agriculture.

10 An early warning system should be provided to alert rural communities, especially those in the remote mountainous regions, to impending floods, typhoons and other natural disasters.

11 The necessary steps should be taken to protect the rights and interests of minorities, and to provide both capital and know-how for developing and upgrading cottage industries, especially in backward rural areas. Greater resources should be allocated to rural areas, and every effort should be made to develop income-generating projects to benefit local communities and to expand the market for locally produced goods.

12 Work should be undertaken in the rural communities to gather reliable socio-economic statistical data on literacy rate, infant mortality rate, birth rate, etc., because such data are an essential prerequisite of good planning.

13 Ecologically sound urbanisation of rural areas should be undertaken, and essential infrastructure should be provided without damaging the environment.

14 Every effort should be made to ensure the widest possible public access to at least primary health care, especially family planning advice and immunisation of children against childhood diseases.

BOX 10.1   RURAL AREAS IN NORTH VIETNAM

The north-west mountain area, upstream of the Da river basin, is the largest of Vietnam's high regions. The landscape is characterised by steep slopes. The area covers 3,610,140 ha of nature and includes the Fanxipan mountain (3.143 m). Huge variations in the relief make it most difficult to use the area for agriculture or other developments. As a consequence, this area has the lowest development level in Vietnam.

The climate in this area is tropical. The region has two distinct seasons. The summer or rainy season is very hot and wet and lasts from May till September. During this period there are 80 to 100 days of rain providing 2,100 mm of rainfall upstream of the Da river. The winter is cold and dry. The frost obstructs agricultural and forestry activities.

Agriculture and forestry
In this region there are, however, cultivated crops and forest plants used as medicine or food:

- The arable land area in the north-west covers 323,295 ha which represents 9 per cent of the area.
- There are 476,544 ha of forest land which occupies 13.2 per cent of the region.
- The open areas cover 2,464,326 ha (68.2 per cent).
- The remaining space of 345,975 ha occupies 9.6 per cent of the region.

Community
In the region there are 30 communities with 2,051,700 inhabitants. Of these, 965,050 constitute the active part of the population. They are employed in the following sectors:

- Agriculture – forestry labour: 76.6 per cent.
- Service – handicraft industry: 5.2 per cent.
- Government employees: 18 per cent.

Social conditions
Education in the region is very limited in comparison with other regions in Vietnam. Of the active population, 49.2 per cent are women. Most villages have no medical station. The road density is 0.1 km/km^2. In 1997, 64 out of the 526 villages had no roads.

Production is falling back. Most of the communities consist of nomads who only produce to support themselves. People of 304 villages, spread over 1,697 groups (461,367 inhabitants) still live as nomads. They basically perform agriculture and forestry on the sloped land.

BOX 10.2   AGRICULTURAL INDUSTRIALISATION IN JAPAN

Japan is a OECD country and has an industrialised agricultural and forestry sector.

Land use in Japan (1993)
The total surface of 37,652,000 ha is used as follows:

- Annual arable land and permanent crops land: 4,463,000 ha (12 per cent).
- Pasture land: 661,000 ha (2 per cent).
- Forest and woodland: 25,100,000 ha (66 per cent).
- Other land (including urban sites): 7,428,000 ha (20 per cent).

During the period 1983–1993 farmland in Japan has been reduced by 343,000 ha (0.9 per cent).

	1983	1993
Agricultural land (ha)	4,806,000	4,463,000
Agricultural land as a percentage of total land	12.8%	11.9%

Farmers in Japan
During the ten years 1984–1994, the population in Japan increased by 4,771,000 people, while the population in the agricultural sector decreased by 4,093,000 people (or a 3.6 per cent decrease of the agricultural group).

	1984	1994
Total population	120,044,000	124,815,000
Agriculture population	10,260,000	6,167,000
Percentage of total	8.5%	4.9%

The agricultural land decreased by 7.1 per cent while the agricultural population decreased by 39.9 per cent. Industrial development and urbanisation had a negative effect on employment in the rural sector. So Japan lost not only agricultural land, but also farmers.

Agricultural input indicators
During the ten years 1983–1993, mineral fertiliser use decreased substantially. In 1983, Japan used 2,098,000 tons of fertiliser, while 1,817,000 tons were used in 1993. This coincides with a decrease from 436.5 kg/ha (1983) to 407.1 kg/ha in 1993.
   Also the number of tractors is indicative of the increasing mechanisation of the sector. In 1983 there were 1,584,300 tractors in Japan. Their number increased to 2,041,000 in 1993.

Agricultural output
Substantially fewer farmers produce, however, slightly increasing amounts of rice, pork, meat and aquaculture products.

	1984	1994
Rice paddy production (1,000 t)	14,848	14,976
Rice paddy: yield (kg/ha)	6,414	6,770
Cereals: total production (1,000 t)	16,012	15,787
Pig population (1,000 heads)	10,423	10,621
Meat: total production (1,000 mt)	3,280	3,334

	1983	1993
Total fisheries production (1,000 t)	11,254.7	8,128.1
Aquaculture production (1,000 t)	1,163.1	1431.9

→

In conclusion, the agricultural area in Japan shrank slightly and the number of farmers diminished substantially, while the overall production remained constant and subject to the offer and demand logic. More recently these conditions have been realised with lower inputs of energy and resources. This evolution is characteristic of the agricultural sector in most industrial (OECD) countries and is indicative of the fast industrialisation of the agricultural sector. This evolution is most different in developing countries.

CONCLUSIONS

Rural environments, which face environmental problems associated with agriculture, are essentially different from urban environments. Since agriculture is the mainstay of rural socio-economic life, especially in developing countries, it is clear that the socio-economic development of rural communities is contingent upon agricultural development in the first instance. This implies increasing agricultural productivity, improving the quality of agricultural products, practising efficient animal husbandry, prudent use of water resources and their proper management, efficient marketing of agricultural products and so on. But to achieve sustainability in the long term, all these activities must be underpinned by sound ecological considerations, so that the integrity of rural ecosystems is preserved. To secure sustainability in the long term would necessitate the provision of appropriate programmes to upgrade farmers' knowledge and understanding of how best to perform their activities; it is equally important to raise public awareness of environmental issues, problems and best practices.

Appropriate mechanisms should be put in place, especially in the rural communities of developing countries, aimed at detaching them from superstitions, habits and practices that are ultimately damaging to them.

Given that by far the majority of people in the world live in rural communities, the importance of educating and informing them about modern developments, especially with regard to health, family planning and sustainable development, can hardly be overstated. Also, it is difficult to see how the deteriorating quality of life of the rural communities can be improved without implementing the above-mentioned policies as a matter of urgency.

REFERENCES

Cleaver, K.M. and Schreiber, G.A. (1992) *The Population, Agriculture and Environment Nexus in Sub-Sahara Africa*, Washington DC, USA: World Bank, XVII.

FAO and IIRR (1995) *Resource Management for Upland Area in Southeast Asia. An Information Kit*, FARM Field document No. 2, Bangkok and Silang, Philippines.

Hoang Niem (1995) 'An assessment status use of water resource by conception ecology and sustainable development', *Environmental Conservation and Sustainable Development, 1995*, International scientific conference, Hanoi, Vietnam, Represented Papers, Vols I, II (in Vietnamese).

Ngoc Chau (1996) 'The starving is threatening our planet', *New World* 181, Hanoi, Vietnam (in Vietnamese).

Ngoc Khanh (1992) *Strategy for Scientific and Technological Development of Vietnam. Based on Geographical Conception*, Hanoi, Vietnam: National Center for Natural Sciences and Technology (in Vietnamese).

Nguyen Thuong Hung, Pham Hoang Hai and Nguyen Ngoc Khanh (1995) *Environmental Impact Assessment of Hoabinh Hydro-Electrics Plant (in North-western Region of North Vietnam)*, Governmental Programme on Environmental Conservation (in Vietnamese).

Pacey, A. (1980) *Rural Sanitation*, London, UK: Intermediate Technology Publications.

Pescod M.B. (1992) *Waste Water Treatment and Use in Agriculture*, Rome, Italy: FAO, XIV.

Statistical Yearbook (1995) Hanoi, Vietnam: Statistical Publications, pp. 395–404.

Tran, D. (1995) *Family's Farm in Vietnam and on the World*, Hanoi, Vietnam: Chinh tri–Quoc gia Press (in Vietnamese).

UNEP/IRSK (1990) 'Global assessment of soil degradation', UNEP/IRSK, Nairobi, *Caring for the Earth. A Strategy for Sustainable Living*, Hanoi, Vietnam, 1993: Khoa hoc & Ky thuat Publishers (in Vietnamese).

WRI (1988) 'World Resources 1988–89, an assessment of the resources base that supports the Global Economy', *Caring for the Earth. A Strategy for Sustainable Living*, Hanoi, Vietnam, 1993: Khoa hoc & Ky thuat Publishers (in Vietnamese).

SUGGESTED READING

Agriculture, Environment and Society (1992) Crow Nest, New Zealand: Macmillan Co., XIV.

Dudley, N., Madeley, J. and Stolton, S. (1992) 'Land is life', *Land Reform and Sustainable Agriculture*, Intermediate Technology Publication, pp. 43–53.

FAO (1993a) *Agro-Ecological Assessment for National Planning, the Example of Kenya*, Rome, Italy: FAO, Chapters 1 and 8.

FAO (1993b) *Women in Rural Saving and Finance*, Bangkok, Vol. 19.

FAO (1994a) *Farm and Community Information Use for Agricultural Programmes and Policies*, Rome, Italy: FAO, pp. 31–47.

FAO (1994b) 'Sustainable agriculture and rural development (SARD) new directions for agriculture, forestry and fisheries', *Strategies for Sustainable Development*, Rome, Italy: FAO, pp. 5–15.

FAO (1994c) 'Policies for sustainable development, four essays', *Economical and Social Development Paper* 121, Rome, Italy: FAO, pp. 39–52.

Sainteny, G. (1992) 'La crise du monde rural, la nature et l'import', *Futuribles* 170: 21–38 (in French).

SELF-ASSESSMENT QUESTIONS

1 Which development tendencies are different in rural areas in industrialised and developing countries?
   (a) Amount of fertilisers used.
   (b) Number of farmers.
   (c) Land surface used for agriculture.
   (d) Quality of the agricultural yield.

2 Indicate the causes that degenerate or pollute land in rural regions.
   (a) Industrial agricultural practices resulting in erosion.
   (b) Industrialisation and urbanisation.
   (c) Socio-economic development without environmental planning.
   (d) Any development increasing the production.

3 Indicate the correct statements on water use in rural areas.
   (a) Excessive irrigation can cause problems because it refills irrigation water reserves.
   (b) Deforestation can result in a loss of water retention in the soil.
   (c) In some developing countries the level of nitrification of water is comparable to the levels of nitrate pollution found in OECD countries with an industrialised agriculture.
   (d) Deep groundwater reserves are virtually unlimited.

4 Indicate the correct statement(s). The development of rural transportation systems
   (a) marginally interferes with the economic development;
   (b) pollutes NO_x and hydrocarbons and is therefore a main contributor to high tropospheric ozone concentrations in rural areas;
   (c) uses neglectable amounts of land;
   (d) is as a rule well accompanied by mitigating measures.

5 The potential of rural tourism is influenced by
   (a) the scenic beauty of a region;
   (b) the presence of natural springs;
   (c) a long-term management plan.

6 Indicate elements of stimulating environmental management in rural areas.
   (a) Increased use of fertilisers to enhance the production.
   (b) Forest planting to reduce soil erosion.
   (c) Preferentially plant species that are well selected for their disease resistance.
   (d) Adopt biological pest control strategies.

I I

URBAN ENVIRONMENTS

Dimitri Devuyst

SUMMARY

The city is a very complex human environment for which many definitions exist. Contemporary urban areas are integral centres of production and communication in a highly interdependent global network. Cities are growing rapidly these days, and environmental problems are often the result. No two cities are alike – major differences exist between North and South, East and West – with the result that there is no prescribed set of measures for environmental management. Traditional environmental problems discussed in this chapter are air, water, noise, solid waste disposal and land contamination. In addition to these problems we shall also consider the built-up areas. In the past, several traditions were established to solve urban problems by way of planning schemes. Alternatives to traditional planning lead to concepts of healthy cities, liveable cities and sustainable communities, which can be considered as the most comprehensive way to urban environmental management. The major goal of these approaches is to improve the quality of life for the urban resident.

ACADEMIC OBJECTIVES

The aim of this chapter is to discuss the complex urban environment, to give an overview of urban environmental problems and to introduce concepts of urban planning in addition to urban environmental management.

On completion of this chapter the reader should be able to:

- discuss differences in definitions of urban areas;
- give an overview of the complexity of the urban environment;
- discuss differences between less developed and more developed regions in the context of urban growth and urban planning;
- recognise aspects of urban environmental pollution;
- identify alternatives to traditional planning;
- introduce urban environmental management.

DEFINITIONS OF URBAN ENVIRONMENTS

Cities are most frequently defined as 'places where residents are primarily engaging in non-agricultural activities' (Dogan and Kasarda, 1988)

The city is a very complex human environment. A little research indicates that there are as many definitions of the urban environment as there are authors, as confirmed by the fact that there is no single definition of the word 'city' that scientists can agree upon. At one

level, the city is a collection of buildings and roads and their associated transport, communication, water and sewage systems – the hard infrastructure. However, this is more a description of an archaeological site than a city. Clearly, a city is more than bricks and mortar. The common elements that describe a city are permanent residents and a large and heterogeneous population living at high densities. To an economist, a city is a place 'where the local inhabitants satisfy an economically substantial part of their daily wants in the local market'. To an anthropologist, a city may only exist 'where there are cultural ingredients considered essential to urban life – the fine arts, exact science and writing'. A sociologist would focus upon the interactions between the inhabitants of the city, while to a political scientist, a city is a legally and politically defined entity with clear boundaries and jurisdictions (World Health Organization, 1988).

Park *et al.* (1974) describe the city as something more than a congeries of individuals and social conveniences – streets, buildings, electric lights, tramways and telephones; something more, also, than a mere constellation of institutions and administrative devices – courts, hospitals, schools, police and civil functionaries of various kinds. The city is rather a state of mind, a body of customs and traditions. The city is not, in other words, merely a physical mechanism and an artificial construction. It is involved in the vital processes of the people who compose it; it is a product of nature and particularly of human nature.

Next to the more traditional definitions of urban environments, one should also consider more innovative ways of describing cities as ecosystems or organisms. Ecosystems are complex self-sustaining systems that consist of organisms and the physical and chemical phenomena associated with them. Most importantly, they include interactions binding the living and non-living components into stable systems. These interactions include those between organisms, the relationships between organisms and their non-living (abiotic) environment and the various phases of the abiotic environment that mould its own change. The term 'ecosystem' generally brings to mind the idea of nature, and most studies of ecosystems concern natural ecosystems. But the principles of biological ecology can also be applied to ecosystems where humans play a major role, and these may be described as human ecosystems. They are unlike natural ecosystems in some important respects because human activity makes human ecosystems what they are. In ecological terms, humans are the dominant species. Human activity is rooted in social systems and is oriented towards goals whose bases are social, not biological (Clapham, 1981). The ecosystems view of cities is quite controversial, because one can argue that the 'ecosystem of the city' stretches far beyond the city itself. Are cities not dependent upon agricultural and wilderness areas for food, water, recreation, etc?

Figure 11.1 shows the energy and material flows involved in the 'metabolism' of the city. The city is not separate from its surroundings: it imports, among other things energy, water, oxygen and solar radiation. It exports products and materials, as well as solid, liquid and gaseous waste. Human activity inside the city causes noise and the release of carbon dioxide and heat. Urbanisation can cause air and water pollution and deterioration of the landscape.

Is it reasonable to use the ecological approach in the case of urban systems? Can we describe the city as an organism and, if so, is it 'parasitic' on the natural environment as its host? Furthermore, is it an organism within an environment or an entire ecosystem in its own right?

It is the complexity of the urban environment that makes it almost impossible to give a clear definition of a city. Figure 11.2 illustrates this complexity by way of a network: it shows the many conflicts that exist between various components of the urban environment; between, for instance, the demand for new housing and the demand for more office space in the city. Urban planners may, for example, be faced with an increase in the number of households at the same time as a changing production and employment situation creates a need for new office space. The consequence is often the loss of housing and the demolition of historic centres. Since urban developers find it more profitable to build offices than housing, families move out to the suburbs, which leads to a division between living space and work space. In turn, this division causes an increase in the use of private transport, traffic congestion, air pollution and noise. It should be clear that although Figure 11.2 looks complex, it is only an example, and a simplification of the real situation, in which many more interactions take place and relationships change; moreover, it reflects a dynamic that is typical only for certain cities in Europe.

THE GROWTH OF URBAN AREAS

Cities are a relatively recent phenomenon, occurring several millennia after the emergence of agriculture some twelve thousand years ago. Agricultural surplus,

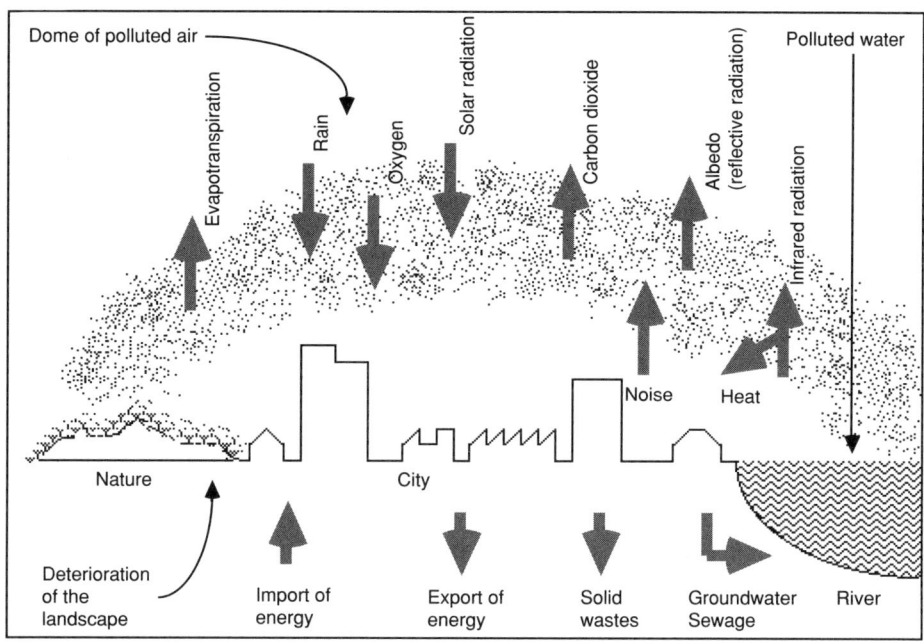

Figure 11.1 A qualitative input–output model showing the energy and material flows involved in the metabolism of the city of Barcelona, Spain
Source: Man and the Biosphere, 1988

expanding populations and a sense of common interests among peoples of a region fostered the initial growth of urban areas. The first known cities evolved about five thousand years ago on the Nile, Tigris and Euphrates rivers when traditionally nomadic peoples began to cultivate crops. Food surplus resulting from successive agricultural advances enabled farmers to support nascent villages and towns. Diversification of trade and the production of a wider array of goods encouraged the continued development of human settlements. Advances in science and the arts seem to have depended on the dynamics of a 'human implosion' as the population density of ancient cities speeded the exchange of ideas and innovations (Brown and Jacobson, 1987).

Despite the importance of cities in past social and economic development, their history merely foreshadowed the dominant role cities now play. Contemporary urban areas are integral centres of production and communication in a highly interdependent global network. But where urban growth is most rapid – and hence particularly in the Third

World – the economic gains normally attributed to cities are being offset by increasingly inefficient use of human and natural resources as a result of uncontrolled urban expansion (Brown and Jacobson, 1987).

In the mid-1700s, only 3 per cent of the world's inhabitants lived in urban areas. By 1950, urban areas held 29 per cent of the population and just 35 years later, they contained 41 per cent. It is expected that by the year 2025, 60 per cent of the world's population will live in and around cities (World Resources Institute, 1990).

Table 11.1 shows that the world-wide projected urban population in the year 2025 is almost seven times as large as it was in 1950. Of the 5,119 million urban dwellers, it is predicted that 79 per cent will be living in less developed countries, while 21 per cent will be in the more developed regions. By comparison, of the 733 million urban dwellers in 1950, 39 per cent were living in less developed regions and 61 per cent in the more developed regions. In other words, urban growth is much greater in the less developed regions of the world.

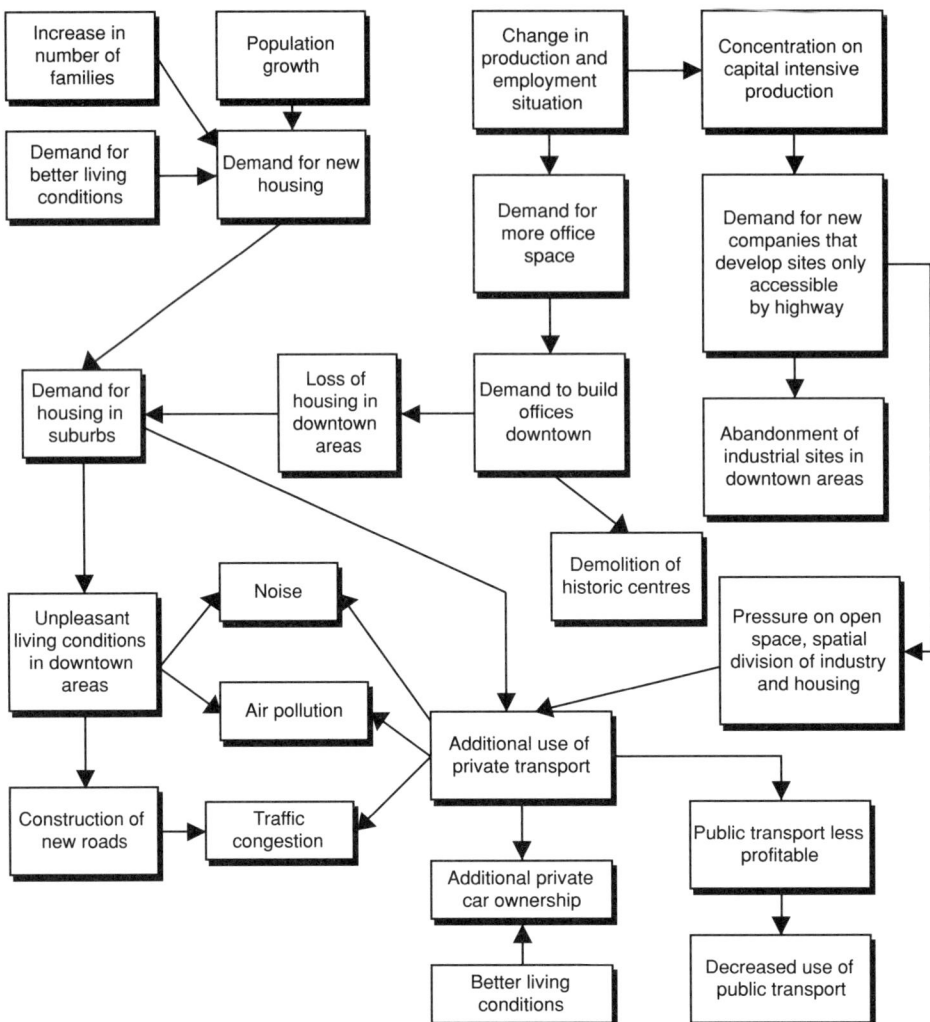

Figure 11.2 A few examples of relations within the urban system
Source: CEC, 1990

Figure 11.3 shows the world's 25 largest cities as projected for the year 2000. Here again, it is clear that cities in less developed countries will grow more rapidly than those in the more developed regions.

Statistics relating to urbanisation should be approached with the appropriate caution. United Nations statistics showing the majority of the world as urbanised are commonly misused to sound a doomsday alarm against the metropolis. United Nations data rest, however, on definitions of 'urban'

established by each individual country. Some countries define any settlement with a population of more than 2,000 as urban, while others use 100,000 as the mark. Some countries include every provincial capital regardless of its population (Angotti, 1993). Angotti (1993) wants to propose an alternative to the anti-urban doomsday theories. The best available data indicate that just 20 per cent of the world's population live in metropolitan areas, and only about 33 per cent live in cities over 100,000. While Angotti

Table 11.1 Actual and projected urban population from 1950 to 2025 (in millions)

	1950	1960	1970	1980	1990	2000	2010	2025
World total	733	1,030	1,374	1,770	2,261	2,917	3,737	5,119
More developed regions	448	571	699	798	876	945	1,004	1,068
Less developed regions	285	459	675	972	1,385	1,972	2,733	4,051
Africa	33	51	83	135	223	361	552	914
Latin America	69	107	163	237	324	417	509	645
Asia	226	359	503	688	931	1,292	1,772	2,589
Europe	221	259	307	340	364	387	405	422
Oceania	8	10	14	16	19	21	24	29
USSR	71	105	138	167	195	217	237	260

Source: World Resources Institute, 1990, from Department of International Economic and Social Affairs, United Nations, Prospects of World Urbanisation, 1988

agrees that the population in urban areas is growing, he does not see this as necessarily constituting a major problem.

The growth of urban areas in the developing countries will parallel anticipated global population trends (Dogan and Kasarda, 1988) and is partly attributable to rural–urban migration. Dogan and Kasarda (1988) conclude that despite the hardships greeting new urban arrivals, they consider themselves better off in the city than in rural areas where chances of economic success are slim. Not only do the fast growing cities offer better employment prospects, but they also provide cultural amenities, stimuli and some very basic services lacking in most rural regions.

DIFFERENCES OF ENVIRONMENTAL PROBLEMS BETWEEN DEVELOPING COUNTRIES AND INDUSTRIALISED COUNTRIES

The scale and types of environmental pollution and associated problems found in the cities of less developed countries are different from those in more developed countries. To begin with, the problems in less developed countries are more acute and more intense. The mega-cities developing today are, however, much more than expanding slums and squatter settlements fraught with environmental problems and social pathologies. Even a superficial analysis reveals that many of the largest cities in the developing world are important global hubs for manufacturing, trade, finance and administration. But it is also clear that as mega-cities expand, the physical inconveniences of dense urban living, ineffective environmental protection policies and adverse patterns of industrial and commercial development are manifesting themselves in increasing air and water pollution, depletion of natural resources and deterioration in the quality of urban life (Kasarda and Rondinelli, 1990).

The growth of giant metropolises in Asia, Latin America and other developing regions is resulting in a multitude of problems. These include high rates of unemployment and underemployment as urban

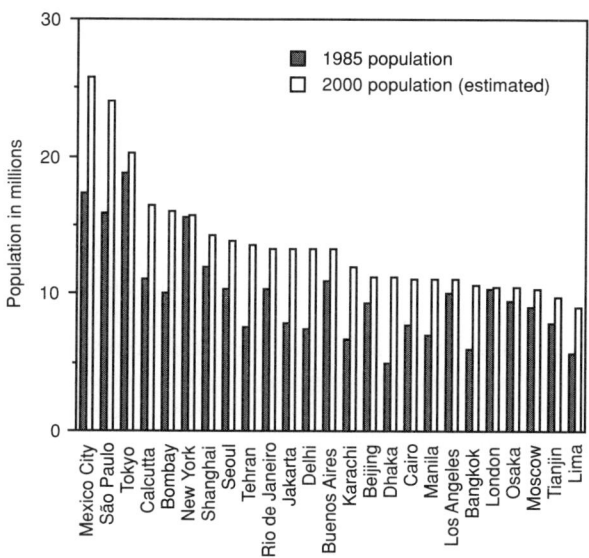

Figure 11.3 The world's 25 largest cities in the year 2000
Source: Lean *et al.*, 1990

labour markets are unable to absorb the expanding numbers of urban job keepers; insufficient housing and shelter; health and nutrition problems; inadequate sanitation and water supplies; overloaded and congested transportation systems; air, water and noise pollution; municipal budget crises; rising crime and other social malaise; and a general deterioration of the perceived quality of urban life (Dogan and Kasarda, 1988).

Again, this information should be approached with the necessary caution. Examples can be given of places where density does not make for major urban problems. Angotti (1993) indicates, for example, that contrary to the doomsday theory, higher urban densities do not necessarily produce crime and alienation. One need only compare the high crime rates in the sprawled low-density metropolitan regions of the USA with the low crime rates in Europe's and Japan's densely populated metropolises. Another example is the use of energy. Despite the reckless waste of resources in today's metropolis, it is generally the most energy-efficient human settlement form (Angotti, 1993).

The most pressing urban environmental problems in more developed regions can be summarised as air pollution, solid waste disposal, water pollution, noise pollution, land contamination and indoor pollution. Although these problems may be very serious, they are not comparable with those in less developed regions, for example, a life-threatening situation where drinking water is contaminated by human waste is much more serious than noise pollution in a city in Western Europe.

When considering issues in urban planning, one should also be aware of local differences. Urban planners in less developed countries should be mostly concerned with the provision of basic needs and services, where planning for water and sanitation services, food, electricity or other energy sources, basic transportation services, waste management and housing is a priority. Informal housing is often built on economically or ecologically undesirable land such as flood-prone, low-lying areas or unstable hillsides. In these cities, planning for basic services should be a priority.

Angotti (1993) describes the metropolis in the developing world as characterised by urban inequalities. Despite the fact that an overwhelming majority of the population lives in poverty, it is also clear that there is a minority that has a decent and stable living environment.

Although environmental problems in developing countries might be different from those in industrialised countries, it could be argued that these cities are merely at different stages in a common evolution. In less developed countries governments are trying to meet a population's bio-physical needs with essential services such as drinking water or sewage systems. In more developed countries, the challenge is no longer to provide running water, but to minimise water loss and improve water filtration, wastewater treatment and waste management techniques.

Running a metropolis calls for a delicate balancing act, demanding a long-term vision, in which the city is viewed as a whole. Cities around the world will continue to attract new residents and society must therefore ensure that urban growth is orderly and planned.

SOME URBAN ENVIRONMENTAL PROBLEMS

This section will focus on the following environmental problems in urban areas: air pollution, solid waste disposal, water pollution and water supply, noise pollution, land contamination, energy use and problems with the built-up areas.

Air pollution

Table 11.2 shows the three main sources of air pollution in cities: industry, motor vehicles and the burning of fuels for heating and power generation. Pollution from industrial sources varies from one city to another – depending on the kind of industry, its concentration in the area, existing environmental regulations and its exact location. Industrial air pollution is not a problem specific to urban areas, in contrast to air pollution caused by urban transport and the heating of buildings which are inherently linked to the functioning of cities. In Europe, the emission of SO_2 caused by the heating of buildings has been reduced by using alternative fuels and is replaced today by air pollution caused by the increased use of motor vehicles. Car exhaust contains NO_X, CO, CO_2, volatile

Table 11.2 Sources of air pollution

Sector	CO_2 (%)	SO_2 (%)	NO_x (%)
Energy	37.5	71.3	28.1
(electricity)	(29.3)	(61.5)	(24.6)
Industry	18.6	15.4	7.9
Road traffic	22.0	4.0	57.7
Others	21.9	9.3	6.3
Total	100.0	100.0	100.0

Source: Commission of the European Communities, 1990

organic substances, particles and lead (Commission of the European Communities, 1990).

Generally speaking, we can distinguish two types of urban air pollution. Older industrial cities, such as New York, belong to a group of grey-air cities; newer, relatively non-industrialised cities, such as Los Angeles, belong to a group of brown-air cities. Grey-air cities are generally located in cold, moist climates. The major pollutants are sulphur oxides and particulates, which combine with atmospheric moisture to form a greyish haze called smog. The grey-air cities depend greatly on coal and oil and are usually heavily industrialised. The air pollution is especially severe during cold, wet winters when the demand for home heating oil and electricity is heavy and atmospheric moisture content is high.

Brown-air cities are typically located in warm, dry, sunny climates and are generally newer cities with fewer polluting industries. The major sources of pollution in these cities are automobiles and electrical power plants, the primary pollutants being carbon monoxide, hydrocarbons and nitrogen oxides. In these cities atmospheric hydrocarbons and nitrogen oxides react in the presence of sunlight to form a number of secondary pollutants such as ozone, formaldehyde and peroxy acetylnitrate by photochemical reaction. The brownish-orange shroud of air pollution is called photochemical smog. Today, the distinction between grey- and brown-air cities is rapidly disappearing and most cities have brown-air in the summer and grey-air in the winter (Chiras, 1988).

Solid waste disposal

The solid waste problem in cities is often the result of high population density, the nature of western society, its culture of consumerism and the diversity of economic activities. Waste generation is mostly a problem of highly developed regions. Though garbage is not unique to rich countries, it is generated there on a different scale. Residents of New York City, for example, throw away three or more times as much as people in Calcutta and Manila (Young, 1991). In developing countries, waste is a luxury only available to a wealthy minority.

Dumping of solid waste in controlled landfills is becoming increasingly difficult because of the lack of open space and protests by neighbourhood communities, while the alternative of burning waste in incinerators has its own drawbacks such as the release of toxins into the air. Prevention and recycling programmes for urban solid waste are being developed. Table 11.3 gives the amount of household waste produced and recycled in different European cities. It shows that both the amount of waste per person and the proportion recycled varies considerably for the different cities.

Water pollution and water supply

The inadequate treatment of domestic and industrial waste prior to emission into water systems is a growing concern, mostly in relation to the health of the public at large. For example, in Budapest (Hungary) bacterial pollution of the Danube is seen to rise substantially downstream of the city. Only a small percentage of its sewage water is treated and bacterial concentrations are five to twenty times higher downstream than upstream of the city. Chemical

Table 11.3 Amount of household waste produced and recycled in different European cities, 1988–1989

	Total amount (kg/person/y)	Recycled (kg/person/y)	Proportion recycled (%)
Paris	512.30	18.1	3.5
Copenhagen	259.05	29.0	11.1
Vienna	347.01	41.6	12.0
Stockholm	397.93	56.8	14.3
Munich	440.67	67.4	15.3
Zurich	522.06	106.7	20.5
Dunkirk	363.00	86.5	24.0

Source: Waste Management Congress, Vienna, October 1989

parameters, though, are generally less dramatically altered (Hoang Anh Phuong *et al.*, 1992).

Large cities use enormous amounts of water which has to be discharged after use: for instance, every day Paris (France) is supplied with over 800,000 cubic metres of water and currently has an accessible sewer system some 2,100 km long. These large amounts of water are supplied to the city from neighbouring regions. New York City, for example, obtains its water from watersheds in the Catskill Mountains, 150 km to the north, which demonstrates the interdependence between the city and the countryside. However, watershed land that was once undeveloped countryside has of late been subject to rapid residential development. If this pattern of environmentally insensitive growth in the watershed continues, the quality of the drinking water for urban residents will soon be degraded.

Noise pollution

Noise, which has an influence on the health of urban populations and their quality of life, is caused by different types of transport: cars, motorcycles, airplanes and trains. Although new technologies and increasingly strict noise limits might reduce noise pollution in the future, the predicted increase in road and air traffic is expected to make the problem of noise more acute. The European Community Programme of Policy and Action in relation to the Environment and Sustainable Development states that more than 16 per cent of the population of EU countries suffer from nighttime noise levels of over leq 65 dB(A) which causes serious health risks. Reducing noise levels in urban environments is therefore one of the main objectives of the European Union (European Community, 1992).

Land contamination

The pollution of soil by past industrial activities and indiscriminate tipping of toxic substances is now recognised as a major concern in urban areas. While there is an increasing interest in reclaiming derelict land, risk of contamination by previous users prevents the re-use of such lands for either residential development or new activities. But derelict industrial land

is not the only type of land that is difficult to recover. In cities the soil is sealed by buildings and roads. Such land is lost as a water reservoir and as a system to sustain life. Roads contaminate soils, watercourses and groundwater through the materials from which they are constructed, by the discharges from traffic (for example, spilt oil, hydrocarbons, heavy metals, lost materials from hazardous goods transport) and as a result of the salt used to keep roads clear of ice in the winter.

Energy use

Production, building processes and transport all use a great deal of energy in cities. Although energy use patterns vary with differences in climate, degree of urbanisation and industrial structure, in most West European countries high levels of energy are used (Deelstra, 1992). Urban energy budgets increase as cities expand their boundaries, pushing back the countryside and lengthening supply lines. The amount of energy needed for households, industry and transportation is closely related to the structure of urban social and economic activity. The efficiency with which energy is used depends less on the size of urban population than on choices regarding land use, transportation, food and waste systems. Considerably higher levels of energy are required where settlement patterns are highly dispersed than where people live in close proximity to jobs and markets (Brown and Jacobson, 1987).

The ravenous appetite for energy of many cities also threatens the environment with consequences such as atmospheric warming and acid rain. New energy strategies based on conservation, innovation and alternative energy sources, with due thought to the environmental consequences of energy use, need to be formulated.

The built-up areas

Every city's built environment is unique. The collection of buildings, spaces, monuments, corridors and edges forms a valuable resource. The identity of these elements contributes to a sense of place, helping to distinguish each part of the city. Through contact with these elements, the resident builds up a mental map of the city, in which the relationships with places of work, education, friends and home are established (Lynch, 1960). Many unique historic city centres in Europe are threatened by the pressure to build new, large-scale development projects which can upset the established social and economic framework. Historic buildings are also threatened by air pollution, with building materials deteriorating as a result of acid rain and encrustation from airborne particles.

Road systems to accommodate the explosion in the use of private cars have also severely eroded the character of certain towns and cities. Historic parts of European cities are ill-equipped to accommodate cars.

Functional exactness and increasing scale destroys the flexibility of the city; buildings conceived as architectural objects are unable to adapt to changing conditions and therefore prevent the city from functioning as a dynamic, organic whole (Jacobs, 1961). This separation of functions leads to increased pressure on transport networks and the use of motor cars.

Urban areas also generate an urban climate characterised by distinct fluctuations in the local thermal balance. The modified surface structure or surface features, as well as the concentration of various substances in the atmosphere affect not only the radiation but also the conduction of heat in the ground and in the atmosphere. This also affects evaporation of the earth's surface (Horbert *et al.*, 1980) and creates local 'dry' areas with a low groundwater table.

Specific urban surface features, such as the large amount of paved areas, also modify the hydrology of the city. When rain falls in a rural area, the water penetrates into the soil, in contrast with the urban area, where it cannot penetrate and is discharged into the sewage system.

ENVIRONMENTAL MANAGEMENT OF URBAN AREAS

Several traditional measures, which are described in Volume 2 of this book, can be proposed to prevent the problems of urban environmental pollution. It is, however, most important to integrate all these measures and adapt them to urban areas in one comprehensive planning scheme. Alternatives to urban planning traditions are considered the most efficient way to manage the environment of urban

areas. In the following sections, healthy and liveable cities and sustainable cities are therefore discussed.

Healthy and liveable cities

Today, urban environmental problems exist in spite of several decades of urban planning traditions which are summarised in Table 11.4. As a consequence, some researchers have proposed alternatives to traditional planning. The most relevant concepts are 'healthy cities' and 'liveable cities'.

The concept of 'healthy cities', which was developed by the World Health Organization (WHO), offers interesting new perspectives. The WHO Healthy Cities Project promotes the idea that health is a result of the complex interactions of people with each other and their physical and social environments. There are, of course, many different opinions as to what is a healthy city. However, commonly shared parameters of a healthy city include a clean, safe, high-quality physical environment and a sustainable ecosystem; a strong, supportive and participatory community; provision of basic needs; access to a wide variety of experiences and resources; a diverse, vital and innovative economy; a sense of historical, biological and cultural connectedness; a city form that makes these possible; and a high health status with appropriate, high-quality and accessible public health and sick care services (World Health Organization, 1988). Vienna, Liege, Montpellier, Munich, Barcelona and Liverpool are a few examples of cities that are part of the WHO Healthy Cities Project. These cities meet regularly to discuss progress with specific projects and to exchange ideas.

Box 11.1 gives an example of the application of the Healthy Cities Concept. To be able to reach the goals set forward, several more detailed measures have to be taken such as, for example, proposals for reducing CO_2 emissions.

The 'liveable city concept' is very similar to the 'healthy city concept', the major goal of which is to improve the quality of life for urban residents. The city should be a place in which it is pleasant to live and raise children. This requires a friendly and quiet environment, areas with a neighbourhood identity, green spaces and urban spaces adapted to the daily needs of people.

Table 11.4 Pre-World War II urban planning traditions

Tradition	Definition	Examples, advocates
Authoritarian	Tended to follow the views of one man or an elite.	Haussmann (Paris)
Utilitarian	Product of *laissez-faire* economics and the rise of nineteenth-century capitalism. Urban developments are functional and subservient to the demands of an industrial economy born of free enterprise.	The unregulated spread of suburban housing, often in ribbons along arterial routes, e.g. in the London suburbs.
Utopian	Idealism in urban planning returned as a response to the social and environmental conditions produced in the early years of the industrial revolution.	Ebenezer Howard (1898), the Garden City.
Romantic	Direct result of the feeling of horror, on the one hand, at the industrial revolution's impact on urban form and, on the other hand, of the authoritarians' impact on the beautiful capitals of Europe.	Camillo Sitte (1889) emphasised town planning as a creative art and dragged urban planning out of the narrow confines of authoritarianism and unco-ordinated utilitarian policies.
Technocratic utopian	Technological tradition has formed the basis of this tradition.	In one of his earliest schemes in 1773, Ledoux placed the factory at the centre of the industrial town. Fourier, Garnier and Godin were also supporters of the technocratic utopian tradition and it was Le Corbusier who translated their vision into modern terms. La Ville Contemporaine and La Ville Radieuse represent the zenith of the technocratic tradition, sometimes also called functionalism.
Organic	The main emphasis was on the achievement of an intelligible order in cities.	Around 1915 Geddes stressed the need to make the city a living organism.
Socialist	This tradition has arisen from the works of Marx, Engels and their interpreters, although the socialist philosophers did not produce plans for cities.	Former Soviet urban planning.

Source: Burtenshaw *et al.,* 1991

Following is a list of characteristics of a vision of shaping cities in a way that will contribute positively towards the liveability of an urban place (Devuyst, 1994):

- Diversity and a mixture of compatible functions within one area.
- Diversity and a mixture of compatible functions within one building.
- Dominance of the residential function.
- Creation of public spaces for which people feel responsible.
- Presence of shops for daily necessities.
- Presence of a multitude of shopping possibilities.
- A mixture of retail trade and a limited number of superstores.
- Small-scale buildings and small-scale development projects.
- Buildings with a limited number of floors.
- A compact urban structure.
- Easy access to a large number of people, food, employment, recreation, education, etc.
- Dominance of pedestrians and bicycles.
- A street design that reduces the speed of motorised traffic and protects pedestrians and bicycles.
- Easy access by public transportation.
- Well-kept pavements.
- Presence of artwork with which people can interact.
- A specific identity by which the place can be distinguished from the rest of the city.

BOX 11.1   APPLICATION OF THE HEALTHY CITIES CONCEPT IN TORONTO, CANADA AND THE CITY OF TORONTO PROPOSAL FOR REDUCTION OF CARBON DIOXIDE EMISSIONS

Healthy Toronto 2000

The City of Toronto translated the healthy cities concept into a practical vision. For the Toronto city government a healthy city is a city with a wholesome environment, free of pollution – quiet, green, with fresh air, clean water, etc. It is also a city of neighbourhoods built to human scale, less high-rise, more low-rise and restored old buildings. It envisaged parks, gardens, trees, grass, community gardens, vegetable plots, sunshine, outdoor recreational facilities and community shopping areas. A healthy city has affordable housing for all, work available to all who want it, a transport system that encourages pedestrians, bicycles and public transportation. It is a caring and sharing community which sees itself as a family with support groups, volunteers and community networks.

The City of Toronto proposal for reduction of carbon dioxide emissions

(a) Set a target of a 20 per cent reduction in CO_2 emissions by the year 2005 (compared with the 1985 level), at a time when the population is expected to rise by 20 per cent.

(b) Set targets to reduce energy use for the gas and electrical companies, with a service role to increase efficiency, rather than the traditional production role. Tariff structures to be changed to encourage efficiency. Set up an Energy Efficiency Office to do life-cycle analysis of all new projects and conversions.

(c) Water to be drawn from the lower levels of Lake Ontario to provide cooling for an extended climate system in the downtown core.

(d) Encourage the use of public transport, bicycles and walking by increasing the price of parking and decreasing its availability. Increase land-use density to provide support for public transport; establish bus lanes.

(e) Plant trees in the City to provide shade and shelter. Plant trees in southern Ontario to increase CO_2 uptake and provide 'migration corridors' for plants and animals. Plant trees in the tropics to promote sustainable development.

Source: Toronto City Government, 1987; City of Toronto, 1991; Whyte, 1994

- Presence of striking or interesting views.
- Organisation of neighbourhood activities.
- Presence of green spaces, water and other natural elements.
- Presence of sunshine.
- Visual and personal access to buildings and open spaces.
- Presence of people regardless of age, nationality or lifestyle.
- Social activity on the street.
- Safety by way of informal control.

Characteristics of cities that are considered to contribute negatively towards the liveability of an urban place can be summarised as follows (Devuyst, 1994):

- Monofunctional places.
- Absence of the residential function.
- Absence of shops for daily necessities.
- Large-scale buildings or large-scale development projects that disrupt the close-grained urban structure.

- Lack of occupancy of buildings.
- Dominance of cars and high speed traffic.
- Large numbers of trucks and buses.
- Signs of disorder.
- Presence of fallow land and construction sites.
- Presence of billboards and commercial signs of different sizes, shapes and colours.
- Presence of wires and poles.
- Presence of services for cars such as petrol stations and parking spaces.
- Presence of barriers.

These or similar lists of characteristics can be used to develop urban management plans which result in an amelioration of urban environmental problems.

Sustainable development in cities

'Sustainability' implies different solutions for different places. It implies that the use of energy and materials in urban areas be in balance with what the region can continuously supply through natural processes such as photosynthesis, biological decomposition and the biochemical processes that support life. The immediate implications of this principle are a vastly reduced energy budget for cities and a smaller, more compact urban pattern interspersed with productive areas to collect energy, grow crops for food, fibre and energy and recycle wastes. New urban technologies will be less dependent on fossil fuels and will rely more on information and a careful integration with biological processes. This will mean cities of far greater design diversity than we have today, with each region developing unique urban forms based on regional characteristics that have long been overridden by cheap energy – the great leveller of regional diversity and unique character of place. A sustainable community exacts less from its inhabitants in time, wealth and maintenance. It demands less of its environment for land, water, soil and fuel (Van der Ryn and Calthorpe, 1986).

In 1992, at the UN Conference on Environment and Development in Rio, over one hundred and fifty nations endorsed a 500-page document, Agenda 21, which sets out how both developed and developing countries can work towards sustainable development. Local Agenda 21 is the Agenda 21 for the local level. It explains the process of developing local policies for sustainable development and building partnerships between local authorities and other sectors to implement these policies. It is a crucial part of the move towards sustainability. Table 11.5 shows the major steps of the process described in Local Agenda 21.

An example of urban environmental management that leads to more sustainable development and in which the population is involved is the installation of sanitation systems in Orangi, Karachi, Pakistan. Orangi is an unauthorised settlement with some 700,000 inhabitants in Karachi. Most inhabitants built their own houses, without public provision for sanitation. A local organisation called the Orangi Pilot Project (OPP) was sure that if local residents were fully involved, a cheap sanitation system could be installed. Over the years, households in Orangi have constructed many thousands of sanitary pour-flush latrines in their homes plus sewage lines and secondary drains – using their own funds and under their own management. OPP's research concentrated on the extent to which

Table 11.5 The major steps of the process described in Local Agenda 21 by the Local Government Management Board of the UK

1 Managing and improving the local authority's own environmental performance:
- corporate commitment;
- staff training and awareness training;
- environmental management systems;
- environmental budgeting;
- policy integration across sectors.
2 Integrating sustainable development aims into the local authority's policies and activities:
- green housekeeping;
- land-use planning;
- transport policies and programmes;
- tendering and purchaser/provider splits;
- housing services;
- tourism and visitor strategies;
- health strategies;
- welfare, equal opportunities and poverty strategies;
- explicitly 'environmental' services.
3 Awareness-raising and education:
- support for environmental education;
- awareness-raising events;
- visits and talks;
- support for voluntary groups;
- publication of local information;
- press releases;
- initiatives to encourage behaviour change and practical action.
4 Consulting and involving the general public:
- public consultation processes;
- fora;
- focus groups;
- 'planning for real';
- parish maps;
- feedback mechanisms.
5 Partnerships:
- meetings, workshops and conferences;
- working groups/advisory groups;
- round tables;
- environment city model;
- partnership initiatives;
- developing world partnerships and support.
6 Measuring, monitoring and reporting on progress towards sustainability:
- environmental monitoring;
- local state of the environment reporting;
- sustainability indicators;
- targets;
- environmental impact assessment;
- strategic environmental assessment.

Source: Burtenshaw *et al.*, 1991

the cost of sanitary latrines and sewerage lines could be lowered to the point where poor households could afford to pay for them. Simplified designs, the use of standardised steel moulds and the elimination of profits of the contractor reduced the costs to less than one-quarter of the contractors' rates (Mitlin and Satterthwaite, 1994).

Another example of alternative urban planning resulting in environmental management is the case of Curitiba (Brazil), outlined in Box 11.2. Again, this example shows that urban planning and urban environmental management go hand in hand. The Curitiba case has taught us that solutions to urban problems are not specific and isolated but rather interconnected. Any plan should involve partnerships among the private sector, non-governmental organisations, municipal agencies, utilities, neighbourhood associations, community groups and individuals. Creative and labour-intensive ideas can often be substituted for conventional capital-intensive technologies (Rabinovitch and Leitman, 1996).

CONCLUSIONS

Urban environments are complex networks characterised by high concentrations of people primarily

---

BOX 11.2 URBAN AND ENVIRONMENTAL MANAGEMENT IN CURITIBA, BRAZIL

Curitiba is located in south-eastern Brazil and has a fast growing population (from 300,000 citizens in 1950 to 2.1 million in 1990). Although its poverty and income profile is typical of the region, it has significantly less pollution, a slightly lower crime rate and a higher educational level among its citizens. City administrations have turned Curitiba into a laboratory for a style of urban development based on a preference for public transportation, working with the environment, appropriate rather than high-tech solutions, and citizen participation. This philosophy was introduced in the late 1960s and officially adopted in 1971. The following are examples of successful measures taken in Curitiba:

- *Flood control.* Instead of building drainage canals, the city set aside strips of land for drainage and put certain low-lying areas off-limits for building. Artificial lakes were constructed to contain floodwaters. These low-lying areas have been turned into parks and recreation areas, extensively planted with trees.
- *Innovative public transportation system.* Express buses on exclusive busways are the most important feature of the system. There is full integration between express buses, interdistrict buses and conventional (feeder) buses. Direct express buses have fewer stops and passengers pay before boarding the buses in special raised tubular stations. These new stations (with platforms at the same height as bus floors) cut boarding and deboarding times. To build a subway system would have cost roughly US$ 60 million to US$ 70 million per kilometre, the express bus highways cost US$ 200,000 per kilometre. Although the city has more than 500,000 private cars three-quarters of all commuters – more than 1.3 million passengers a day – take the bus. This has meant savings of up to 25 per cent of fuel consumption city-wide. Curitiba's public transportation system is a major reason for the city having one of the lowest levels of ambient air pollution and one of the lowest accident rates in Brazil. The implementation of the public transport system allowed the development of low-income housing with 40,000 new dwellings.
- *Land development.* Growth in Curitiba is only allowed along prescribed structural axes, close to the express bus lanes.
- *Involving the public and rejecting high-tech solutions.* Owners of historic buildings in the city's historic district are compensated for preservation measures. Incentives and systems for encouraging beneficial behaviour are also introduced at the individual level. Courses and programmes for adults and children are organised. Curitiba has repeatedly rejected proposals that emphasise technologically sophisticated solutions to urban problems. For example, instead of introducing an expensive mechanical garbage-separation plant, Curitiba introduced the 'Garbage that is not Garbage' initiative. This has resulted in more than 70 per cent of households sorting recyclable materials for collection. In the 'Garbage Purchase' programme poor families can exchange filled garbage bags for bus tokens, parcels of surplus food and children's school notebooks. The 'All Clean' initiative hires retired or unemployed people to clean up specific areas where litter has accumulated. These innovations rely on public participation and labour-intensive approaches.

Source: Rabinovitch and Leitman, 1996

working in non-agricultural sectors. The positive effects of these concentrations on the quality of life are often overshadowed by major environmental problems. The lack of housing, drinking water or food may become life-threatening in expanding cities in developing countries. Managing the complex urban systems becomes increasingly difficult because of rapid population growth and a lack of financial resources.

Among the alternatives to traditional planning schemes that are being developed today are the concepts of 'healthy cities', 'liveable cities' and 'sustainable communities'. Putting these new concepts into practice, and making use of advanced instruments for environmental management, offers the solution for many urban environmental problems. It should be clear that while cities contribute to environmental problems, they can also be, indeed must be, part of the solution.

REFERENCES

Angotti, T. (1993) *Metropolis 2000. Planning, Poverty and Politics*, London, UK: Routledge.

Brown, L.R. and Jacobson, J.L. (1987) 'The future of urbanisation: facing the ecological and economic constraints', *Worldwatch Paper 77*, Washington DC, USA: Worldwatch Institute.

Burtenshaw, D., Bateman, M. and Ashworth, G.J. (1991) *The European City. A Western Perspective*, New York, USA, Toronto, Canada: Halsted Press.

Chiras, D.D. (1988) *Environmental Science. A Framework for Decision Making*, 2nd edition, Redwood City, Calif., USA: Benjamin/Cummings Publishing Co., pp. 328–330.

City of Toronto (1991) *The Changing Atmosphere: Strategies for Reducing Carbon Dioxide Emissions*, Toronto, Canada: Special Advisory Committee on the Environment.

Clapham, W.B. (1981) *Human Ecosystems*, London, New York: Macmillan Publishing Co., pp. 1–2.

Commission of the European Communities (CEC) (1990) *Green Book on the Urban Environment*, Brussels, Belgium: CEC.

Deelstra, T. (1992) 'Western Europe', in R. Stren, R. White and J. Whitney (eds) *Sustainable Cities. Urbanisation and the Environment in International Perspective*, Boulder, Col., USA: Westview Press.

Devuyst, D. (1994) 'Instruments for the evaluation of environmental impact assessment', Ph.D. thesis, Brussels, Belgium: Human Ecology Department, Vrije Universiteit Brussel.

Dogan, M. and Kasarda, J.D. (1988) *The Metropolis Era. A World of Giant Cities*, vols 1 and 2, Thousand Oaks, Calif., USA and London, UK: Sage Publications.

European Community (1992) *Towards Sustainability. A European Community Programme of Policy and Action in Relation to the Environment and Sustainable Development*, Brussels, Belgium: EC.

Hoang Anh Phuong, Gyarmati, G., Janossy, L., Regenyi, P. and Rohrbacher, L. (1992) 'Industrial environment of Szazhalombatta. Case study', Paper submitted for the European Master's Degree Course in Environmental Science and Engineering, Budapest, Hungary.

Horbert, M., Blume, H.P., Elvers, H. and Sukopp, H. (1980) 'Ecological contributions to urban planning', in R. Bornkamm, J.A. Lee and M.R.D. Seaward (eds) *Urban Ecology*, 2nd European Ecological Symposium, Oxford, UK: Blackwell Scientific Publications, pp. 255–275.

Jacobs, J. (1961) *The Death and Life of Great American Cities*, New York, USA: Vintage Books.

Kasarda, J.D. and Rondinelli, D.A. (1990) 'Mega-cities, the environment, and private enterprise: toward ecologically sustainable urbanisation', *Environmental Impact Assessment Review* 10, 4: 393–404.

Lean, G., Hinrichsen, D. and Markham, A. (1990) *WWF Atlas of the Environment*, London: Arrow Books, pp. 22–23.

Local Government Management Board (1994) *Local Agenda 21 Principles and Process. A Step by Step Guide*, Luton, UK.

Lowe, M.D. (1991) 'Rethinking urban transport', in L. Brown (ed.) *State of the World 1991*, New York, USA, London, UK: W.W. Norton & Co, pp. 56–73.

Lynch, K. (1960) *The Image of the City*, Cambridge, Mass., USA: MIT Press.

Man and the Biosphere (1988) *Man belongs to the Earth*, Paris, France: UNESCO, pp. 36–44.

Mitlin, D. and Satterthwaite D. (1994) *Cities and Sustainable Development*, Background Paper Prepared for Global Forum '94.

Park, R.E., Burgess, E.W. and McKenzie, R.D. (1974) *The City*, Chicago, Ill., USA: University of Chicago Press.

Rabinovitch, J. and Leitman, J. (1996) 'Urban planning in Curitiba', *Scientific American* 274, 3: 26–33.

Toronto City Government (1987) *Healthy Toronto 2000. A Discussion Paper*, City of Toronto, Canada: Healthy Toronto 2000 Subcommittee, Board of Health.

Van der Ryn, S. and Calthorpe, P. (1986) *Sustainable Communities. A New Design Synthesis for Cities, Suburbs, and Towns*, San Francisco, Calif., USA: Sierra Club Books.

Whyte, R.R. (1994) Urban Environmental Management. Environmental Change and Urban Design, Chichester, UK: Wiley.

World Health Organization (1988) 'Promoting health in the urban context', *WHO Healthy Cities Papers No. 1*, Copenhagen, Denmark: FADL.

World Resources Institute (1988) *World Resources 1988–1989*, World Resources Institute in collaboration with the International Institute for Environment and Development and the United Nations Environment Programme, New York, USA: Basic Books, pp. 35–50.

World Resources Institute (1990) *World Resources 1990–91*, Oxford, UK: Oxford University Press.

Young, J.E. (1991) 'Reducing waste, saving materials', in L.R. Brown (ed.) *State of the World 1991*, Worldwatch Institute, New York, USA: W.W. Norton & Co.

SUGGESTED READING

Angotti, T. (1993) *Metropolis 2000. Planning, Poverty and Politics*, London, UK: Routledge.

Hall, P. (1992) *Urban and Regional Planning*, 3rd edition, London, UK: Routledge.

Jacobs, J. (1961) *The Death and Life of Great American Cities*, New York, USA: Vintage Books.

Lowe, M.D. (1992) 'Shaping cities', in L. Brown (ed.) *State of the World 1992*, New York, USA: W.W. Norton & Co, pp. 119–137.

Rabinovitch, J. and Leitman, J. (1996) 'Urban planning in Curitiba', *Scientific American* 274, 3: 26–33.

Robson, B. (1987) *Managing the City. The Aims and Impacts of Urban Policy*, London, UK, Sidney, Australia: Croom Helm.

SELF-ASSESSMENT QUESTIONS

1 Today, air pollution in urban areas in Western Europe is generally caused by:
   (a) the rise in SO_2 caused by heating of buildings;
   (b) the rise in NO_x, CO, particles, and lead caused by the increase in the use of motor vehicles;
   (c) inner city industrial facilities.

2 The utilitarian tradition in urban planning is:
   (a) a product of *laissez-faire* economics and the rise of nineteenth-century capitalism resulting in urban developments which were predominantly functional and subservient to the demands of the economy;
   (b) a response to the social and environmental conditions produced in the early years of the industrial revolution;
   (c) a result of the rise of electronic engineering and its application to a variety of secondary and tertiary employment which has opened up a whole range of potential scenarios for urban change.

3 The healthy cities concept is:
   (a) a tradition that has arisen from the works of Marx, Engels and their interpreters;
   (b) a concept developed by WHO which stresses a clean, safe, high-quality physical environment and a sustainable ecosystem, a place with a high quality of life for its inhabitants;
   (c) the provision of hospitals and medical services in urban areas.

4 Public transport in urban areas:
   (a) should be promoted because it can reduce fuel consumption, emissions and the amount of space required;
   (b) should be discouraged because it is inefficient and noisy and cannot solve traffic congestion;
   (c) should become an alternative to walking or biking to work because it is safer to use public transport.

5 Building office towers in downtown areas often causes:
   (a) the loss of historic urban features, historic buildings, open space, and housing leading to a more mono-functional region;
   (b) better housing conditions and reduced traffic congestion;
   (c) an explosion of the urban population.

THE ARCHAEOLOGICAL HERITAGE IN ENVIRONMENTAL MANAGEMENT

Jozef Buys

SUMMARY

Environmental management should be considered as a means for attaining sustainable development. However, the direction of this development should not be solely determined by a purely ecological environment, but must include the human aspect, as expressed by cultural diversity. Although present-day indigenous populations may still display original characteristics that illustrate their unique ways of adapting to the natural as well as to the human environment, often the most authentic expressions of knowledge and know-how are hardly remembered and lie hidden in the archaeological heritage. These remains of the past may hold a huge potential for understanding how we became what we are and what lessons past successes and failures may teach us, but they are, at the same time, prone to destruction for multiple reasons. Current protection measures don't work well, partly because of a lack of efficient enforcement, and partly because of the great complexity of the problem.

All development projects in archaeologically interesting areas should therefore include an archaeological impact assessment (AIA), and foresee appropriate measures for protecting ancient remains. This means that political will, at all levels, must exist to include AIA in the project cycle and to execute it properly, after which monitoring of the archaeological heritage must be organised. Hence the need for awareness-raising and capacity-building at all levels. A possible way of reaching this objective could be to include the cultural dimension in existing legally binding environmental agreements on the international level, or to negotiate a convention *sui generis* that combines both aspects.

ACADEMIC OBJECTIVES

This chapter aims at establishing the links between sustainable development, environmental management and the safeguarding of the cultural heritage, with an emphasis on archaeological patrimony. After providing some basic definitions, ways and means of integrating the archaeological heritage into environmental management are suggested, both on the specific project basis and on the broader level of anticipatory programming.

Upon completion of this chapter, the reader should be able to:

- understand the importance and role of cultural and archaeological heritage in development plans, as an integral part of the environment;
- identify the levels at which appropriate decisions can be made for integrating this heritage into the project cycle;
- turn to the right authorities in order to assess correctly the archaeological impact of development projects;
- evaluate the quality of these assessments and of the suggested mitigating measures that derive from it; and
- be acquainted with the general principles that should safeguard the archaeological heritage against future destruction, by implementing a system of protection, a process of monitoring and campaigns of awareness-raising and capacity-building at both the national and international levels.

INTRODUCTION

If sustainable development means that we can continue to use all kinds of resources for our own development, while guaranteeing that future generations can also establish and enjoy a good quality of life, and if, furthermore, we accept that sustainability comprises aspects of natural environment, social welfare and economic well-being, then the very concept of sustainable development must be broad enough to encompass the multiple facets of human society.

In this chapter we argue that the omnipresent, though mostly unconsidered, cultural characteristics of human society in general make up a very important aspect of the driving forces that ultimately determine the direction of development. However, human society varies widely according to historical background and geographical setting. Hence the need to take into account the cultural specifics of countries, or regions, when designing development plans.

Regardless of their diversity, cultures share general components (social, economic, political, religious) which relate to their earlier stages of development or to foreign inputs, and show the ways in which they managed to interact with, and adapt to, their environment, both natural and cultural. Many clues to this understanding can be found in the cultural heritage in general, but particular aspects may actually lie buried in the archaeological heritage.

Since this book focuses on the practice of environmental management, and the cultural dimension therein is not currently taken into account, we shall divide this chapter into a first part that gives some basic definitions, a second section with more practical points about cultural and archaeological heritage, environment and development, including examples in the form of case studies, and a final part suggesting some ideas for the future. Since the author's experience in this field mainly lies in South America, the topic under discussion is written against this background and with examples from Ecuador. When transposing the same problems to other regions, some degree of adjustment may be necessary.

BROADENING ENVIRONMENTAL MANAGEMENT

The objective

First of all, we consider environmental management to be a means for sustainable development, not an end in itself. Sustainable development must be the objective for human society as a whole, be it in the developed world or in the developing world. The first needs to become a healthier society, while the latter must catch up in a sound way. As stated in the Brundtland report:

sustainable development is not a fixed state of harmony, but rather a process of change in which the exploitation of resources, the direction of investments, the orientation of technological development, and institutional change are made consistent with future as well as present needs.
(WCED, 1987: 8–9)

This means that, for the time being, we must limit the use of the environment if we don't want to exhaust our natural resources. Simultaneously, we need to help poor countries develop in a sound way, which, in turn, demands an extra amount of resources. In order to resolve the apparent contradiction, we must combine environmental policy with development policy. In doing so, we should be able to reduce poverty, increase democracy and control consumption and demography.

However, care must be taken not to establish a one-sided, narrow-minded concept of sustainable development, which would then tend to impose itself in a rigid way everywhere. It would almost certainly be a western point of view, based on exclusively western assumptions about development and sustainability. This would only be a repetition of past errors and be doomed for quick failure. On the contrary, a lot of input is needed from the quiet majority, the ones that traditionally have been recipients and have been directed towards targets that were not their own. If we are serious about sustainable development, we need to accept that there is not one but several facets to sustainability, reflecting the many actors who will be participating in the drama of the future.

Culture and sustainable development

When speaking about development, culture usually is not an issue. When considering development assistance, cultural aspects are hardly ever included in plans, programmes or projects. A possible explanation is that development assistance has always been regarded as a purely technical matter, while culture has been viewed rather in the opposite way. A more important and, at the same time, more negative distinction between both has to do with feelings of solidarity with broad population levels when talking about development assistance while thinking in terms of distance, abstraction and being somehow elite-oriented when mentioning culture.

Culture can be defined as learned behaviour, transmitted from generation to generation, for coping with both the natural and the human environment. It is therefore based on tradition, but also on innovation, and is characteristic of a geographical region at a certain point in time. Cultures can be very long-lived, evolving slowly in one place, or expanding their influence in an imperialistic way. They can disappear suddenly, due to disaster, or be conquered and replaced by other human groups. Wherever cultures are located and whenever they existed, the natural environment has always been of key importance, by the mere fact that human development has been a process of progressive domination of nature in order to ensure survival. They have been managing their environment in their own right, in the same way as we are trying to do now. In this continuum, our ideas may replace earlier ones or combine past and present knowledge.

Although culture dynamics can ultimately teach us a lot about the success stories or failures of extant or extinct social systems, the careful observation of particular culture traits may give insights into their unique ways of adapting to specific environments. These ways often include ideal solutions to particular problems, examples of adapted technology that are generally more environment-friendly than present-day practices. As Chapter 26 of Agenda 21 states, indigenous people often have a unique knowledge of their environment and its natural characteristics. In combination with modern scientific research, this knowledge can produce new opportunities for sustainable development. Therefore, indigenous people should be given more rights of self-determination, should administer their own natural resources and participate in the decision-making. Finally, indigenous people should be protected against dominant societies that threaten their natural resources, culture, prosperity and way of life (Johnson, 1993).

First, the many different, specific cultural answers to environmental conditions can form an inventory of possible alternatives for good environmental practices in development policy. Therefore, thought should always be given to local knowledge when planning development actions, seeking direct involvement of the local population. In this regard, recent trends in development assistance and environment, calling for more bottom-up and participatory approaches, must be warmly welcomed.

Second, it will soon become clear that not only cultural specifics, that is technical solutions, may play an important role in finding the right solutions, but the whole of society targeted by the development activities must have an input into the process. Only then can development be socially integrated, guaranteeing not only positive direct results, but also long-term improvements and multiplicator effects.

In the long run, repeating the exercise over and over again will enhance cultural sensitivity on both sides of the development co-operation, making for better results, closer relations and mutual understanding of each other's needs and deeds.

Cultural heritage

Cultural heritage encompasses all possible expressions of a society. We generally think of material phenomena such as art, technology, architecture, etc., but whatever particular cultural trait we look at, no abstraction can be made of the ideas and beliefs embedded in it. One of the main problems in the present-day cultural reality of certain societies is the degree to which the original traits have been altered by the importing of foreign ('modern and progress-oriented') elements, such as technology, economic systems, models and the like. A good example is South America, where Iberian conquest willingly changed all aspects of pre-Columbian society. By now, 500 years later, a lot of knowledge would have been lost forever, were it not that historical and archaeological research is recuperating it.

The developed countries also have an important share of the responsibility in the cultural realm. Haven't we imposed our culture on many societies in different latitudes, thereby inflicting serious damage or actually destroying unique local cultures? The principle of 'the polluter pays' may work in the material dimension, provided there is still a way to mend the damage and that nature can regenerate, but on the cultural level, being entirely man-made, the destruction may have been terminal. If a local population has been exterminated, who is going to bring it back to life? At most, it can be re-created to serve as a showcase in some ethnographic museum, but the living thing, the dynamic society, has stopped existing. Need it be said that the developed world was a cultural depredator long before turning to destroying nature? If the origins of the latter can be tracked back to the industrial revolution, did not the first start shortly after the Neolithic revolution, several millennia earlier?

As rich countries, with the extra resources and knowledge to take care of cultural heritage, it is our duty to be guardians not just of our own cultural heritage and that of the countries we are related to, but also of world heritage, so as to ensure that future generations can know and understand the rich cultural diversity on our planet. In order to do so we must decisively undertake actions to preserve and protect the cultural heritage. This means that we should invest in the safeguarding of culture, for we can only protect and preserve those cultural elements we know. Through anthropological research in the field, a lot can be learned about important aspects such as local political and social organisation. Present-day knowledge and technology can be studied through direct contact with people, but certain traditional features may be harder to detect by simple observation or interview. Some of them may actually be forgotten, buried in the minds and in the soil. Archaeological research may therefore not only recover the origins of the local population, explain the reason for their presence and actual situation, but additionally discover fine examples of ecologically well-adapted technology, dating from centuries ago (Box 12.1).

Several other examples could be mentioned, but entering into the details would lead us too far. There are, however, interesting elements on the level of infrastructure, such as condensing water from daily mist and using it for irrigation in the desert west coast of South America, or increasing agricultural areas and preventing soil erosion by a wide array of terracing and channelling techniques, of which many good examples exist all over the world. On a smaller scale, and still awaiting further exploration, ancient technology continually startles researchers with many examples of simple and very logical technical solutions for a wide variety of problems in metallurgy, ceramics, architecture, food

---

BOX 12.1 RECUPERATING ANCIENT TECHNOLOGY

In many regions of South America, a traditional agricultural system known as 'raised fields' has been reported since ancient times (Denevan et al., 1987). It is composed of raised parcels of land, limited by small, shallow channels, which are interconnected and feed into bigger watercourses, while being interspersed with dammed areas for stocking water reserves. This way a dynamic system of drainage and irrigation is created. When located in tropical lowlands, this is its main function. However, identical agricultural systems also occur in the interandean valleys or high altitude plains of South America. Here, research has proven that during the dry period in the Andes the water in the channels creates a microclimate by absorbing sun heat during the day that is radiated at night, providing the extra degree or tenths of a degree Celsius that may make the difference between conserving the crop or losing it all by freezing (Knapp, 1984). As these raised fields are located on the valley floor or lower slopes, drainage/irrigation may also be a function.

Archaeological investigations of the raised field complex to the North of Guayaquil, Ecuador, have dated the earliest phase of this system as far back in time as 2000 BC (Parsons and Shlemon, 1987), increasing its extension and complexity over time. During the last pre-Columbian period, that is AD 500–1500, vast areas of the Guayas plain were covered with raised fields, articulated around a hierarchy of settlements (Buys and Muse, 1987), providing the economic basis for chiefdoms that reigned over the tropical wetlands behind the pacific coastline.

Not only did the raised fields yield different kinds of crops, planted higher or lower on the slope of the field according to the need for more or less water, but the channels too provided an important part of the diet, through the fish and molluscs they contained, and the birds they attracted. Maintenance of the system required periodic cleaning of the channels by removing the deposits and spreading them on the fields, as a natural fertiliser. This way the raised field system became an interestingly diversified source for subsistence and surplus production.

Excavations of raised fields and settlements associated with them have given a clear picture of how adapted technology could sustain a large population and its social development for many centuries, failing only at the moment when the Spanish Conquest disrupted the indigenous society. Experiments with planting crops on selected fields have proven the potential of the system even today (Muse and Quintero, 1987) – an excellent example of lessons to be learned from the past and possibilities for applying them to the future.

storage, etc. They may not serve industrial purposes, but they can still be applicable on the artisan level.

Archaeology

Archaeology is the scientific discipline that studies past cultures, and their development, based on the material remains that these cultures have left behind. It does so through surveying the landscape, registering all kinds of visible remains and by excavating sites where those remains are buried.

The remains of the past are not only limited to what we traditionally consider as such, that is objects and buildings, but include all things produced, used or slightly modified by man, comprising a variety of elements – bone, seeds, soils, stone, wood, shell, metal, ceramics, etc. All these elements, together with the specific context in which they are encountered, and its structure, allow the archaeologist to build a sequence of cultural events, reconstruct the way of life at different points in time and infer the broad historical processes. In order to fulfil its objectives, archaeology seeks the help of many other disciplines: anthropology, history, geology, physics, chemistry, biology, etc.

For the present publication the past is defined in a very broad way, be it fifty or a thousand years ago. Archaeology discerns many different periods, based on geographical, chronological, cultural or evolutionary parameters. There may be South American, prehistoric, Roman or Iron Age archaeology, each of them – and many more – with its own specifics. They may be important at the time of actually conducting archaeological research, but not for this kind of general discussion. Age is not necessarily a factor in determining the importance of archaeological remains.

The standing remains of the past may have a fair chance of surviving, at least partially, if the population is respectful of them, through identifying themselves with the ruins of their ancestors. Unfortunately, architectural remains often serve as quarries for cheap building materials. The finest Inca masonry embellishes modern chimneys in Peru, and Indian carpentry becomes incorporated as decorative elements of houses in the west. These are just two simple examples of a much wider problem. In a way, it may be a minor problem, because of the size, weight or visibility of the objects being moved. On the other hand, there are hundreds of thousands of less spectacular, though even more destructive, pillages going on. Most of these concentrate on the buried remains of the past, with a marked preference for ancient graves. Millions of funerary objects can be found on the shelves of museums and private collections, be they ceramic pots, textiles, basketry, metals or any other artefact that accompanied the dead in the afterlife. The great majority is the product of illegal excavations and trade between developing countries and the rich nations. On both sides, unscrupulous people have established the necessary channels to plunder the archaeological heritage for the sole sake of economic profit, thereby destroying beyond repair what may be unique contexts.

Although archaeological cemeteries may be particularly vulnerable to this greed, a lot of other situations are disastrous for the buried archaeological patrimony. Most big building projects, such as roads and dams, don't take into account the existence of archaeological sites, and numerous controversies have arisen where ancient remains have been detected in time, holding up progress and causing huge economic losses. But a lot of damage can also occur on the level of the individual citizen. It covers the wide range from innocent sherd-picking on archaeological sites to organised treasure hunting, from souvenir buying to illegal traffic in antiquities. All these destructions seriously limit the potential for discovering lost knowledge that may be of use today.

Quantifying the amount of trade in cultural heritage will always be difficult, since an important part of it occurs clandestinely. However, recent estimates speak of US$ 1 billion worth of commerce, a great part of it made up by archaeological patrimony. This makes it second only to drug trafficking on world level. Some figures may illustrate the seriousness of the situation:

- In 1973 some 1 per cent of the economically active population of Costa Rica was estimated to be involved in illegal excavation or commerce of the results of it, representing about US$ 0.5 million annually (Walden, 1991).
- Of all thefts of cultural property, and in the case of illegal excavation probably less, only 12 per cent

is ever recovered, meaning that over 88 per cent enters the illegal trade network (ibid.).

- One of the incentives to loot and sell the products can be found in the dazzling increase in prices for art and cultural property. The most important auction houses report incredible growths in revenue over the last ten years. Sothebys increased 800 per cent, Christies 700 per cent (UNESCO Sources, 1991: 12–13).

Archaeological heritage

We look at the archaeological heritage as an important resource for sustainable development, a resource that is both instructive and finite, and thus irreplaceable. We therefore urgently need to consider the cultural aspect of our environment as essential as the ecological one. We must work out ways to protect the cultural environment from becoming as deteriorated as the natural environment now is. We suggest that one way of doing this is to define our environment in an integrated way, meaning that we are talking of a cultural-ecological environment. A few general reasons underlie this thinking:

- The natural environment doesn't have a lot of meaning if it doesn't include all of its components, a key player being man. The mere fact that the human element is now jeopardising its natural environment, makes it all the more interesting to investigate how such a situation was arrived at and, more importantly, how it used to be, when man and nature had a more symbiotic way of life. Maybe history can show us similar problems in the past and teach us how they were solved.
- Ecological preservation has hit the news for decades now, but it used to be completely different. We remember the times when almost anything was allowed in order to attain a higher level of development, including destroying the natural environment, resulting in the situation we have now. So let us not commit the same error on the cultural level, as we need it for assuring the right quality of life of future generations.
- There is a perceived need for safeguarding biological diversity, the rich natural variety of living species, which should be transposed to the cultural

level. The many different cultural expressions represent the variety of adaptive responses to equally different situations, both natural and social, that can, ultimately, teach us solutions to present-day problems. So preserving the remains that illustrate the cultural diversity can have modern applications.

- Cultural identity is a fundamental need of populations that are undergoing major processes of change, as is the case of developing countries trying to catch up. Their feeling of proper identity must be strong enough to assimilate foreign inputs and to retain their own, while being constantly influenced and measured by international, universal standards that have no meaning whatsoever in their native world-view. Historical background, reflected by the archaeological remains, represents a key factor in cultural identity.

These reasons notwithstanding, modern-day measures for protecting the archaeological heritage leave much to be desired. Although legislation exists in almost every country, enforcement is usually very low key on both sides of development. If Italy still struggles with its tombaroli (Etruscan) grave looters, one can imagine what the situation is in developing countries. Alarming, and often very sad, reports come from everywhere: primarily Middle and South America and Africa, but also South-East Asia. In all these regions, countries are facing the ever-increasing pillaging of their archaeological sites (Brent, 1994). Manifestly the most efficient protection measure is expropriation of the archaeological site, turning it into an archaeological park with appropriate infrastructure for tourism or recreation. But only very few sites qualify for this kind of treatment, because of the high costs involved and the often very complicated legal aspects.

INTEGRATING ARCHAEOLOGICAL HERITAGE IN ENVIRONMENTAL MANAGEMENT

Levels of decision-making

Once the importance of the cultural heritage in general, and the archaeological patrimony in particular, has been accepted, one must ensure that it will be taken into account when designing development plans. First of all, of course, the political will must exist to redefine the environment as a dual concept that comprises both natural and cultural aspects. This battle against traditional theories and conservative bureaucracies may not be easy, but the concept of sustainability in all its facets should be of paramount usefulness. Its emphasis on social and economic aspects, besides the ecological one, should allow for broadening the scope of the definition and, as a corollary, the scope of action.

We must beware of being overoptimistic: the systematic inclusion of cultural aspects, let alone the archaeological heritage, in the broad field of development planning will indeed be a very hard task. A comparison with ecological conservation can be instructive. Even now, after decades of warnings, the systematic screening of development projects on their environmental consequences is only slowly finding its way into the higher levels of decision-making. The discussion about extending the environmental impact assessment (EIA), on the project level, to the strategic environmental assessment (SEA) on the policy level, has only just been initiated and promises to be lengthy. We can only hope that the new ground being broken there will provide a path for our purposes.

At the highest level then, policy-makers need to be convinced, but regional authorities and local people are equally important. Indeed, many times already the initiatives that eventually lead towards the execution of development projects have originated within the population that, later, will become the target group of those projects. Development should always happen this way. Problems should be recognised by those who suffer their consequences, and be communicated to the next level of hierarchy, which should take action to solve the problems. Too many times in the past, and even now, existing problems were perceived at a high level of decision-makers who then turned to external sources for help, both economic and technical, without a thorough knowledge of all the aspects of the problem in question. As a result the solution offered often failed to meet the expectations and the real needs of the population involved.

The input of the local population is especially indispensable when considering cultural aspects and archaeological heritage. High-level policy-makers may belong to an entirely different cultural group, or may simply ignore the existence of archaeological remains. Many reasons may exist for not taking these elements into account, most commonly lack of interest or fear of complicating the action under consideration. This does not mean that the central authority cannot have good ideas for developing local groups; it does mean that close co-operation between the local population and the authorities at all levels, from the very beginning of development planning, will undoubtedly lead to easier execution and positive results. External donors should be very much aware of this mechanism and, without intending to dictate to the developing countries, should definitely try to promote local involvement in decision-making (see Box 12.2).

When speaking of local authorities, a wider range than the official authorities should be envisaged. Sometimes, government officials at the local level are people brought in from outside the region and may lack sympathy with the local group and its historical roots. In parallel, there may be an ethnic local government, according to traditional social organisation. It is not uncommon to have several tribal chiefs, ethnic lords, or whatever, living under the authority of a representative from the central government. These chiefs themselves head a complex social structure, not necessarily compatible with the official government structure, and are probably acting as buffers between the official and the ethnic sides. When it comes to implementing development plans successfully, the local ethnic players may prove to be of decisive importance. Their familiarity and identification with the region, both ecologically and historically, their knowledge of ancient technology and archaeological sites will all be assets in designing and implementing integrated development plans.

The intervention cycle

If we are serious about giving consideration to the cultural aspects of a society, and among them the historical testimonies of its identity, then we need to do so from the outset. Any intervention in a country, region or human group should evaluate the possible consequences for the archaeological heritage of the

BOX 12.2   INVOLVING LOCAL PEOPLE

The community of Agua Blanca, located in the valley of Rio Buena Vista, lies in the southern part of the Province of Manabi, on the coast of Ecuador. This small community, composed of some 40-odd families, living in a small hamlet and some scattered houses in the valley, subsisted traditionally on a mixed economy of fishing, some cattle-herding and limited agriculture.

Being situated in the heartland of the pre-Columbian Manteño culture (AD 500–1500), surrounded by a fair quantity of architectural remains, an important complementary income was generated by pot-hunting in the archaeological sites.

When the area was incorporated into the Machalilla National Park in 1979 and serious restrictions were imposed on the exploitation of natural resources, archaeological looting practically became the main revenue for the local population, in spite of the fact that one of the park's objectives was to preserve the archaeological resources. Regularly, recognisable Manteño objects from the region invaded the antiques market, a practice that continued for several years.

It wasn't until the 1980s, when a British archaeologist became interested in the area and started designing a research project (McEwan, 1982), that the local people became aware of the damage they were inflicting on their heritage. By explaining the importance of the ancient remains, showing their extension in the valley and beyond, and inferring the economic and social situation, he was able to raise the locals' interest in their 'ancestors'.

When starting his research, the archaeologist hired the local people to participate in the survey and excavations. During fieldwork frequent comments were heard about similar finds the locals had made during their looting campaigns and the archaeologist could gather a lot of additional information. He was even allowed to see several personal collections and learned about others.

Over time, becoming ever more acquainted with the local population and its problems, the archaeologist managed to finance certain improvements in the infrastructure of the valley and took effective steps to promote the archaeological heritage, so that tourism started to increase.

After the fieldwork and the processing of the archaeological finds, the local people became involved in the maintenance of the site and finally ended up attending the visitors. In order to present a better view of the history and the archaeological heritage of the region, a museum was built and was equipped with a few objects stemming from the excavations and . . . a lot of artefacts donated by the local people. Nowadays, the daily care of site and museum is in the hands of the community. The income from entrance fees and from the selling of handicrafts has completely replaced that from archaeological looting. Additionally, the Agua Blanca community is known as an important defender of the archaeological heritage, organising meetings with other communities in order to raise awareness through communicating their own experience.

target population, in exactly the same way that the impact on the natural environment is being assessed.

This means that an archaeological impact assessment should be considered, as an integral part of EIA. Ideally, this exercise should begin as early as the identification phase of an intervention, but should continue throughout the rest of the cycle. Indeed, not everything can be foreseen when it comes to evaluating the archaeological heritage of an impact area, given that many elements may be buried, and trying to detect the totality of them beforehand may be beyond the overall scope of the intervention. Flexibility during the whole project cycle thus becomes a very important characteristic. Flexibility in general should always be a prime concern when designing and executing development plans.

The screening of interventions for their eventual archaeological impact should be integrated into the systematic environmental screening, by including specific questions that seek general information on visible or known remains from the past. Further key questions relate to the amount of subsurface intervention that is planned, as it may affect buried elements. The quantity and quality of the remains must then be evaluated and safeguarding priorities established. Not everything can be preserved but a representative sample should be protected by changing the intervention modalities or localities. A few exceptional sites may be worth saving from destructive large-scale projects in the way UNESCO did at the time of the building of the Aswan dam.

If, at an early stage, the projected activities of the intervention suggest that part of the archaeological heritage will be affected in a negative way, specific actions need to be taken in order to avoid or minimise impact. If no danger exists, the intervention can go forward. If there are uncertainties about the type and extent of the archaeological heritage that will be affected, then further study is warranted. Ultimately, an entire project could be cancelled because of unjustifiable destruction. Before such a harsh decision is taken, however, alternatives to the original project should be sought or redesigning of the project should be intended.

Archaeological impact assessment (AIA) as part of EIA

As soon as the archaeological elements of a region are considered to be an integral part of the environment, then any AIA should not be separated from the general EIA. It would just be another dimension of EIA, to be handled by an archaeologist within the EIA team, since only a professional can have the accurate knowledge of both the existing information and the techniques actively to discover ancient remains. Its incorporation should follow the general scheme of EIA, much along the lines expressed in Volume 1, Chapter 10 of this book.

The first step is getting acquainted with the intervention area and the objectives of the project. This should give initial indications about a potential impact on the archaeological heritage. Many interventions may be rapidly qualified as non-destructive, so that no further action is required. Projects in sectors such as public health and education rarely jeopardise archaeological remains. In general, only interventions that modify the landscape, including any form of digging, should be closely watched.

During the second step, unless the archaeologist in charge is completely familiar with the region, time will be needed to study the specialised literature. Information may be found in old maps, travel accounts, early geographical descriptions, historical archives, former official documents, excavation reports, anthropological studies, etc. All elements thus discovered should be mapped provisionally and their existence and state of conservation verified in the field. Interviewing people who know the area and its history well may increase the preliminary inventory of archaeological remains. At this point, a better idea of the potential impact can emerge. Once the verification and final mapping is completed, certain patterns

in the ancient remains, or the lack of them, may show up and a decision on the need for further research must be taken.

Up to this point the assessment may be considered as a standard screening and scoping procedure, identical to environmental screening and resulting in a broad classification of the kind 'no impact', 'low impact', 'high impact'. In addition, however, there may often be insufficient information available. Indeed, lack of information in archaeology doesn't automatically mean that there are no remains; usually it means that they haven't been found yet, mostly because they are buried and thus invisible to the naked eye. The decision on whether or not to continue investigating will be difficult, since any further research will be more specialised, longer term and more expensive (Figure 12.1).

The next step represents the active search for archaeological remains through a wide variety of surveying techniques. Where available and applicable, remote sensing is a great tool for having easy access

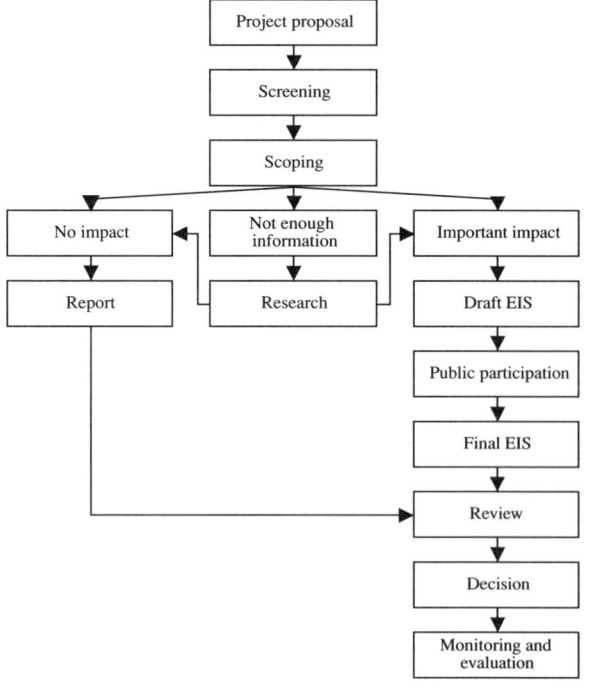

Figure 12.1 Framework for an archaeological impact assessment

to extended areas. Aerial photography, originally used for national cartographic purposes, is an important source for detecting monumental remains. Satellite images can do the same with an ever-increasing degree of detail and resolution. Low altitude aerial photography can even show buried remains through soil marks, crop marks or shadow marks. But no survey would be complete or accurate without a pedestrian phase, walking the area while scrutinising the surface for remains of buried sites. In order to limit time and expenses, several sampling techniques can be applied. If a realistic image of the archaeological heritage is being pursued, then test excavations must be conducted. Only in this way can the depth and the characteristics of archaeological components of the region be established.

Now the archaeological picture should be sufficiently clear to evaluate the real impact of the planned intervention. The archaeological impact assessment in itself can be considered to be finished, but mitigating actions may be warranted. Setting aside an eventual follow-up, it should be clear that, at this stage, already a lot of energy and resources will have been allocated while the possibility of cancelling the intervention still lures. This may result in very tough decisions to take, much in the same way as applies to the ecological environment, but sooner or later this is the point we should reach. If planning starts a long time ahead and if it is integral, that is integrating all aspects of sustainable development, then a lot of surprises can be avoided and enough time will remain to either conduct the necessary research or look for alternatives. Ultimately, the real anticipatory approach, as stipulated by SEA, should prevail, thereby transcending the narrow and restrictive project-by-project approach.

Following up on AIA

Once the archaeological picture is crystal clear, decisions with regard to the project must be taken. Several options may be present and we shall run through them briefly, from no special action required to project cancelled.

After an exhaustive study of the project impact area there may be no evidence at all of standing or buried archaeological remains. Obviously no special action is then required and the planned project can go ahead unmodified.

The archaeological study may have come across remains of the past that will be impacted by the project activities as originally set out. Different possibilities for the conservation of the cultural elements arise:

1 They may be left untouched by changing the project.
2 They may be safeguarded against the effects of the project.
3 They may be saved by removing them from their original location.
4 They may be destroyed.

In order to arrive at the best solution, many factors must be studied, compared and weighed for each option and a lot will depend on the kind of project activity and the type of archaeological remain. In order to illustrate these options let us suppose that a development project will inundate part of a valley with the aim of providing drinking water, irrigation and electric power to the populations downstream. After having studied the area archaeologically, the project may be approved on the condition that certain modifications be introduced and complementary activities be taken up. The modified version of the project may reduce the height of the dam, limiting the flooding so that archaeological sites may be left untouched (1). The most important remains may be surrounded by a protective wall, creating sunken islands (2), or they may be dismantled, transported to higher grounds and reassembled (3). A few sites, considered less important because they are repetitious or have been studied thoroughly, could be sacrificed (4). However, a site should never be destroyed without being properly registered, meaning that its location and characteristics are carefully described. Because each archaeological site has a story to be told, destruction should be the last resort, conservation being the prime target. Nevertheless, trying to preserve every single testimony of the past is utopian and undesirable. It would turn the whole world into a showcase, denying the right to progress. The ultimate goal of conservation is sustainable development, not stagnation by exaltation of antiquity.

It should be clear by now that any form of AIA and its follow-up will inevitably cause additional time and money requirements on the part of the project. These incremental costs should be considered in the same way as the present environmental impact consequences of development projects. Funding facilities should be available for cultural/archaeological impact mitigation, much in the manner they operate for the impact on the ecological environment. In this sense it is a promising perspective that the World Bank encourages borrowers to incorporate into their environmental action plans any major cross-sectoral issue, including cultural and natural heritage. The World Bank considers the conservation of important cultural property to be part of the sustainable development process because the important cultural values of a society or social group are frequently reflected in cultural property (UNDP/UNESCO Regional Project for Cultural, Urban and Environmental Heritage, 1995: 49–60). Also UNDP has expressed and practised the linkage between its global urban development co-operation strategy and the cultural preservation of historic cities, especially in Latin America and the Caribbean.

IS THERE A FUTURE FOR THE PAST?

Protecting sites for conservation

Given the importance of the archaeological heritage as a key factor in the identity of people and as a potentially rich source for traditional knowledge and know-how, given also the possible omnipresence of the archaeological patrimony and therefore its proneness to destruction, given finally the ever-increasing lack of already scarce resources for research and *ad hoc* conservation, mechanisms should be developed to go beyond the project-oriented approach, that is to anticipate possible problems during project preparation, much in the way that the EIA approach is now trying to move towards an SEA approach.

The strategic approach in archaeological conservation would involve vast programmes of archaeological reconnaissance and survey, in order to establish the extension and characteristics of the heritage, followed by effective and efficient measures to assure its long-term conservation. Indeed, if we want to protect our common patrimony, we first need to know precisely where it is located and what the most immediate dangers are. We therefore suggest we start out around big cities or other places that are expanding rapidly so that once specific urbanisation plans appear we already know how to adapt them to the situation of the archaeological heritage (see Box 12.3).

Important archaeological sites or areas may eventually become real archaeological parks, that is well-protected areas where any kind of disturbing activity is banned or controlled by law. The number of these parks, at any time, will probably be limited and reserved for very special monuments, but a fair amount of archaeological sites may be combined with existing ecological reserves or natural parks, thereby enhancing these with the human aspect, while profiting from the protective ruling. Where the previous solutions are not feasible, archaeological sites should be turned into green areas within the urban perimeter. The kind of activities allowed in these places will depend on the characteristics of the site. Standing remains should be protected against direct contact. Buried sites could serve as a recreation area as long as interference with the subsurface archaeological layers is avoided. Sometimes adding a protective cover of topsoil may create a perfect archaeological reserve, while allowing some social function.

Monitoring

Once a clear picture of the archaeological heritage exists and the different degrees of conservation categories, with their specific conservation measures, have been established, a matching system of monitoring should be devised. Monitoring is here understood to be the periodic evaluation of the state of conservation of an archaeological site.

According to the kind of site and the protection measures that apply to it, monitoring can range from a simple inspection visit in the case of public green areas, to a specialised and integral monitoring mission when checking on important monuments (Buys, 1995). It should be carried out by a specialised national institution, normally the same one that establishes and assigns the conservation categories.

In any case, special care must be given to an adequate level of reporting of the monitoring exercise,

---

BOX 12.3   MANAGING URBAN EXPANSION

Quito, the capital of the Republic of Ecuador, is located on a step-like terrace of the interandean valley, that runs North–South along the Western Cordillera. A long-stretched city of a little over 30 km long by 5 or 6 km wide, at 2,800 m above sea level, and almost exactly on the Equator, it is endowed with year-round spring weather and a rich variety of ecological niches in the immediate surroundings.

No wonder human occupation in and around Quito is among the oldest in Ecuador, dating back as far as 7000 BC with well-established settlements in what is now the city from 1500 BC onwards. The Neolithic village of some 26 hectares and 700 inhabitants has now become a capital that occupies approximately 20,000 hectares and houses 1.1 million people, and is growing rapidly (Ilustre Municipio de Quito, 1991).

The Municipal Planning Unit has analysed the historical information on urban growth over the last 85 years, establishing a decreasing growing rate, but still projecting a doubling of the urban area to over 40,000 hectares by 2020, and a tripling of the population to almost 3.5 million. On the other hand, recent archaeological research in the Quito Valley by the Belgian Agency for Development Co-operation has established a density of one archaeological site per 2 km^2, based on a systematic regional survey by sampling (Buys, 1994).

The conclusion from both types of information can only be that over the next 20 years a total of 100 archaeological sites will be lost, due exclusively to the urban expansion of the capital. Therefore, a project was suggested, and will hopefully be implemented some day, that would develop the capacity of the municipal services to anticipate the emergency situations that will occur once effective urbanisation starts.

The project would basically aim at the organisation of a municipal archaeological service which, in close co-ordination with the office for urban planning, would design a survey programme systematically to cover all the areas that will eventually become incorporated into the city. The presence and characteristics of ancient sites would be registered and evaluated in order to assign conservation categories and priority degrees for their protection. According to this system, urbanisation plans will have to take into account the archaeological heritage and adapt to it. The project would also contemplate campaigns of awareness-raising in the other municipal services and at the level of the general public.

Although the basic research preceding this conservation classification, i.e. survey and excavations, will require specific human and financial resources, positive outputs may be expected in terms of understanding the region's archaeology, and in terms of significant economic gains at the time of urbanisation, by avoiding costly delays in work already started.

---

developing a standardised reporting system so that the information gathered over time may be comparable. That information should include aspects other than the mere physical condition of the site, covering its use and abuse, its social and economic functions or potentials, paying special attention to the informal economy that tends to build up around tourist places.

Awareness-raising and capacity-building

In a field that is as new as environmental management and archaeology, much awareness-raising and capacity-building is required. In many developing countries a lot has still to be done on the level of basic archaeology teaching. Indeed, with a few notorious exceptions, professional training leaves much to be desired and when formal teaching exists, it is often concentrated on the theoretical level, leaving the students with hardly any field experience by the time

they graduate. Whenever these young professionals get involved in emergency situations or regular research, their work clearly shows those training deficiencies.

This situation has originated a kind of intellectual or cultural colonialism on the part of foreign scholars or graduate students, who are eager to find new fields of study in the developing countries. They used to invest time and money to explore these lesser-known regions and become experts and founders of new archaeology departments at home. Fortunately, many of the early explorers have taught local people in the field or have provided grants to study abroad. However, even today we must ascertain that there still are, mostly at the doctoral dissertation level, European and American graduates who exploit intellectually the archaeological heritage of a developing country by conducting their research without feeding back its results to the country that generously hosted them.

In order to take up the archaeological heritage in the environmental management, capacity-building should be applied to different levels. At the highest level, political will must be created by informing the policy-makers about the importance of the cultural and archaeological heritage in identity-building and sustainable development, emphasising its educational role and taking into account its economic potential. Ministries of culture and education, national institutes for the cultural heritage or national museums for history and archaeology should take the lead in this awareness-building, seeking examples of proven good practices in other countries. Enhancing the capacity to cope with the more practical action plans on the national level may require specific assistance from outside.

In the professional realm, a quality basic training in archaeology at the national university level should be assured, complemented with specialisation possibilities abroad, through grant agreements with foreign academic or technical institutions. Emphasis should be put on the interdisciplinary approach of archaeological research and on the social function the discipline should pursue (see above), beyond the pure scientific immediate results. General notions about the objectives and functions of archaeology should also be introduced in related disciplines, especially those that are frequently in the field, that is natural sciences but also, and very importantly, urban planning and general development.

At the service level, much archaeological information is lost through lack of reporting or destruction by ignorance, due to the fact that fieldworkers of companies or institutions that take care of electricity and telephone distribution, of water or sewer infrastructure, etc. have not the slightest clue to the importance of the remains they come across. These people especially are in constant contact with subsurface layers in the urban and rural area, and countless stories have been heard of the many fortuitous finds they have discovered during work. Workers and field supervisors, as well as the higher administrative levels of these companies and institutions should therefore receive some basic knowledge of how the past may appear in their daily work, who they should report to and what emergency measures eventually could be taken.

Finally, at the level of general education, starting in primary school and from there up, the young should not only learn about their country's history on a textbook basis, but should be confronted directly with whatever remains of that history are left. Raising awareness about the importance of conserving the evidence of past cultures should be developed together with building up the ecological conscience. Basic education probably represents the most important level on which to work for this purpose, since it means both investing in the future – this youth will soon become the main actor in sustainable development – and creating a multiplicator effect through the influence youngsters exert on their parents.

International co-operation

Our self-imposed programme for the future, Agenda 21, covers a wide array of hopes and aspirations for a better future, intrinsically relating the environment to the economics and social aspects of development. Several legally binding offsprings of the Rio Conference are well underway in taking effective and efficient measures to put ideas into action. However promising and necessary all this is, the cultural dimension of development is still missing. Little regard is given to the uniqueness of the social systems that managed the environment in ways that were much more sustainable than our 'modern' alternatives. At the most, brief references are made to the need for recuperating local know-how, but only very rarely is mention made of the preservation of indigenous culture. Since no precise definitions are usually given, the question of how culture in this context is exactly conceived remains wide open. One strongly senses that it is rather limited to folk culture, the present 'exotic' aspects of foreign people, not the rich totality of traditional beliefs, knowledge, art, technology, etc. that has its most genuine expression in the archaeological heritage.

One of the reasons for this lack of cultural dimension may be due to the fact that other bodies have been involved in these problems at an earlier stage and, at the same time, from a more technical point of view. The international organisation that first comes to mind is UNESCO, which adopted the World Heritage Convention in 1972 – the same year of the Stockholm Conference on the Environment – in order to protect the *natural and cultural* heritage against increasing threats of destruction by natural and human causes. Earlier conventions deal with the protection of cultural heritage in armed conflict (The Hague, 1954) and with the measures to prevent illegal import, export and transfer of cultural heritage (Paris, 1970). So, apparently, several international agreements are taking care of the cultural and archaeological heritage. In effect, a lot has been done, but past and present efforts were not able to, and still can't, keep up with the ever-increasing pace of development and the subsequent destruction of ancient remains. UNESCO alone can't handle the situation, although it has been building partnerships on a regional or national basis everywhere.

Clearly, synergy must be developed if we want sustainable development to become a truly integral approach. In the same way that the mixed sites, that is natural and cultural heritage sites, on the World Heritage List make up only 3.86 per cent of the total (Mujica, 1995), cultural considerations within the purely environmental realm are equally insignificant. Chapter 26 of Agenda 21 refers to the ILO Convention 169 on indigenous and tribal peoples (Tomei and Swepston, 1995). However, neither Agenda 21 nor the ILO Convention contain any direct reference to the safeguarding of the indigenous cultural heritage, although both admit the importance of traditional knowledge and know-how concerning the management of natural resources. The same recognition of local knowledge, and appeal to use it, can be seen in the Convention to Combat Desertification, while the Convention on Biological Diversity shows only very indirect reference.

It seems then that a better co-ordination between the existing international agreements is warranted. Probably a combination of environmental conventions and cultural agreements, as a legally binding mechanism, could do the job. The right formal framework and practical organisation of such a co-operative effort should be established as soon as possible. The conclusions and guidelines that might result from this negotiating process should be incorporated into the overall sustainable development policy. Only if there is a concerted exercise at the international level with enough technical and moral authority can we expect the individual countries to start effective actions on behalf of the cultural and archaeological heritage. The environment will then have broadened its definition, beyond the aspect of mere natural resources, to the integral concept of environment with an emphasis on the interaction of man and his natural environment through time, the historical experiences of this relationship and lessons learned therefrom for their application in the future.

CONCLUSIONS

Environmental management, as a means for attaining sustainable development, must encompass all aspects of the human environment, including natural and cultural elements. When this objective is clear to all stakeholders, recognising the fact that certain cultural elements and archaeological remains contain knowledge and know-how that is applicable today as an alternative to current, negative practices regarding the natural environment, then the cultural/archaeological heritage should become an integral part of any form of environmental management. Consideration of this heritage should become fully integrated into the initial screening and scoping of development projects on their environmental effects and, if necessary, an archaeological impact assessment (AIA) should be fully incorporated into the EIA procedure. As a result mitigating actions or simple disapproval of a project may occur, but monitoring should be ongoing. Needless to say, for this suggested course of action awareness must be raised at all levels and local capacities must be systematically increased. It seems logical that this objective could best be reached within the framework of existing, international and legally binding environmental agreements, but if this proves to become difficult, then a special convention may be warranted. The concept of global benefit also applies to the archaeological aspect of the environment, since this heritage represents our collective memory, locally but also universally, and the loss of any culture, past or present, is tantamount to the loss of a biological species. It would be ridiculous to preserve biological diversity while disregarding the diversity of our own species.

REFERENCES

Brent, M. (1994) 'Pillaging Archaeological Sites', *Interpol Review* 448/449: 25–36.

Buys, J. (1994) *Investigación arqueológica en la Provincia de Pichincha*, Quito, Ecuador: Ediciones Libri Mundi (in Spanish).

Buys, J. (1995) 'Keeping an Eye on the Past: Monitoring Archaeological Sites as a Conservation Tool', in UNDP/UNESCO Regional Project for Cultural, Urban and Environmental Heritage, *Systematic Monitoring Exercise. World Heritage Sites Latin America, the Caribbean and Mozambique. Findings and International Perspectives, Report 1991/1994*, Lima, Peru, pp. 85–88.

Buys, J. and Muse, M. (1987) 'Arqueología de Asentamientos Asociados a los Campos Elevados de Peñón del Río, Guayas, Ecuador/chapter 11', in W. Denevan, K. Mathewson and G. Knapp (eds) *Pre-Hispanic Agricultural Fields in the Andean Region*, Oxford: BAR International Series 359 (ii), pp. 225–248 (in Spanish).

Denevan, W., Mathewson, K and Knapp, G. (eds) (1987) *Pre-Hispanic Agricultural Fields in the Andean Region*, Oxford: BAR International Series 359 (i) and (ii).

Ilustre Municipio de Quito (1991) *Quito Actual, Fase I, Medio Ambiente y Población*, Quito, Ecuador: Dirección de Planificación, Plan Distrito Metropolitano (in Spanish).

Johnson, S.P. (1993) *The Earth Summit. The United Nations Conference on Environment and Development (UNCED)*, London, UK: Graham and Trotman.

Knapp, G. (1984) 'Soil, Slope, and Water in the Equatorial Andes: A Study of Prehistoric Agricultural Adaptation', Ph.D. Dissertation, Department of Geography, Madison, Wis., USA: University of Wisconsin.

McEwan, C. (1982) 'Asientos de Poder: evolución sociocultural de Manabì, Costa Central del Ecuador', paper presented at the 44th Congress of Americanists, Manchester (in Spanish).

Mujica, E. (1995) 'Mixed Site Monitor: The Machu Picchu Experience where the Works of Man and Nature seem Contemporaneous', in UNDP/UNESCO Regional Project for Cultural, Urban and Environmental Heritage, *Systematic Monitoring Exercise. World Heritage Sites Latin America, the Caribbean and Mozambique. Findings and International Perspectives, Report 1991/1994*, Lima, Peru, pp. 69–73.

Muse, M. and Quintero, F. (1987) 'Experimentos de Reactivación de Campos Elevados, Peñón del Río, Guayas, Ecuador/chapter 12', in W. Denevan, K. Mathewson and G. Knapp (eds) *Pre-Hispanic Agricultural Fields in the Andean Region*, Oxford: BAR International Series 359 (ii), pp. 249–266 (in Spanish).

Parsons, J. and Shlemon, R. (1987) 'Mapping and Dating the Prehistoric Raised Fields of the Guayas Basin, Ecuador/chapter 9', in W. Denevan, K. Mathewson and G. Knapp (eds) *Pre-Hispanic Agricultural Fields in the Andean Region*, Oxford: BAR International Series 359 (ii), pp. 207–216.

Tomei, M. and Swepston, L. (1995) *A Guide to ILO Convention No. 169*, Geneva: International Labour Organisation.

UNDP/UNESCO Regional Project for Cultural, Urban and Environmental Heritage (1995) *Systematic Monitoring Exercise. World Heritage Sites Latin America, The Caribbean and Mozambique. Findings and International Perspectives. Report 1991/1994*, Lima, Peru.

UNESCO (1991) *UNESCO Sources* 28, July/August: 12–13.

Walden, D. (1991) 'Raiders of the Cultural Ark', *UNESCO Sources* 28, July/August: 6–8.

World Commission on Environment and Development (1987) *Our Common Future*, New York, USA: Oxford University Press.

SUGGESTED READING

INTERPOL (1994) *International Criminal Police Review*, 448/449, Lyon, France: International Criminal Police Organization.

UNDP/UNESCO Regional Project for Cultural, Urban and Environmental Heritage (1995) *Systematic Monitoring Exercise. World Heritage Sites Latin America, the Caribbean and Mozambique. Findings and International Perspectives. Report 1991/1994*, Lima, Peru.

SELF-ASSESSMENT QUESTIONS

1 Is the ultimate objective of integrating the archaeological heritage into environmental management:
   (a) to establish ancient knowledge;
   (b) to learn about history;
   (c) to attain sustainable development.

2 Would you define culture as:
   (a) the material expressions of a society, like art, technology or architecture, and the ideas and beliefs embedded in it;
   (b) learned behaviour, transmitted from generation to generation, for coping with both the natural and human environment;
   (c) behaviour developed by indigenous people to adapt to disrupting interventions by foreign populations.

3 Archaeology may be defined as the scientific discipline that studies past cultures based on their material remains, in order to:
   (a) explain the evolution of the interaction between populations and the natural environment, with a special emphasis on adapted technology;
   (b) understand the development of standing remains and objects, including all things produced, used or modified by man;
   (c) build a sequence of cultural events, reconstruct past ways of life and infer broad historical processes.

4 Suppose the initial screening of the impact area of a development project reveals that no archaeological remains exist. Would you:
   (a) ask for additional research;
   (b) study the possibilities for changing the project;
   (c) clear the project for approval.

5 A fast growing city is advancing rapidly into areas that are archaeologically unexplored. Being in charge of the urban planning you:
   (a) start raising funds for emergency excavations whenever archaeological remains may appear;
   (b) make arrangements for an archaeological service to be set up and become operational when urbanisation advances;
   (c) initiate a programme of systematic archaeological survey in order to determine where sites are located and how urban expansion should adapt to them.

6 You are the head of a very successful National Institute for Cultural Heritage, having established the national map of archaeological sites, and receive a considerable budget every year that you can dispose of freely. You decide to:
   (a) produce numerous publications on the basis of the archaeological inventory in order to promote tourism in your country;
   (b) design a vast programme for monitoring the archaeological sites, excavating a representative sample of them and organising national campaigns of awareness-raising and capacity-building;
   (c) organise international conferences on archaeological heritage in order to communicate your experience and good practices to colleagues who are still struggling back home.

7 Do you think there is a future for the past by:
   (a) trying to combine international agreements on both the natural and cultural dimension into new legally binding instruments that would promote integral sustainable development;
   (b) pressing international cultural organisations to increase activities in their specialised fields of action;
   (c) putting more emphasis on the recuperation of local knowledge and know-how in the existing, legally binding, international environmental conventions.

13

ENVIRONMENTAL MANAGEMENT OF LANDSCAPES
Landscape Ecology

Bostjan Anko

SUMMARY

Traditionally the conservation of the environment has been species- or site-oriented, and the concept of the 'ecosystem' management approach has only recently begun to gain momentum. Now, however, the fragility of whole landscapes is becoming an issue. Extending conservation endeavours to the level of landscapes has significantly added to the complexity of conservation planning and management processes.

In this context, cultural landscape has to be considered not as an arbitrarily chosen part of the Earth's surface but as an ecological system with a distinct structure, functioning and trends of change. At the same time cultural landscape should also be viewed as a social system.

In the last decade particularly, understanding the natural and social complexity of landscape has been made easier by means of new tools such as GIS and similar systems, modelling, environmental impact assessment, etc. These tools might be also instrumental in bridging the gaps between individual sciences/professions active in landscape study and management.

A thorough understanding of landscape as an ecological system as provided by landscape ecology is essential for attaining two basic environmental management goals: (dynamic) sustainability and biodiversity – at all levels. To achieve this, landscape ecology has to operate as a cross-sectional discipline, strictly demanding that the functioning of landscape structure and changes be fully understood also in terms of social, economic and cultural givens and processes.

As a living and social system landscape is constantly changing. From the management point of view, understanding such change – its causes and directions – is essential. Landscapes have to be evaluated constantly for their natural and socio-economic parameters as well as for their non-material values. Due to the complexity and diversity of landscapes and their (natural and social) environments it is not possible to develop a universally valid set of planning and management procedures. Nevertheless, European experience does offer some principles and approaches that seem to be applicable in most situations.

ACADEMIC OBJECTIVES

The aim of this chapter is to introduce the cultural landscape as an ecological and social system and to discuss some issues related to environmental planning and management at the landscape level.

After having read this chapter the reader should be able to:

- distinguish among abiotic, biotic and noospheric layers of any landscape;
- identify relevant aspects of landscape structure and functioning;
- identify the moving forces in the landscape social sphere;
- recognise the present and future processes of landscape changes, and their causes;
- prepare (in very general terms) a landscape evaluation scheme for parameters relevant from the management point of view;
- appreciate the complexity of landscape planning and management;
- outline (in broad terms) the approaches relevant for a concrete landscape planning and management procedure.

INTRODUCTION

> Managing a river includes also management of the human uses of the surrounding drainage basin in the context of changing institutional and jurisdictional arrangements ... in order to foster the health of the river ecosystem as a self-organising system and to avoid its degradation.
>
> (Tolba *et al.*, 1992)

The case of integrated river management is a good illustration of the complexity entailed in such an approach. It points out clearly that:

- the river is to be considered a multi-purpose ecosystem and not just a flow of water in space and time;
- the river as ecosystem interacts with many other ecosystems;
- the utility of the river to people ought to be optimised rather than maximised, by taking into consideration the health and viability of the river and of the other ecosystems interacting with it.

Similarly, integrated landscape management should also consider that:

- landscapes are ecological systems and not just four-dimensional phenomena;
- cultural landscapes in particular are heavily dependent on human factors;
- the sustainability of cultural landscapes can only be secured as a compromise between natural givens and human requirements.

Ecosystem management has been making rather slow progress, not only because of its complexity, but also because of some radical changes in thinking and in attitudes. The same applies to landscape integrated management. An ecosystem – a river, field or forest, for example – is relatively easy to define and to grasp conceptually. A landscape, however, is much more complex, but this should not prevent us from trying to deal with landscapes in the context of environmental management. Several ground-breaking political documents, notably the Pan-European Biological and Landscape Diversity Strategy as endorsed at the Ministerial conference in Sofia, in October 1995, are clearly marking future developments in this area of environmental management as well as in landscape ecology.

LANDSCAPE AS AN ECOLOGICAL SYSTEM

Environmental management at the landscape level requires accepting landscapes as 'living systems (those containing life)', exhibiting structure, function and change (Forman, 1995).

However, very few landscapes we see today could be considered natural (primeval) ecological systems, that is without any traces of human presence or activities. The palaeolithic hunters who required 10 square kilometres per person to feed themselves (McHale, 1972) hardly left any lasting impacts upon the primeval landscape. Subsequently, landscapes marked by herding, agriculture, industry, urbanisation or leisure increasingly bear the imprint of a given social sphere, of a certain culture. Thus, such cultural landscapes can be considered as consisting of three closely interwoven layers: the abiotic, the biotic and the social.

There are many definitions of landscape, which vary according to the authors' background and the given purpose of defining landscapes. In the early nineteenth century, the geographer A. von Humboldt defined landscape as the 'total character' of an Earth region. The biogeographer Troll, who coined the expression 'landscape ecology' in the late 1930s, defined landscape (in 1971) as the 'total spatial and visual entity' of human living space, integrating the geosphere with the biosphere and its noospheric man-made artefacts (Naveh and Lieberman, 1994). Landscape ecologists Forman and Godron (1986) defined landscape as a 'heterogeneous land area composed of a cluster of interacting ecosystems that are repeated in a similar form throughout'.

Most of the current definitions of landscape seem to converge towards defining landscape as an ecological system, that is the spatial and temporal expression (resultant) of interacting ecosystems and their natural and social environment, characterised by specific structure, functioning and developmental trends, all of which make it distinctly different from adjacent landscapes.

Thus, a cultural landscape can be viewed as a system of ecosystems – most of which are controlled to some degree by 'human intelligence' (Vink, 1975) through land utilisation and management. Accordingly, a landscape can be placed in the ecological organisational hierarchy between ecosystems and biomes (Odum, 1989) or ecosystems and regions (Forman, 1995). 'Landscapes vary in size, down to a few kilometres in diameter' (Forman and Godron, 1986).

Such statements should give an idea of the size and complexity of landscapes. However, in seeking to understand a landscape as an ecological system we should not merely ask 'how big is a landscape?', but also 'how small can a cluster of ecosystems be, and what attributes should it have, for us to be able to call it a landscape?'

People have very different perceptions of landscapes. Most laypeople will claim that the landscape is the area surrounding them at the moment, which they can perceive through their senses. Artists, particularly poets, painters and musicians tell us of their own visions and perceptions of landscapes. Is it possible to argue with Lorca's pictures of the Spanish countryside, with Mickiewicz's rendering of Crimean landscapes, or with the poetic paintings by Heine or Robert Frost? Did not the Dutch landscape painters of the fifteenth century onwards, or the Canadian 'Group of Seven' leave landscapes behind? And is not the purest rendering of the essence of landscape what we hear in the music of Sibelius' *Finlandia* or Smetana's symphonic cycle *My Fatherland*?

It most certainly is; it is not the only possible way of grasping the landscape, but it is an important one. The aesthetic and spiritual dimensions of landscapes should be considered in any kind of dealings with them. Enhancing the landscape's general well-being – and not just its efficiency and effectiveness – is also one of the imperatives of landscape management.

Landscape management requires identification of as many meaningful, quantifiable landscape parameters as possible. Determining these parameters still seems to be one of the major obstacles in developing a general landscape management theory based on ecological principles.

If we truly view a landscape as an ecological system, the following parallel should not be difficult to accept. By clearly defining what is to be understood under ecosystem structure and functioning, Odum (1965) greatly enhanced ecosystem studies – and their comparability. Thus he significantly contributed to the development of ecosystem theory.

Similarly, it should be possible to contribute to the progress in theoretical and practical dealings with landscapes by adjusting Odum's scheme for ecosystem structure and function to deal with landscapes (Anko, 1986).

'Reading' landscapes means assessing three of their characteristic features, that is structure, function and change. No matter how we look at landscapes, these three landscape characteristics are always the focus: we always want either to change or to preserve them. In addition they lend themselves to landscape studies (and management measures) because they are mostly quantifiable.

Landscape structure

The structure of any ecological system is relatively easy to observe, quantify and express in one way or another. It is a reflection of the natural givens and their amplitude, and also (in the case of anthropogenic systems) of the level and mode of human interference.

Structure is a result of past and present givens and processes. At the same time it determines the mode and scope of functioning of the system, at present and in the future.

There are various possible approaches to dealing with landscape structure: landscape architects, for example, emphasise the visual characteristics and qualities of landscape and distribution patterns of landscape components. Forman and Godron (1986) and Forman (1995), on the other hand, interpret landscape structure in terms of matrices, patches, corridors and networks, which is the most useful interpretation for landscape planning and management. From a purely landscape-ecological point of view, however, it might be interesting to draw a parallel between Odum's interpretation of ecosystem structure and that of a landscape.

At the ecosystem level, Odum dealt with various aspects of structure in terms of populations. At the landscape level, however, the role of structural units will be assumed by ecosystems.

The comparison of an ecosystem and landscape structure study scheme is shown in Table 13.1.

'Reading' landscapes begins with reading and interpreting their structure. Noting and correctly

Table 13.1 Structure of the ecosystem and of the landscape

Ecosystem	Landscape
1 Composition of community with regard to population: species; numbers; biomass; life history; distribution in space. 2 Quantity and distribution of abiotic matter, e.g.: nutrients; water. 3 Range or gradient of conditions of existence, e.g.: temperature; light.	1 Relief 2 Composition of landscape with regard to ecosystem: kinds; numbers; life history; distribution in space. 3 Abiotic matter needed for primary production: kinds; quantities; distribution (spatial, temporal); availability. 4 Energy conditions: temperature (vegetation period, extremes); insolation; other sources of energy. 5 Quantity and structure of plant biomass: in individual ecosystems; in the entire landscape (comparison between primeval and actual landscape). 6. General observations of human influence upon cultural landscape: genesis; stability; non-material values; direction of changes.

interpreting certain landscape-ecological features is a prerequisite for environmentally correct dealing with landscapes. Therefore some brief explanations of and comments on the items in Table 13.1 may be needed.

In most ecosystems relief is a homogeneous feature. Change of relief most frequently results also in change of ecosystems. In landscapes, however, relief characteristics or changes (e.g. watershed divides, basins, altitudinal belts, floodplains, escarpments, etc.) often do not just represent landscape boundaries but also influence ecosystem (land-use) distribution patterns. Relief energy, reflecting the altitudinal differences per unit area and expressed as potential energy (Kerényi, 1981), is an interesting measure of erodibility, additional transportation energy costs, or severity of disturbances caused, for example, by road construction or any similar intervention.

Listing all the ecosystems in a given landscape yields an 'ecosystem inventory'. Ecosystems can be arranged with regard to their increasing distance from the natural state – either in terms of necessary artificial energy or material inputs required to maintain them or as a ratio of total (natural and artificially introduced) energy flow needed to maintain a unit of biomass. At all events, the presence, prevalence or absence of given ecosystems (e.g. vineyards, rice-paddies, wetlands or forests) does give an approximate idea of the type of landscape in question.

In natural landscape, a greater number of ecosystems present indicates a greater natural diversity

of the abiotic sphere. In cultural landscapes where the common trend seems to be toward homogenisation, a greater number of ecosystems (land uses) can mean greater pressure upon land, or a greater degree of self-sustainability of the local population. In all cases special attention is to be given to the numbers of the so-called minority ecosystems (barren rock, springs, small wetlands, abandoned quarries or gravel pits, ponds, etc.). They may not have any immediate economic utility, but they are most important for the conservation of biodiversity.

Like populations, ecosystems too have life cycles of various duration. Thus a landscape may be viewed as a well-orchestrated blend of ecosystems with various lengths of life cycles. Even in cultural landscapes it is important for their ecological stability and for the richness of life-forms that they consist of the whole range of ecosystems – from those that have an extremely long life cycle (such as a forest) to those whose life cycle is completed in a matter of months (such as a cornfield).

In natural landscapes the distribution of ecosystems most frequently follows some gradient (water or nutrient availability, insolation, elevation, etc.). In addition to natural gradient distribution some other factors influencing the (cultural) ecosystem distribution patterns can be observed in cultural landscapes, such as proximity to rural settlements or large urban-industrial centres, land division patterns among individual owners, vicinity of transportation corridors, and so forth.

Gathering information on the abiotic matter needed for primary production usually requires lengthier and more demanding studies. In old cultural landscapes the presence of fields may also indicate a relative abundance of nutrients, while the absence of fields or settlements in seemingly suitable localities may indicate more or less frequent flooding (overabundance of water) – or drought. In many cases we have to resort to soil or hydrological maps – if these are available. Some basic knowledge of biogeochemical cycles (particularly for the critical elements) is also needed to understand the factors influencing the (un)availability of certain nutrients.

Energy plays an important role in landscape structuring. Temperature conditions expressed in terms of the length of vegetation period, dates of late and early frost or by means of extreme values are certainly a factor that is relevant from the agricultural point of view, but can also significantly influence settlement patterns, living conditions, or the development of tourism, for example.

Many land uses have relatively well-defined insolation niches (Anko, 1986): in the temperate zone, for instance, too high an insolation may mean too short a snow cover – and no skiing area as a result; on the other hand, particularly with the growing elevation or geographic latitude, people will be searching for south-facing slopes for vineyards or fields. Forests have been often left on extreme sites – with the lowest or highest insolation (causing drought) – and so forth.

Other natural sources of energy such as water, wind or thermal energy have often been instrumental in shaping the settlement patterns and changing cultural landscapes. The energy of water streams gave rise to the milling industry, ironworks, water traffic, etc. (Electricity generating) windmills, indicating the presence of relatively steady winds, are as much a part of the modern Danish landscape as they traditionally used to be in Holland. The discovery of thermal springs often brings radical changes in landscape structure due to development of tourism, which may change the traditional land-use patterns.

In any ecological system the quantity of biomass is a good indicator of its potential for life. In ecological analysis of biomass at the landscape level we generally limit ourselves to plant biomass alone, for purely practical reasons: it is more predictable, easier to estimate and is also more abundant by several orders of magnitude than the animal biomass. However, it is not just the quantity of biomass that is relevant in the analysis of landscape structure. Only a reasonably high permanent stock of diversely structured plant biomass guarantees the desirable diversity of life at all four trophic levels.

By their very nature, not all the ecosystems within a landscape can meet such a requirement. Therefore in landscape management special attention should be given to the share, appropriate spatial distribution and general conservation of the ecosystems with a permanently high stock of biomass, structured in a well-balanced manner including ecosystems containing a high content of dead biomass such as forests or some types of wetlands.

Cultural landscape is an ever-changing artefact. This very fact makes landscape ecology a bridge connecting the natural and social sciences. The slow but unavoidable metamorphosis of natural into cultural landscape made the original ecological system 'landscape' cultural (social) also. In this context the word 'cultural' (landscape) is to be taken in its original meaning, as derived from the Latin verb 'colere', meaning to till or cultivate. Through innumerable acts of tilling and cultivation which filled the nature with artificial ecosystems and other works of the human mind and hands, the landscape itself became an artefact, faithfully reflecting the society that created it, or was trying to change it. Thus it is important to understand whether the cultural landscape we are dealing with is an ancient or relatively recent one.

No primeval landscape would have been colonised in the same manner and form by two different cultures. For that reason it is interesting to know about the beginnings of continuous settlement and traces in landscapes that individual periods, that is cultural layers, left behind. Their presence is a testimony to the harmony between natural and social structures and functions. If we perceive the origins of cultural landscapes in this manner, we may see that the structural instability of any given landscape is most frequently brought about by instability (disturbances, gradual or radical changes) in the social sphere. Thus, predictable developments in the latter (from global to local level) are the best means of assessing the present and future stability and direction of changes of a given landscape.

Landscape functioning

The functioning of landscapes is a sum of interactions among the landscape components, that is ecosystems and their environment. These interactions can be expressed in the form of flows of matter, energy and species. In the case of cultural landscapes, flows of information *sensu* de Groot (1992) can be considered important too.

The above-mentioned flows are much more difficult to assess than the structural characteristics of landscapes. Although there is a clear correlation between landscape structure and functioning, not much attention has been given so far to the study of the latter, even though energy and material flows in landscapes are often targets of human intervention.

Changes of cultural landscape structure occurring as a result of human (un)planned interferences inevitably bring about changes in flows which may not seem to be directly related to them. They may occur at a distant place and/or at a much later time. They cause new structural changes, and are often perceived as 'environmental problems'.

Flows of matter, energy, species or information are seldom studied at landscape level. We may have fairly good ideas about the cycling of major chemical elements on the global or continental level (Butcher *et al.*, 1994). However, energy flows (particularly artificial ones) are more frequently studied within

various administrative boundaries and not in conjunction with the natural energy flows that characterise landscape systems.

Neither the structure nor the change of landscape can be understood without some basic understanding of its function. Studying landscape functioning may be difficult and may not promise immediate returns. Yet it does require different approaches to landscape as an ecological system, and forces us to ask sometimes quite unconventional questions such as 'What is the energy cost of maintaining (or preserving) a given cultural landscape?', or 'What is the material input needed to support the change of a traditional agricultural landscape into a modern one?'

As in the case of the change of landscape structure, landscape functioning too can be studied following the E. Odum scheme of ecosystem function (Odum, 1965) as shown in Table 13.2.

It takes a great deal of skill and practice to be able to read the functional specifics of any given landscape. Tracing flows (either energy or material ones) opens up a fascinating perspective of the landscape as ecological system.

The study of energy flows in landscapes is not limited to biological flows but includes all kinds of energy, powering various processes: those of living and non-living nature as well as those that are by and large manipulated by humans.

It is not just the size and quantity of these flows that concern us in this regard, but also their qualitative characteristics such as their origin (natural vs. introduced by humans), speed, predictability, etc. From the system point of view it is interesting to be

Table 13.2 Functioning of the ecosystem and of the landscape

Ecosystem	Landscape
1 Rate of biological energy flow through the ecosystem, i.e. rates of production (of populations and of community); rates of respiration (of populations and community). 2 Rate of material or nutrient cycles. 3 Biological or ecological regulation: of organisms by environment; of environment by organisms.	1 Rate and quality of all energy flows: energy inputs, outputs, balance; energy efficiency; energy flows among ecosystems. 2 Characteristics of material cycles: translocation of matter by means of (in)organic processes; accumulation of matter (sinks); pulsating of biomass; openness. 3 Flows of species. 4 Mutual regulating of landscape and ecosystems with the respective environment: self-regulation and maintenance; adjustment to environment; influences upon environment.

able to assess energy inputs (natural and artificial ones) into a given landscape, the respective outputs and their short-term balance, as well as the energy efficiency (i.e. the total amount of energy required to maintain a unit of biomass) of individual ecosystems and of a landscape as a whole.

Studies of energy flows among ecosystems reveal that, just like populations of a given ecosystem, the ecosystems of a given landscape can also be divided into those providing energy and those that are dependent on such energy inflows – in addition to human-manipulated energy inputs. This broad scope of ecosystem interdependence ranging from, for example, forests to agricultural ecosystems does, in fact, resemble the trophic organisation of populations into producers, consumers, predators, etc.

Reading the general characteristics of the material flows in a given landscape may be just as difficult as following the invisible energy flows. However, it should not be difficult – with some practice – to read and identify certain areas in (natural) landscapes where such flows take place, as, for example, areas that are losing matter due to erosion, valley alluvium terrains where sediments are accumulating, avalanche paths and screes, tidal flat terrains, landslide areas, etc. Similarly it is possible to identify areas where artificial manipulations with material flows (harvesting biomass in any form, irrigation, drainage, use of fertilisers, etc.) significantly change the original (natural) flow patterns.

The above-mentioned processes may result in a more or less even distribution of matter in a given landscape: in certain cases, however, certain kinds of matter tend to accumulate in relatively easily definable areas that we consider sinks: we know, for example, that forests act as CO_2 sinks, thus reducing the 'greenhouse effect'; that many wetlands due to their filtering capacity serve as sinks for many pollutants, for example, heavy metals; that riparian vegetation often intercepts a lot of agricultural pollution (pesticides, fertilisers) and prevents it from entering the streams, etc. In a broader sense we can consider sinks also as material deposits (e.g. from some past processes that influence the natural givens and related human activities in a given area.

The pulsating of biomass in a landscape (and in its components) is particularly important from the biodiversity point of view. Ecosystems (as well as landscapes) having low and/or intensively pulsating biomass (see Figures 13.1 and 13.2) are less suitable for maintaining a wealth of life forms. They simply cannot support populations of higher trophic levels. For that reason it is important to maintain in any cultural landscape a reasonably high proportion of ecosystems with a high and stable biomass content that secures some basic level of biodiversity. In this regard there is an obvious lack of scientific knowledge to assess the best kind of management for maintaining a desirable dynamic level of biodiversity in a sustainable way.

The natural health of landscapes depends to a large extent also on the undisturbed cycling of matter between organisms and soils (Mg, K, Ca, P, S, . . .), or organisms and atmosphere (O, C, . . .) or between atmosphere, organisms and soils (e.g. N). Human interference with material cycles frequently causes disturbances in the cycling of elements.

In most cases this cycling exceeds the boundaries of a single ecosystem, or even of a landscape. Usually biogeochemical cycles are studied at the regional, continental or even global level. Nevertheless, from the management point of view it is important to have an idea about the part of such a cycle that runs through a given landscape. Material inputs, speed of flow, creation of sinks (undesirable in the case of pollutants) and the form and quality of outputs are definitely parameters to be considered in landscape management.

The extent to which ecosystems and landscapes are open or closed to material inputs or outputs is another of their fundamental characteristics. In any

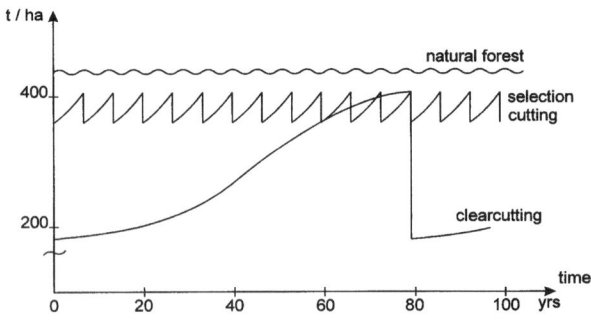

Figure 13.1 Schematic presentation of plant biomass pulsating in primeval forest and in a managed forest
Source: Anko, 1994

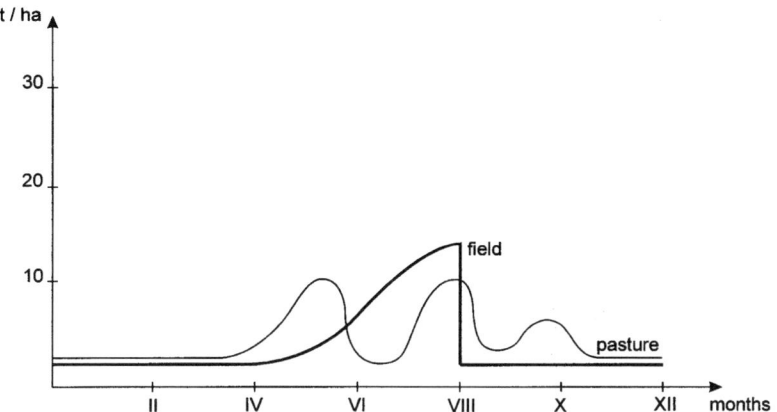

Figure 13.2 Schematic presentation of plant biomass pulsating in the field and in the pasture
Source: Anko, 1994

given landscape, it is interesting to identify the eco-systems that are adapted to huge material (organic or inorganic) exports that have to be compensated for, for example, either by weathering in the case of natural terrestrial ecosystems, or by steady application of fertilisers in the case of agricultural systems. The same applies in the case of landscapes: primeval forest landscape is (was) relatively closed, whereas a contemporary urban-industrial one is open and could not exist/function without considerable steady material inputs.

Some authors (e.g. Forman and Godron, 1986; Forman, 1995, and in a sense also Zonneveld, 1995) consider flows of species as one of the important aspects of landscape functioning. Although these flows do not occur at ecosystem level but involve individual populations only, they can be considered relevant at landscape level as they serve as the vectors of spreading ecosystems. This aspect is particularly interesting from the point of view of biodiversity conservation at the landscape level. The existence of many (particularly animal) populations is threatened by the fragmentation of their natural habitats by means of which their respective remnants become smaller and smaller and ever more isolated. Thus, the ability of such plant or animal populations to communicate at the gene level is often crucial for their survival. The presence or absence of a given species can be considered a piece of ecological information on the quality of a given landscape. Thus, without

going into detail, we may consider vigorous and steady population flows – no matter whether they concern frogs, bears or flows of information – as a measure of the ecological health of a given landscape.

The fourth aspect of landscape functioning is related to the mutual regulating of landscapes (and their components) and their respective environments. It is relevant to be able to assess to what extent the constituent ecosystems and entire landscapes are capable of self-regulation, and to what degree they are dependent on human intervention. Further, we need to know to what degree the ecosystems and entire landscapes are influenced by their immediate (physical and social) environment. Watt (1973), for example, points out that 'more diverse systems or subsystems not only parasitise their less diverse neigh-bours, they actually suppress development of diver-sity in other parts of the system'. This notion indicates certain tensions among variously developed landscape ecosystems and landscapes themselves that result in certain energy costs comparable to the respi-ration losses of individual organisms, or even systems.

On the other hand, it should also be considered that ecosystems and the landscapes they constitute do not just passively adjust to their respective environ-ments, but also exert certain influences upon them. A decrease (or more rarely an increase) in natural eco-systems areas within a given landscape – or within landscapes that are close to a natural condition within a given region – are a good example of this.

Landscape change

Change of landscape is the 'alteration in the structure and function of the ecological mosaic over time' (Forman and Godron, 1986). In natural landscapes, landscape change could be compared in a sense to the ecosystem succession, that is to the 'sequence of biotic communities which tend to succeed one another in a given area over a period of time' (de Groot, 1992), with the end product of such succession known as a climax community.

In natural landscapes the progress towards the 'climax state' is normally slow. Regardless of disturbances such as fire, floods, droughts or storms, which are part of the life of the natural landscape, the general trend is still towards a more natural state, which means increase in biomass and its structuredness, increase in energy efficiency, increase in material closedness, and slowing down of matter and energy flows.

Normally the natural landscape succession also means an increase in biodiversity – particularly in autochthonous species and their genetic diversity as well as in natural biotopes, an increase in the numbers of functional connections, resulting in better resilience toward disturbances, completeness of natural 'trophic webs' among ecosystems and an increase in the size and longevity of individual organisms.

Changes in natural landscapes are relatively predictable. The pattern of postglacial landscape development, for example, still applies in the European Alpine landscapes; in cultural landscape, however, the overall change is usually a result of innumerable single-purpose actions by humans.

From the ecological point of view the transition from natural into cultural landscape retains certain characteristics of the degradation process, characterised by decrease of biomass and its structuredness, decrease of energy efficiency, decrease in closedness (with regard to matter and energy), increase in artificially introduced energy, decrease in natural biodiversity (little room for ecotones, appearance of artificially maintained species and ecosystems), decrease in the numbers of natural functional connections, and invasion of anthropogenic functional connections (not tested from the evolutionary point of view), thus causing functional instability, and finally smaller size and shorter life expectancy of individual organisms.

In brief, the cultural landscape is actually a successional stage marked by increased dependency on people. As a result it is not possible to talk about the 'climax cultural landscape', at least not in purely ecological terms.

Several major forms/stages of transformation of natural into cultural landscape are observed most frequently (see Forman, 1995): deforestation, agricultural intensification, desertification, urbanisation-industrialisation, suburbanisation, construction of infrastructure, etc. On the other hand, land abandonment, reforestation and designation of protected areas for the purpose of nature protection represent an opposite process. Each of these transformations means a different scenario for landscape development – and for its ecological relations.

A real understanding of landscape metamorphosis requires not only identifying the transformation types but also conducting a thorough analysis of the real reasons for their occurrence. In most cases landscape management is concerned with some landscape changes that are to be either introduced or prevented. For this reason, the study of landscape changes should consider that changes of a cultural landscape are not just a given, but can originate either in the abiotic sphere, for example, due to changes in water regime or climatic changes; in the biotic sphere, for example, due to the appearance, overcrowding or extinction of an important species; or in the social sphere, because of changes in human condition (demographic conditions, economics, ownership, policies, value systems, politics, etc.).

Due to all these factors, no direct analogy (as in the case of structure and functioning) is possible between ecosystem and (cultural) landscape change. In order to make the landscape change studies systematic and their results comparable the following points might be considered (see Table 13.3).

While the structure and functioning of landscapes are something immediate, and therefore relatively easy to perceive, the landscape change, due to its temporal scale, seems to be more elusive. As has been pointed out, some changes may be abrupt and cataclysmic while others are slow and hardly noticeable. Both cases seem to offer little ground for the successful

Table 13.3 Change of the cultural landscape (some basic parameters)

1 Origin of the change (abiotic, biotic, social sphere).
2 Character of change (acute, chronic, composite, single-cause disturbance).
3 General place of change (abiotic, biotic, social sphere).
4 Exact place of change (ecological variable, ecosystem).
5 Scope of change (point, ecosystem, landscape, inter-landscape).
6 Dynamics of change (slow, rapid).
7 (Un)predictability of change (its beginning, development/direction, ending).
8 Reactions of landscape homeostatic mechanisms (responses in abiotic, biotic, social sphere).
9 Possibilities of preventing changes (feasibility, desirability, etc. of such measures).
10 Value systems behind landscape change or behind the opposition to it.

management of landscape changes. And yet, cultural landscapes have been changing over the millennia. How much have we learnt from all these processes for the future? The most important lessons for present and future landscape managers might be summed up in three simple statements:

1 If landscape is a living system, then change is innate to it.
2 If landscape is a living system, then it cannot be 'planned' but only steered towards some desirable state.
3 Future change of cultural landscape is not necessarily an extrapolation of its present state.

Understanding landscape change as a continuous process consisting of short- and long-term sub-processes – many of which exceed a normal human lifespan – is perhaps the most important prerequisite for successful landscape management.

LANDSCAPE EVALUATION

Cultural landscape is in a permanent state of flux. People change it, perceive the resulting changes from the aesthetic, environmental, economic, social or cultural point of view, and base new changes on the ground of these perceptions. In other words, cultural landscapes (in particular) are constantly being re-evaluated.

By definition (see Naveh and Lieberman, 1994), evaluation is 'an examination and judging concerning the worth, quality, significance, amount, degree or condition of something'. Evaluation depends on the point of view and on the purpose of the evaluator. In the noosphere of any cultural landscape, the values as perceived by people have the same role as energy in the biotic sphere: they are its driving force.

The wide scope of factors that created cultural landscapes, and continue to change them, is reflected in the many values attached to them. Identifying landscape values and scaling them can be a rather subjective undertaking. Considering the fact that dealing with the environment on the landscape level is relatively new, it is easy to understand that there is a lack of internationally harmonised approaches to describe, classify and evaluate cultural landscapes (see Stanners and Bourdeau, 1995).

Zonneveld (1995) writes about two basic human attitudes leading to evaluation which are never strictly separated: the attitude 'to use or exploit' (considering suitability and constraints) and the attitude 'to care for' (considering the intrinsic value and the vulnerability of the landscape). In the case of landscapes one might easily add a third attitude – 'to enjoy'. This is the attitude of the indirect user (traveller, visitor, tourist, nature protector), which does not necessarily conform to that of the people who actually live there. According to Spellerberg (1992), the methods used for landscape evaluation have been researched for at least three decades, mainly in Europe and North America, by geographers, policy-makers and planners. However, all these methods have dealt largely with visual qualities and people's preferences, which is certainly not the only landscape quality aspect. An integrated landscape assessment methodology is obviously needed, to include both ecological and perceptual aspects, if the entire evaluation process is to serve as a guide for sustainable landscape planning and management.

While natural landscapes can be evaluated for their pristineness, abiotic and biotic diversity, intrinsic value or spiritual values of wilderness to modern mankind, and so forth, cultural landscapes should be evaluated on additional grounds that stress the man–landscape relationships.

In seeking common denominators, that is of landscape aspects that could be objectively evaluated in any cultural landscape, Stanners and Bourdeau (1995) suggest the following: (1) sustainable use of natural resources; (2) wildlife habitats; (3) economic benefits; (4) open spaces and scenery; and (5) cultural heritage.

If cultural landscapes are to be dealt with as ecological and social systems the above scheme could be slightly modified and enlarged to include (1) natural givens; (2) social givens; (3) functional harmony; (4) biodiversity; (5) spiritual values; (6) sustainability.

1 Natural givens

If cultural landscape is to be considered a result of the interaction between nature and people, its past, present and future can be understood only if we have a reasonable appreciation of its natural givens, that is geological, climatic, hydrological, soil, vegetation, wildlife, etc. characteristics and processes. Early accessibility to natural resources gave rise to the oldest cultural landscapes. Changes in natural resource availability – either new discoveries thereof or exhausting them – radically rewrote the development scenarios of many cultural landscapes. The notions of carrying capacity and of sustainability are based on the evaluation of natural givens.

2 Social givens

Evaluating the human community interacting with a given landscape is just as important as evaluating its natural givens. People are an important factor in the shaping of cultural landscapes. The demographic and social parameters of a given human community are probably one of the best indicators of the future of a given cultural landscape. Not only the numbers of people and their structures by gender, age, education, occupation, etc. are to be considered, but also their value systems, aspirations, lifestyle, social structuring, land ownership and its patterns, etc. It is also important to identify and evaluate the influences of the most powerful beneficiaries of the values of a given landscape: the conflict between local and outside interests is often an element of landscape instability. And finally, understanding traditional ways and relevant legislation, past and present, is also essential for evaluation of the social sphere of a given landscape.

3 Functional harmony

In evaluating functional harmony at least three levels should be taken into account:

(a) Compatibility with human uses of landscape with its natural functioning and its natural carrying capacities.
(b) Compatibility between the old and the new landscape uses.
(c) Compatibility between local and other land(scape) use interests.

Searching for balanced human aspirations within the boundaries of natural possibilities is the beginning of sustainable management. Landscape uses have to be optimised rather than maximised. Harmony between natural givens and many traditional, essentially sustainable land(scape) uses often creates the specific aesthetic quality of traditional landscapes that makes them particularly valuable.

4 Biodiversity

Shaping cultural landscapes in terms of human needs and scales, in addition to general tendencies towards landscape homogenisation – for example, by urbanisation, water management, macro-agriculture (Forman, 1995) or plantation forestry – is by no means friendly to the natural richness of life. Thus, considering the conservation of biodiversity in landscape management, particularly at the ecosystem (biotope) level, is a guarantee of ecologically viable landscapes that can support a great number of existences – including that of humans. The practical evaluation of landscape biodiversity can be done in three ways: (a) via landscape ecological pattern and function; (b) via selected single species potential habitat models; and (c) via multi-species gap analysis modelling (Steinitz, 1995).

In the context of evaluation, particular attention should be given to the protection of minority ecosystems, to fragmentation of relatively natural ecosystems, to their restoration, and to preserving a diversity of cultural landscapes themselves which, in fact, represents the fourth level of biodiversity conservation.

5 Spiritual values

The notion that a landscape is not just a place of many objects of the natural and cultural heritage, but is itself a heritage, that is a value by itself, is relatively new to the landscape manager in particular (Vos and Stortelder, 1992; Stanners and Bourdeau, 1995; CE et al., 1996). Traditional cultural landscapes can be viewed as rich records of past human existence. As such they represent ties with ancestors, create a sense of belonging and identity for local people, and an aesthetically pleasing (even if not understood) scenery for the casual visitor, or for numbers of tourists. As such these landscapes are a source of inspiration – not only to artists but to most people who encounter them. Thus, assessing their non-material values is by no means the least important aspect of the evaluation.

6 Sustainability

The question 'Is this (cultural) landscape sustainable?' has to be answered from three points of view: ecological, social and economic. First, one should consider its ecological sustainability, which depends on its distance from the natural (primeval) state and on the state of preservation of its homeostatic mechanisms. Its biological sustainability depends also on the willingness and ability of the people to maintain it in its present form. Second, from the social point of view a landscape can be considered sustainable when it meets the needs of its users. If this is not the case the landscape will change. In appreciating the social aspect of landscape sustainability attention ought to be given to the sustainable land-use patterns that required social adaptations just as much as considering the natural carrying capacities. Third, economic sustainability usually depends not only on the locally available natural resources but also on factors that lie far beyond the boundaries of a given landscape. Self-sustaining agrarian economies were usually a guarantee of landscape economic sustainability. The opening of (new) markets for a certain agricultural product – or the disappearance of those markets – frequently caused radical landscape changes.

Similarly we can trace profound landscape changes to the economic policies of some transnational associations, for example, the Common Agricultural Policy of the European Union.

Biological, social and economic diversity in a given landscape seems to be the best guarantee of its sustainability – or still better, of its gradual change in line with the changes within the human community.

LANDSCAPE PLANNING AND MANAGEMENT

Planning and management of most natural resources began only when people became aware of the fact that these resources are valuable, because they are scarce, endangered or already destroyed.

Land is one of the most obviously limited natural resources, and land planning of one form or another has a long tradition. Yet traditional land planning and management is not to be confused with landscape planning and management on a landscape-ecological basis. In many cases, designating certain planning or management as 'landscape' simply indicated a level of complexity or spatial dimension that was higher than the ecosystem but lower than the regional or national one. Germany, for example, has a long and rich tradition of such landscape planning, and yet Buchwald and Engelhardt (1980) state that it is 'surprising that the landscape planning obviously does not have a generally accepted set of instruments'; and Pflug (cit. Buchwald and Engelhardt, 1980) states: 'Up until this moment there is no conception in the Federal Republic of Germany about the development of our landscapes on an ecological basis.'

Some of the most obvious explanations for this situation might be the following:

1 Conceptual difficulties in landscape designation and delimiting (in comparison to ecosystems).
2 Non-congruity of natural and social landscape boundaries (e.g. most planning is done within administrative rather than landscape-ecological boundaries, and these do not overlap).
3 Failing to accept landscape-ecological doctrine as the first among equals when trying to find a balance among various interests. Ecological sustainability is, in fact, the basis of social and economic sustainability.
4 The non-holistic approach in past landscape planning and management, which stressed the suitability of a given landscape for a given use rather than looking at the question the other way around.

Landscape planning and management in theory

Modern landscape planning and management considers landscape as a whole, as a system whose existence depends on natural, social and economic conditions. It views the landscape as a carrier of environmental health, and accepts landscape sustainability and conservation of landscape biodiversity as supreme planning and management imperatives.

All this explains, at least in part, why the conservation of landscapes as a special planning and management category has been chosen as one of the action themes of the Pan-European Biological and Landscape Diversity Strategy Action Plan 1996–2000 (CE *et al.*, 1996). According to this, the following challenges – in most cases relevant world-wide – are to be addressed in landscape planning and management:

- Prevent further deterioration of the landscapes and their associated cultural and geological heritage.
- Redress the lack of integrated perception of landscapes as a unique mosaic of cultural, natural and geological features.
- Establish a better public and policy-maker awareness and more suitable protection status for these features.

Meeting these challenges may require some further clarification of landscape as a subject of planning and management processes.

In an ideal cultural landscape, people should be capable of meeting their social and economic needs without compromising the landscape's natural health. 'Health' here can be understood also as the presence of some level of homeostatic mechanisms that will guarantee a certain degree of ecological flexibility. This should render possible further changes introduced by humans without jeopardising a certain level of biodiversity. This, in fact, could also be considered a definition of sustainable landscape.

In the planning-management context, landscape has to be considered as an undetermined, open system whose functioning or changes can be planned or managed only with a limited degree of predictability due to the many factors influencing the system, such as natural complexity, large temporal and spatial scales that reach beyond human lifespan and beyond actual landscape boundaries, of the multiplicity of competing and often incompatible interests. In addition to all this, a cultural landscape has to be viewed as both an ecological and a social system at the same time.

Different societies shape different landscapes – considering the environmental conditions, constraints and possibilities and reflecting their civilisation level, value system, tradition, experience, even political system, and so forth. Regardless of all this, there seem to be some common landscape ecological principles, applicable in one form or another to almost any kind of landscape. They are summarised in Forman's aggregate-with-outliers principle (Forman, 1995) which states that one should 'aggregate land uses, yet maintain corridors and small patches of nature throughout developed areas, as well as outliers of human activity spatially arranged along major boundaries'. According to Forman, the following mainly landscape-ecological attributes are incorporated into or solved by this principle: (1) large patches of natural vegetation; (2) grain size (patch texture); (3) risk spreading; (4) genetic variation; (5) boundary zone; (6) small patches of natural vegetation; and (7) corridors (Forman, 1995).

In (cultural) landscape planning and management, these ecological attributes are confronted by functional aspects, concerning landscape as a social entity having to fulfil a number of functions: safety, settlement, primary production, mining, water management, industry, infrastructure (e.g. traffic, energy), recreation, tourism, natural heritage protection, and so forth.

There are obvious conflicts between the two sets of aspects/objectives. The gap between the two can be bridged by the idea and practice of landscape tending (in German: Pflege) which, according to Gildemeister (1976), contains planning (and management) measures for: (1) protection; (2) maintenance; (3) restoration; and (4) development. Depending on the situation such tending can be a different blend of these measures: in the case of a natural protected area it can consist of protection alone, in the case of a well balanced, stable agrarian landscape a blend of protection and maintenance will be sufficient, while a landscape that has been overloaded from the environmental point of view may need all four kinds of considerations. This inevitably leads to conflicts between the various interests – protection and development, for example – and moves the planning process from the drawing board into the political arena.

From the landscape management point of view the non-material values of landscapes present a very special set of dilemmas. Management of natural landscapes of outstanding scenic beauty or of exceptional ecological value designated as protected areas of one category or the other frequently requires special approaches and measures. In most cases the very fact that they have remained more or less untouched indicates the relative absence of any major economic interest in the past. However, such interest may be created (e.g. tourism, development) by the very act of designating them protected areas.

In many cultural landscapes it is not just for their natural beauty that they are considered valuable but rather for the harmony between people and nature most obviously expressed in various artefacts, certain of which may be considered as values by some people, while by others they are seen as obstacles to 'progress'. Conflicts of local and other interests over these values may make landscape management in such cases extremely difficult. However, such cases (and they are by no means isolated) testify to the fact that also (non)perceived, non-material landscape values may play a significant role in landscape management.

Landscape planning and management in practice

Due to the complexity and multiplicity of interests, landscape planning is a typically interdisciplinary process – and this is exactly where it is at its weakest. We do not seem to be able to handle its too many variables at the same time. Landscape planning is essentially still sectoral planning within not-so-certainly-set landscape boundaries. Like Buchwald and Engelhard (1980), we still talk about agricultural, urbanisation, highways, water management, or restoration, landscape planning, or even about planning for nature protection (Skage, 1995) or biodiversity (Steinitz, 1995). Real integration of all these aspects still seems to be a long way away. This, however, should be taken as a challenge. The very fact that all these sectors are using the same technology in order to gain control over the complexity of the databases (e.g. GIS and similar information systems), to master the spatial scales (remote sensing techniques), to surmount the time dimension (by means of computer modelling) or to prevent undesirable developments (environmental impact assessments) may indicate that sooner or later those involved will begin to talk a

similar language and to think in an interdisciplinary way.

A truly integrated interdisciplinary planning and management approach can only develop in an evolutionary manner: in recent years, the concept of an integrated ('ecosystem') management approach has gained momentum, particularly in the context of environmental management at the regional level (Tolba and El-Kholy, 1992). In 1994, on Isle au Haut, Maine, USA a charette was convened on ecosystem management for sustainability. It concentrated on the Everglades, south Florida, yet its conclusions (Anon., 1994) (see Box 13.1) seem to be equally applicable from the planning or the management point of view – at the ecosystem level, or (bearing in mind the changes in scale and complexity) at landscape level.

Due to the undetermined character of landscape systems, and because of the sometimes large spatial and temporal scales involved, special attention should be paid to the design of monitoring as early as the planning stage: parameters in the interior as well as exterior (broader) environment that are to be monitored should be identified, and the monitoring techniques and frequencies determined as far as possible.

'The monitoring should include standardised and repeatable surveys, which would allow corrective action to be taken. It would also enable management effectiveness to be evaluated – at any level' (IUCN, 1994) – and from various points of view.

Actual landscape management is just as complex; it is to follow the same basic principles that apply to planning, but the managers (as Forman (1995) vividly puts it) 'do much more than plan. . . . They deal with crises. They capitalise on change. . . . Sometimes they throw plans away, but not the land.'

Who then are these landscape managers? They come from a variety of professions dealing with natural resources in one way or another. In practice, no individual is vested with the power to manage an entire landscape, and very seldom – with the rare exception of protected areas or extreme emergencies of landscapes that need restoration – is an agency given centralised power to make decisions on a given landscape. Landscapes are actually managed by a host of land users, owners, concessionaires, squatters, etc., each pursuing their own ends. An orchestra without

---

BOX 13.1   ECOSYSTEM MANAGEMENT PRINCIPLES APPLICABLE ALSO IN LANDSCAPE PLANNING AND MANAGEMENT

- Use an ecological approach that will recover and maintain the biological diversity, ecological function and defining characteristics of natural ecosystems.
- Recognise that humans are part of ecosystems, that they shape and are shaped by the natural systems; the sustainability of ecological and societal systems is mutually dependent.
- Adopt a management approach that recognises that ecosystems and institutions are characteristically heterogeneous in time and space.
- Integrate sustained economic and community activity into the management of ecosystems.
- Develop a shared vision of desires, human/environmental conditions.
- Provide for ecosystem governance at appropriate ecological and institutional scales.
- Use adaptive management as the mechanism for achieving both desired outcomes and new understandings regarding ecosystem conditions.
- Integrate the best science available into the decision-making process, while continuing scientific research to reduce uncertainties.
- Implement ecosystem management principles through co-ordinated government and non-government plans and activities.

Source: Anon., 1994

---

a conductor? Hardly. In most countries landscape (spatial) planning and management supervision authorities do exist, although there may be vast differences among these structures and the ways in which they operate. The way countries and communities manage their landscapes is, in fact, the litmus test of their land ethics, environmental awareness and social order. In most countries real (ecological) landscape planning is only just starting; for this reason it is most difficult to find examples of such management where landscapes would be managed as ecological and social systems on a sustainable basis, with conservation of biodiversity in mind. In developing countries with relatively preserved landscapes good examples of modern landscape planning are extremely rare. There is a lack of expertise, funds – and political will. Landscape management planning is long-term oriented, and because of this it will never receive priority over short-term issues. In addition, land management planning, which might attempt to moderate the ruinous clashes between traditional and modern land-use patterns and norms, is often considered as interference in politics – and doomed from the start. Thus, the best examples of modern landscape planning (considering the ecological and social aspects) in the developing world might be the planning (= establishing) and management of protected areas such as parks; this requires obtaining international, national and local (!)

support for its values, clear objectives, much (usually imported) expertise, and some basic institutional network – from legislation to supervision, monitoring, research, law enforcement, etc. Thus, founding parks in many cases activates processes and institutions that will also be instrumental in introducing landscape management into other ('normal', 'everyday') landscapes. In this regard it is interesting to consider the wide scope of frequently overlapping management objectives set by the IUCN (1994) for the six categories of protected areas, such as national parks (Category II) or protected landscape/seascape (Category V) (Box 13.2).

Although the objectives are similar, in principle it is obvious that there are significant differences between national park management objectives and those in protected landscape management where the influence and input of the local community is much stronger, and management, as a result, is much more complex and demanding. Another example of landscape management, particularly (but not only) applicable in the tropical countries is the system of agroforestry as described in *Caring for the Earth* (IUCN *et al.*, 1991) – see Box 13.3.

Although 'agroforestry techniques offer exceptional promise for whole landscape rejuvenation' (Forman, 1995), it is notable that the above description of management guidelines makes hardly any reference to the

---

BOX 13.2   OBJECTIVES OF MANAGEMENT OF NATIONAL PARKS AND PROTECTED LANDSCAPES/SEASCAPES

(a) Category II National Parks

- To protect natural and scenic areas of national and international significance for spiritual, scientific, educational, recreational or tourist purposes.
- To preserve, in as natural a state as possible, representative examples of physiographic regions, biotic communities, genetic resources, and species, to provide ecological stability and diversity.
- To manage visitor use for inspirational, educational, cultural purposes at a level which will maintain the area in a natural or near natural state.
- To eliminate and thereafter prevent exploitation or occupation inimical to the purposes of the designation.
- To maintain respect for the ecological, geomorphologic, sacred or aesthetic attributes which warranted designation.
- To take into account the needs of indigenous people, including subsistence resource use, in so far as these will not adversely affect the other objectives of management.

(b) Category V Protected Landscapes/Seascapes

- To maintain the harmonious interaction of nature and culture through the protection of landscapes and/or seascapes and the continuation of traditional land uses, building practices and social and cultural manifestations.
- To support lifestyles and economic activities which are in harmony with nature and the preservation of the social and cultural fabric of the communities concerned.
- To maintain the diversity of landscape and habitat, and of associated species and ecosystems.
- To eliminate where necessary, and thereafter prevent, land uses and activities which are inappropriate in scale and/or character.
- To provide opportunities for public enjoyment through recreation and tourism appropriate in type and scale to the essential qualities of the areas.
- to encourage scientific and educational activities which will contribute to the long-term well-being of resident populations and to the development of public support for the environmental protection of such areas.
- to bring benefits to, and to contribute to the welfare of, the local community through the provision of natural products (such as forest and fisheries products) and services (such as clean water or income derived from sustainable forms of tourism).

Source: IUCN-CNPPA/WCMC, 1994

---

economic and social implications of introducing such landscape management systems, which are definitely a welcome alternative to the wanton destruction of the tropics.

On the other hand, landscape management in developed countries seems to be slow in development for other reasons: it is not necessarily the lack of funds or even expertise but the complexity of values, interests, traditional ties and of social and institutional webs that makes its progress so slow. Sometimes it takes an emergency (e.g. natural disaster) or an extreme degradation of landscape for all these obstacles to give way to a logical and systematic planning and management procedure for the purpose of landscape restoration.

In the following two boxes such examples are given: the first (Saunders *et al.*, 1993 – Box 13.4) is

not unrelated to the idea of agroforestry. It shows the relevance of appropriate farm woodland management for landscape restoration – equally applicable in Australia and Europe. The second case illustrates an example of landscape restoration by means of river restoration in Florida (Box 13.5). The case of the Kissimmee river illustrates how it is possible to restore a landscape by re-regulating a single ecological variable – in this case water flow. Ecological (and social) restoration or rehabilitation of whole landscapes and regions is, indeed, an enormously important frontier.

CONCLUSIONS

Landscape management, understood as a more highly integrated level of ecosystem management –

BOX 13.3   AGROFORESTRY

Agroforestry systems include trees as a main component in a multi-crop production process. The trees protect the soil from raindrop impact and insolation. Some tree species fix atmospheric nitrogen and enrich the soil, while deep-rooted trees prevent nutrient loss from the system and draw nutrients to the surface. The interaction between the trees and other components of the system leads to good soil protection and the conservation of water and nutrients. In that sense, agroforestry systems behave rather like natural multilayered ecosystems. The drawback is that production of the associated crops is generally less than in monoculture systems.

The main types of agroforestry system are:

- alley cropping, where annual crops are grown between lines of trees that produce valuable mulching material;
- mixed growth of permanent crops like coffee or cacao between timber trees;
- growth of crops in fields sheltered by tree windbreaks or hedges;
- orchard systems, where the trees provide edible fruits, medicines and fuelwood, while the ground layer is cropped or grazed;
- growth of scattered trees in pasture land, to improve soil conservation and provide shade, timber or fuelwood;
- plantation systems where the ground layer is grazed by livestock;
- shifting cultivation systems, where small tilled plots are allowed to revert to forest after a number of years of cropping.

Agroforestry should be considered in all formerly forested marginal land that is currently managed for production, regardless of whether the systems are in high- or low-rainfall areas, or at low or high elevation. Agroforestry systems restore tree cover to cleared land but they are not substitutes for forests, which are often more effective in maintaining environmental functions and conserving biological diversity, and may also provide a more sustainable source of income.

Source: IUCN *et al.*, 1991

BOX 13.4   COMPARISON OF PURPOSE AND NATURALNESS FOR FARM WOODLANDS (APPLICABLE TO BOTH AUSTRALIAN AND EUROPEAN RESTORATION PROJECTS)

Purpose	*Naturalness*
Windbreaks	Possibly, if species-rich and wide enough. Connections to existing woodland aid colonisation by native species.
Water/silt/nutrient uptake	Depends on size, species-richness, situation and capacity to regenerate naturally or through invasion by local species via corridors from remnants. After establishment, riparian woods should be allowed to develop naturally to enhance stability and conservation value.
Trees for stock feed	Usually low naturalness but pollarded trees in Europe can have high conservation value for their lichens and insects associated with old trees and woodland continuity. Vegetation belts in the Australian wheatbelt planted as windbreaks are often browsed by stock yet are still valuable habitats attracting birds out into otherwise open paddocks and on to connected remnants.
Trees for timber	Not usually natural, mainly exotics. Even single species stands of native species harvested regularly will not have mature trees. Many new initiatives in Europe aim to increase the area of farm woodland with mixes of native tree species; these woods will have a more natural character if other components of woodland communities can colonise.
Restored natural systems	Yes, when of adequate dimensions, of local species with effective corridors for emigrants from remnants, and able to regenerate naturally.

Most woods in a restoration scheme will be multiple use combining the above and additional functions including recreation, game management and landscape design.

Source: Saunders *et al.*, 1993

BOX 13.5   RESTORATION OF A LANDSCAPE BY MEANS OF RIVER RESTORATION

In Florida the Kissimmee river meandered 165 km from approximately Orlando to Lake Okeechobee. In the twentieth century it was channelled into a wide, deep canal of half that length. The surrounding water table dropped, and most floodplain wetlands dried out. The seasonal pattern of water flow was altered, so constant water levels degraded the remaining wetlands. The river straightening reduced or eliminated plant community complexity, habitat for waterfowl and other wetland wildlife, nursery areas for fish, and interactions between river and floodplain.

However, a recent demonstration project (using weirs) redirected a minuscule flow from the canal into selected stretches of the old river channel. This tiny bit of river burst into life. Almost overnight these channels had more forage fish, game, flushing of muck accumulation, original riparian vegetation, connections among wetland habitats and habitat diversity. Adjacent previously dried out wetlands had more waterfowl and wading birds, export of invertebrate food to channels, and cover by wetland plants. Perhaps these demonstration results will lead to restoration of the longer, several kilometres wide river landscape. If so, it will be the first restored river, and the largest restored wetland world-wide.

Source: Forman, 1995

---

meaning the integration of the entire ecological–technological–social system of a landscape into a single conceptual framework – may still be in its early stages. It may have been unthinkable several decades ago. Now it is becoming not only possible, because of the developments available, but also a necessity if the biosphere is to survive.

It is a logical step in the development of conservation thought – which started by protecting endangered species, went on first to endangered and then to all managed ecosystems, and spread to landscapes: endangered, valuable and 'ordinary' ones in which we live every day. It will move on to the regional, and then to the continental level – this is the only possible way of saving the entire biosphere in a rational, planned manner.

The integrating idea of management philosophy at all these levels is the notion of sustainability, particularly at the landscape level. Why? Because it confronts local or individualised interests, values and aspirations with broader ones on a scale that is conceptually still manageable. Because it teaches environmental responsibility and natural ethics on a spatial and temporal scale which people may not yet have mastered, but are capable of doing so.

Zonneveld (1995) puts the issue succinctly:

Sustainable land(scape) management is supposed to combine technologies, policies and activities with the aim of integrating socio-economic principles with environmental concern. It aims simultaneously to: 1. maintain and enhance (natural) productivity; 2. reduce risks; 3. conserve quality and potential; 4. be economically viable; 5. be socially acceptable.

Land ethics, as A. Leopold (1968) described it, is – in one form or another – an inseparable part of any culture. In saying this we should not forget that the word 'culture' itself has its roots in the Latin verb 'colere', that is to till, to cultivate land. Some nations became rich and powerful not only through wars and commerce but also through dealings with their own land; others faded into oblivion. There is no doubt that there are some common denominators and lessons to be learnt, which are equally applicable at any time and at any place. In this chapter we have seen that landscape management involves not just knowledge and technologies but also natural and human values, traditions, interhuman relations, institutions, written and unwritten laws, conflicts of interests, and that it also requires a certain measure of autonomy in the community, which should plan and manage its land in a sustainable manner.

Because of the natural and social complexity of landscapes and the complexities of managing them, it is virtually impossible to prepare a set of landscape planning and management rules that would be applicable world-wide. We can only look for principles and approaches that might have a general validity.

The European experience might set an example in this regard. Europe, indeed, has a long tradition of man–landscape interplay with many successes and failures, which gave rise to the new landscape ethics and environmental awareness at the landscape level. The knowledge is there, so too are the resources, and there seems to be the political will to tackle the issue of landscape management. The report on the state of the pan-European environment requested by the European environment ministers (Stanners and Bourdeau, 1995) extensively addresses landscape management in the modern context. Its most important findings are, first, that while in the past the approach to conservation has been species- or site-specific, the fragility of landscape is now an issue, second, that numerous countries and local administrations have adopted aspects of landscape planning and management in their policy and management, and third, that the most effective principles include:

- professional landscape planning and management addressing the processes of change rather than the changes themselves;
- working with local people to produce better results; and
- an acceptance that the landscapes will change, because they are a function of human land use depending on social, economic and natural factors (world markets, advances in technology, changes in society, etc.).

And finally the chapter on landscape suggests that the following approaches are important:

- To study, record and monitor European landscapes for their ecological, social, cultural and economic values.
- To understand the processes of change, their causes and consequences.
- To take action to protect outstanding landscapes, using such recognised tools as designated landscapes and seascapes.
- To put in place effective land-use planning mechanisms for use in all areas (for making plans and for regulating what happens on the land).
- To make landscape considerations an important factor in shaping national and regional strategies for sustainable development, that is to treat landscape as an environmental resource in the planning process.

- To recognise the landscape scale as one that is important for strategies addressing the conservation of biodiversity, complementing space- and site-specific approaches.
- To develop a support system for rural communities, and farmers especially, to achieve greater prosperity without the need to destroy landscapes.
- To introduce measures to encourage farmers not only to protect existing landscape features, but also to create new elements in the landscapes.
- To encourage greater public awareness of the value of landscape, locally, nationally and internationally.

To the traditional landscape ecologist, this emphasis on social, cultural and economic issues might seem a bit too strong. However, the document clearly describes the future role of landscape ecology in these processes: 'landscape ecology needs to operate as a cross-sectional discipline by addressing cultural, economic and ecological issues in a scientifically and conceptually creative manner' (Stanners and Bourdeau, 1995). What a challenge!

REFERENCES

Anko, B. (1986) 'Role of the Forest in the Energy Flow of a Mountain Farm', in *18th IUFRO World Congress Proceedings, Div. I., vol. 1*, Ljubljana, Yugoslavia: IUFRO, pp. 19–30.

Anko, B. (1994) 'Application of Landscape Ecology in Forestry', in D. Cattaneo and P. Semenzato (eds) *Atti del XXXI Corso 'Landscape ecology – ecologia del paesaggio'*, S. Vito di Cadore, Italy: Centro Studi per l'Ambiente Alpino.

Anon. (1994) 'Isle au Haut Principles of Ecosystem Management', *U.S. MAB Bulletin*, July, 18: 6.

Buchwald, K. and Engelhardt, W. (eds) (1980) *Handbuch für Planung Gestaltung und Schutz der Umwelt, Vol. 3, Die Bewertung und Planung der Umwelt*, Munich, Germany: BLV Verlagsgesellschaft (in German).

Butcher, S.S., Charlson, R.I., Orians, G.H. and Wolfe, G.V. (eds) (1994) *Global Biogeochemical Cycles*, London, UK: Academic Press.

Council of Europe, United Nations Environment Programme, European Centre for Nature Conservation (1996) *The Pan-European Biological and Landscape Diversity Strategy*, Amsterdam, The Netherlands: CE, UNEP, ECNC.

de Groot, R.S. (1992) *Functions of Nature (s.l.)*, The Netherlands: Wolters-Noordhoff.

Forman, R.T.T. (1995) *Land Mosaics*, Cambridge, UK: Cambridge University Press.

Forman, R.T.T. and Godron, M. (1986) *Landscape Ecology*, New York, USA: John Wiley & Sons.

Gildemeister, R. (1976) 'Gedanken zum Wort Pflege', *Natur und Landschaft* 51, 9: 245–246.

IUCN (1994) *Parks for Life*, Gland, Switzerland: IUCN.

IUCN-CNPPA/WCMC (1994) *Guidelines for Protected Area Management Categories*, Gland, Switzerland: IUCN.

IUCN/UNRP/WWF (1991) *Caring for the Earth*, Gland, Switzerland: IUCN.

Kerényi, A. (1981) 'Relief Energy as Potential Energy', *Acta Geographica Debrecina* 1979–1980, 18–19: 5–30.

Kiemstedt, H. (1995) 'Landschaftsplanung und Eingriffsregelung als Instrumente eines umfassenden Naturschutzes in Deutschland', in D. Ogrin (ed.) *Nature Conservation Outside Protected Areas*, Ljubljana, Slovenia: Office for Physical Planning, Ministry of Environment, and Physical Planning and Institute for Landscape Architecture, Biotechnical Faculty, pp. 119–130 (in German).

Leopold, A. (1968) *A Sand County Almanac*, London, UK: Oxford University Press.

McHale, J. (1972) *World Facts and Trends*, New York, USA: Collier Books.

Naveh, Z. and Lieberman, A.S. (1994) *Landscape Ecology*, New York, USA: Springer-Verlag.

Odum, E. (1965) 'Relationships Between Structure and Function in Ecosystems', in E.J. Kormondy (ed.) *Readings in Ecology*, Englewood Cliffs, NJ, USA: Prentice-Hall, pp. 211–215.

Odum, E.P. (1989) *Ecology and Our Endangered Life-Support Systems*, Sunderland, Mass., USA: Sinauer Associates, Inc.

Saunders, D.A., Hobbs, R.J. and Ehrlich, P.R. (eds) (1993) *Reconstruction of Fragmented Ecosystems*, Chipping Norton, NSW, Australia: Surrey Beatty and Sons Pty Limited.

Skage, O.R. (1995) 'Nature Conservation through Landscape Planning', in D. Ogrin (ed.) *Nature Conservation Outside Protected Areas*, Ljubljana, Slovenia: Office for Physical Planning, Ministry of Environment and Physical Planning and Institute for Landscape Architecture, Biotechnical Faculty, pp. 131–138.

Spellerberg, I. (1992) *Evaluation and Assessment for Conservation*, London, UK: Chapman and Hall.

Stanners, D. and Bourdeau, P. (eds) (1995) *Europe's Environment*, Copenhagen, Denmark: European Environment Agency.

Steinitz, C. (1995) 'Landscape Planning and the Management of Biodiversity', in D. Ogrin (ed.) *Nature Conservation Outside Protected Areas*, Ljubljana, Slovenia: Office for Physical Planning, Ministry of Environment and Physical Planning and Institue for Landscape Architecture, Biotechnical Faculty, pp. 139–152.

Tolba, M.K. and El-Kholy, O.A. (eds) (1992) *The World Environment*, London, UK: Chapman and Hall.

Vink, A.P.A. (1975) *Land Use in Advancing Agriculture*, Berlin, Germany: Springer-Verlag.

Vos, W. and Stortelder, A. (1992) *Vanishing Tuscan Landscapes*, Wageningen, The Netherlands: Pudoc Scientific Publishers.

Watt, K.E.F. (1973) *Principles of Environmental Science*, New York, USA: McGraw-Hill Book Company.

Zonneveld, I.S. (1995) *Land Ecology*, Amsterdam, The Netherlands: APB Academic Publishing.

SUGGESTED READING

Harris, L.D. (1984) *The Fragmented Forest*, Chicago, Ill., USA: The University of Chicago Press.

Hayward, T. (1994) *Ecological Thought*, Cambridge, UK: Polity Press.

IUCN/UNEP/WWF (1991) *Caring for the Earth*, Gland, Switzerland: IUCN.

Leopold, A. (1968) *A Sand County Almanac*, London, UK: Oxford University Press.

Nash, R.F. (1989) *The Rights of Nature*, Madison, Wis., USA: University of Wisconsin Press.

Noss, R.F. and Cooperrider, A.Y. (1994) *Saving Nature's Legacy*, Washington DC, USA: Island Press.

SELF-ASSESSMENT QUESTIONS

1 From the ecological point of view, landscape is:
   (a) an arbitrarily chosen part of the Earth's surface;
   (b) an ecological system with a distinct structure, functioning and change patterns;
   (c) part of the environment that one can perceive through the senses;
   (d) none of the above.

2 Landscape ecology is a scientific discipline dealing with:
   (a) geomorphological specifics of a certain area;
   (b) survival of endangered plant species;
   (c) flows of matter, energy and capital;
   (d) none of the above.

3 In comparison with ecosystem management landscape management is:
  (a) more complex;
  (b) has wider spatial and temporal horizons;
  (c) involves noosphere to a higher degree;
  (d) all of the above.
4 The so-called 'minority ecosystems' (e.g. quarries, ponds, small wetlands, barren land, etc.)
  (a) take up a lot of useful space;
  (b) present an obstacle to efficient land use;
  (c) significantly add to the biodiversity;
  (d) none of the above.
5 Which of the following is not a part of landscape tending (*sensu* Gildemeister)?
  (a) protection;
  (b) maintenance;
  (c) fragmentation;
  (d) restoration.
6 Which of the following is not an objective of land management?
  (a) maximising financial returns;
  (b) attaining sustainability;
  (c) conservation of biodiversity;
  (d) all of the above.

14

DISAPPEARING HUMAN ECOSYSTEMS

Philippe Lefèvre-Witier

SUMMARY

The disappearance of a human ecosystem can happen under different types of constraints. Any human group is at risk, but particularly those populations defined as marginal because of complex adaptations due to climactic conditions, geographical localisations, economic exclusions or political isolations. To foster a good understanding of the process of destructuration, two examples are examined through their recent evolution: the Tuareg, nomadic herders of the Sahara desert, and the Mixtecos Indians of the tropical forest in Pacific coastal Mexico.

The main traditional Tuareg characteristics are described, and their mutations evaluated through the recent history of the northern part of Africa, in particular the consequences of decolonisation. Territorial problems and the effects of drought of the last 30 years appear to be the main factors contributing to the Tuareg sedentarisation and their undertaking of new professional activities. Among the northern tribes the herding of camels and goats has been kept up, but only in very limited areas; among the southern tribes a conflict between nomads and black powers of Mali and Niger is still in progress, disorganising the Tuareg social and economic conditions. As a result the Tuareg no longer appear to be a nomadic people, no longer herders and no longer rulers of the central Sahara.

In view of the development of Mexico, especially before the recent oil crash, the Mixtecos Indians (like many other rural and marginal groups) have found their social and national integration frustrated. They declare their need for better food, better medical assistance, better communications and better education. As a consequence many families resort to migration to the main cities around the Mixteca, including Mexico City. In many villages, 50 per cent of the population has moved away, destroying local systems of food production, of mating exchanges and of religious practices.

These two examples illustrate the effects of rapid transformation and modernisation within and around traditional nomadic or sedentary populations, and demonstrate the threat of their total disappearance.

ACADEMIC OBJECTIVES

This chapter analyses the concepts of fragility and sustainability in human ecosystems through the knowledge of their biocultural and environmental structures.

This can be facilitated by selecting as examples human groups living under severe ecological contraints with elaborate adaptations, to demonstrate clearly the conditions of their recent and rapid evolutions.

Through this approach we offer a better understanding of the sophistication and fragility of human adaptations, whose flexibility, in both biological and cultural terms, appears very limited in many cases.

Our two cases studies will analyse modernisation as a complex and sometimes dangerous process leading to the threat of marginal populations disappearing altogether. They also confirm the rich distribution of human biocultural diversity in remote and difficult environments of the world and the importance of its loss.

INTRODUCTION

In the living world most species discover places favourable to their development and survival. Usually they stay there in a very sustainable way, and disappear only if some lethal mutations appear in their gene pool, or if they suffer an important climatic change.

Humanity, on the other hand, thanks to creative adaptation and management, has built its environment within many ecosystems, fitting in with all biomass on the planet.

In most places, due to good climatic conditions and soil composition, and rich fauna and flora, it was easy for human groups to develop, and their evolution through changing socio-economical strategies saw many socio-technological exchanges and genetic hybridisations.

However, a few ecosystems are characterised by their incredible biocultural adaptations, which reach the limits of species' capacity to adapt. These 'marginal' ecosystems can be observed in extremely cold areas, at high altitude, in harsh deserts, in humid tropical forests and in totally isolated habitats; under these constraints the structure of ecosystems (and their consequent survival) has to be maintained by a high level of education and an autonomous way of life.

For the last 3,000 years (and in some cases recently) most of the Earth's large populations have survived thanks to appropriate and flexible development. Despite wars, famines and epidemics, the destruction of such human groups was almost unknown. During the same period we can assume that many 'marginal' groups, at risk and fragile, have disappeared. We can also assume that the ones we can observe and study now have survived because they have stayed far away from any other populations.

As a consequence the 'marginal' ecosystems were ignorant for a long time of our important technological revolutions, and of the new conditions of life they imposed. Electricity and running water have only recently been experienced by some of them. Some have yet to discover cars and towns, modern disease and medical birth control, and (perhaps) genetic manipulations. But as the first key developments, TV, phone and computer technology will, through the change in mental representations they generate, modify ways of thinking, learning, forms of motivation and behaviour. In large societies, all these evolutionary factors and their effects on relationships between humanity and the environment globally trouble and destabilise people. In 'marginal' ecosystems people are genuinely damaged by the discovery of such modernities, as their adaptive flexibility to the new environment is theoretically very limited. To maintain a sound process of evolution,

modernisation should be introduced to them slowly, step by step, and preserving as far as possible their traditional values and ways of life. Otherwise the increasing complexity of life engendered by aggressive innovation makes their ways of life more and more fragile, destroys their autonomy and ultimately makes them disappear.

Listing the world's disappeared or disappearing human ecosystems would not fit with the general aims of our book. To gain an understanding of the disappearing process it is better to focus on one or two recent cases so as to analyse the main factors of their evolution.

We shall explore this process of increasing fragility through innovations in two well-known 'marginal' ecosystems:

- The Tuareg, in the Central Sahara desert of northern Africa.
- The Mixtecos, Mexican Indians, in the tropical forest of the Sierra Madre.

Both have survived for centuries if not millennia in very different geographical areas: the Tuareg in harsh desert and the Mixtecos in humid mountainous surroundings. Both have been very successful in adapting. Both are at risk of disappearing.

THE TUAREG ECOSYSTEM

The Tuareg live in a very large territory (around 1,000,000 km²) in the Sahara (Figure 14.1); they speak a Berber language ('tamahaq' or 'tamacheq') and are a nomadic pastoral group of more or less 10 million people. Because of their famous characteristic indigo veils, the Tuareg are sometimes known as the 'Blue people' or 'Veiled people'.

Tuareg are divided into four to five confederations of tribes (Figure 14.1). Three of these live in mountainous areas: the Kel Adagh in the south-western Ifoghas mountains, the Kel Aïr to the south-east, near the Aïr mountain, and the northern Tuareg in the mountains of Hoggar and Tassili n'Ajjer. The two others move nomadically near the Niger river: the Tuareg Udalan to the west and Iwellemeden to the east. They have shared many things, but above all the Sahara itself with its severe constraints; most of the tribes nomadise between isohetes of 50 mm and

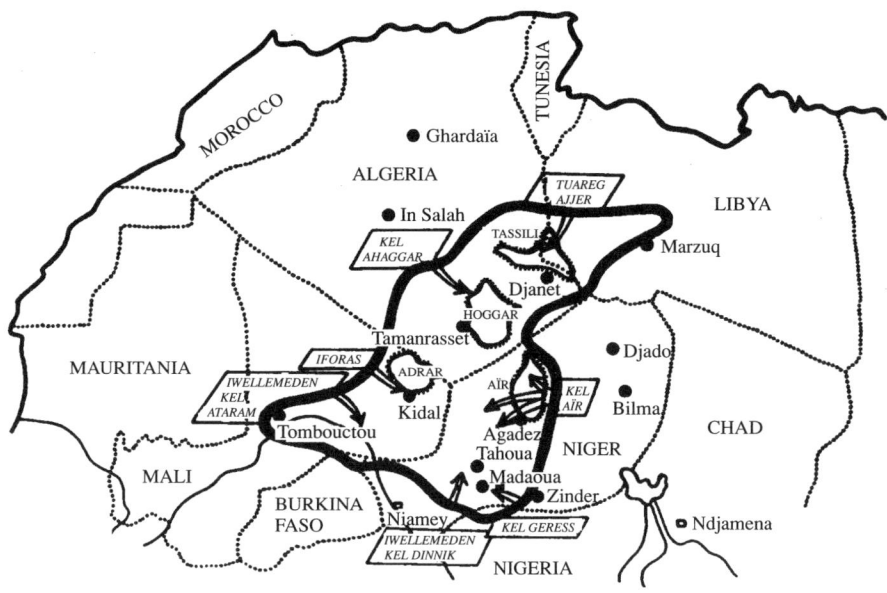

Figure 14.1 The Tuareg tribes in the central Sahara

150 mm (Figure 14.3). In this area rains come roughly every three years. This climatic condition explains why pastoralism at this latitude cannot be a regular transhumance but a permanent nomadisation over short distances (about 100 km), continuously seeking out water and vegetation.

Traditionally Tuareg have also moved over long distances to enrich their diet by bartering salt in the north and south for cereals (millet or wheat) or dates.

Figure 14.2 A view of Atakor, central part of the Hoggar mountain, Algeria

Some of their longer journeys ('rezzou') were also made to steal camels or capture black slaves mostly in the nearby southern populations of Jerma, Haoussa and Fulani.

The social structure of Tuareg is strictly hierarchical, having three tiers (Figure 14.4):

1 Families or tribes of high rank, 'nobles', transmitting power in the confederation through the maternal line.
2 Families or tribes dominated by the above and paying them an annual tribute mainly in livestock or slaves.
3 Black slaves for tending cattle and for domestic labour.

Religious people, blacksmiths and released slaves co-exist at different levels of the structure.

Pastoralism provided the main income of the confederations. Livestock in the north was usually made up of camels and goats. Some northern tribes limited their flocks to goats; these people were called Kel Wulli (Wulli = goat). In the southern, more Sahelian territories, large herds of cattle are kept. These animals need much more water, as they must drink every three days.

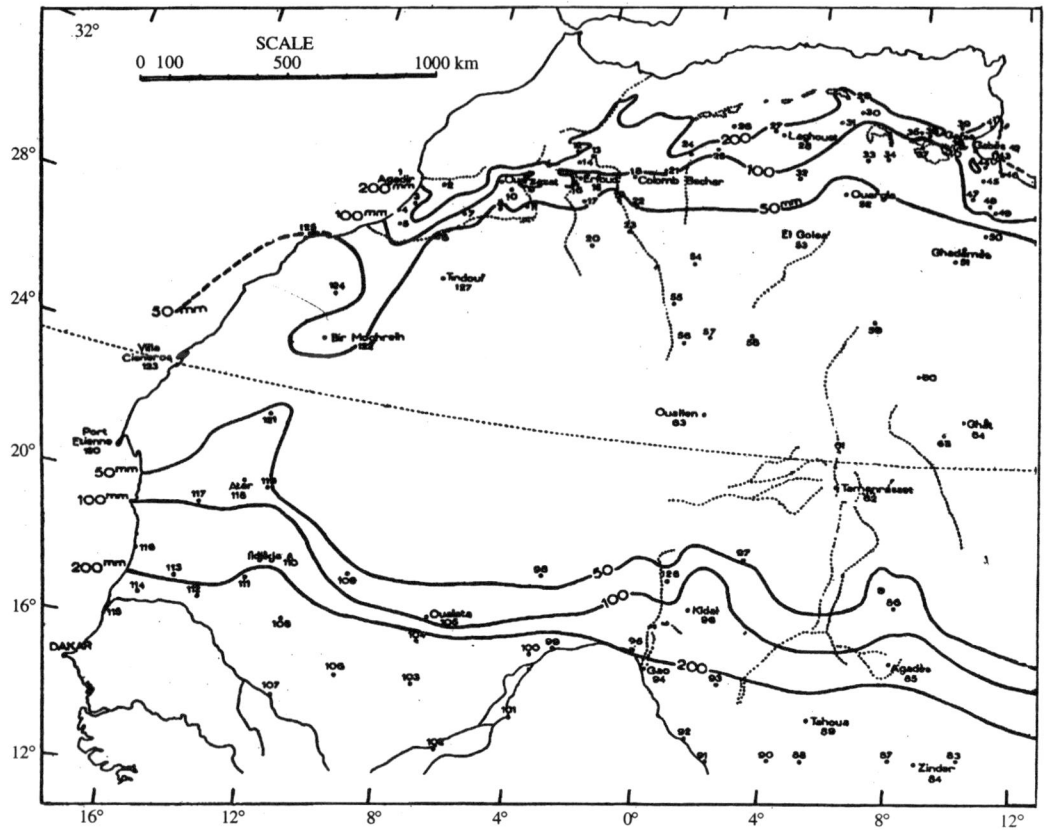

Figure 14.3 Annual mean rainfall lines in the French Sahara, 1931–1945
Source: after Capot-Rey, 1953

As in many large pastoral nomadic people – Masaï and Nuers in Africa, Evenki in Siberia – many adaptation traits are linked to both the quantity and the quality of livestock production, as these provide basic components of nutrition and habitat.

In the Tuareg Sahara, milk is excellent in taste and digestibility; it is low in lipids and contains large quantities of sodium chloride and vitamin C, especially the goats' milk. Milk is also a key nutriment as part of the everyday meal (Gast, 1968). By contrast meat consumption is irregular and occurs mostly at special occasions; however, most of the nomads in transit keep some dry pieces of meat with them to supplement their food. Butter and cheese are used but their preservation is problematic: butter is cooked with aromatic plants, cheese dried to a hard plastic state. As with all pastoral nomads, animal hair, wool and hide are used for making ropes, bags, carpets, tents and huts, cloths and ornaments.

There is a heavy dietary dependence on cereals and dates. The latter provide potassium to these people, whose cardio-vascular system is constantly on overdrive.

In parallel to this brief survey of Tuareg nutrition, let us consider a few more aspects of their biology.

Tuareg cardio-vascular activity is akin to that of athletes; for example, heart frequency is kept almost normal (76 beats per minute) and very regular even during the hottest hours of the day (Lambert, 1968). Thermo-regulation reveals hyperadaptation; the individual has to face soil temperature differences per day of 50 to 60°C, and consequent evapotranspiration holds the 'world record' with 6 metres per year (in Piche evaporimeter), compared to 4 metres in any

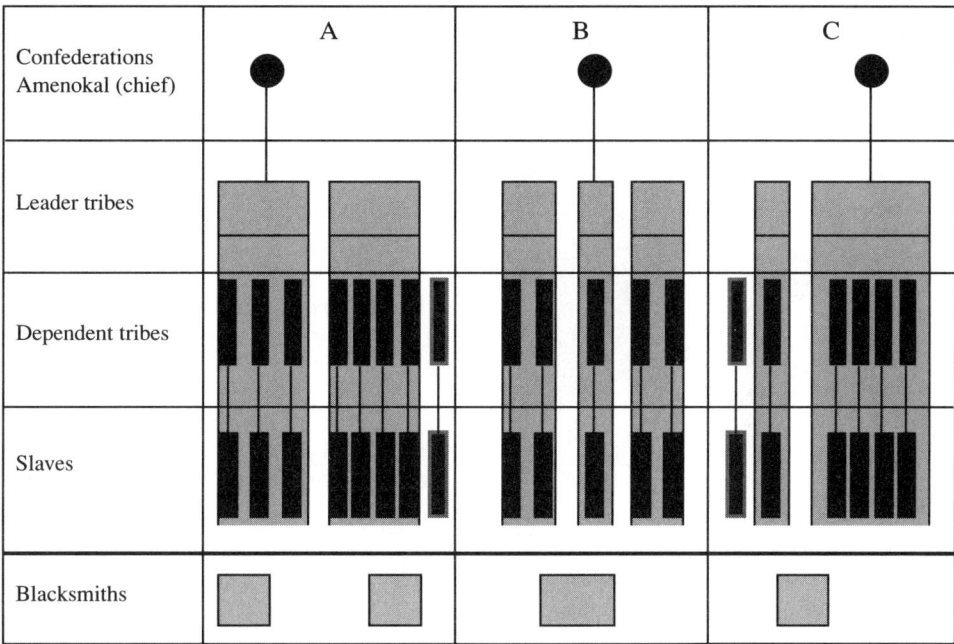

Figure 14.4 Social structure of the Tuareg

other desert. Despite these conditions of heat and dryness, Tuareg show a low transpiration rate (2 to 3 litres per day), with a slight rise in salt concentration, whereas a European in the Sahara sweats 6 to 7 litres per day, with a salt concentration of 7 grams per litre. Tuareg also show a reduction in diuresis,

Figure 14.5 Young Tuareg watering camels at the well, Niger

limiting it to 500 cc or less per day and have a rectal temperature lower than that of Europeans at any time of the day (Lambert, 1968). As a consequence the loss of and need for water and sodium chloride are both very low. Veils and large cloths increase this metabolic economy of water and salts. Tuareg can be active with 3 litres of water or milk per day and 2 grams of sodium chloride whereas the average European drinks 10 to 12 litres and needs 7 grams of sodium chloride per day. As an example, some young Tuareg lost without water have survived because they have been able to keep walking for several days and nights. To these biological 'feats' we must add the psychological attributes of patience and endurance acquired by children through their family education and their experience of nomadic herding.

All these conditions make the Tuareg thin, fast and agile people; a sort of gracility originating partly in the poor nutrition they endure, with around 1,200 to 1,500 calories on a good day, and the hyponutrition they suffer as infants which results in a growth retardation of a specific type, a global and harmonious 'miniaturisation' (Dop *et al.*, 1984).

Figure 14.6 Portrait of a Tuareg of a southern tribe

The genetic structure of the Tuareg shows particular frequencies of mutations common to Mediterranean populations of the Maghreb as well as Negroid groups of the Sahel. This can be explained by an intermediate differentiation specific to these 2–3,000 km of geographical distance in latitude (Lefèvre-Witier, 1996). A process of hybridisation could also be responsible, but it would have been slowed down by the strong traditional antipathies between social and ethnic groups in this part of Africa.

Let us complete this short description of the Tuareg ecosystem by mentioning the true cement of the group: the language 'tamahaq' (in the northern tribes) or 'tamacheq' (in the southern tribes). Following the recent classification of Wilms (1980), this is a true Berber language, and not one of the dialects spoken in different Berber-speaking areas.

Although they record and sustain a rich culture, the technology of pastoralism, literature and poetry, 'tamahaq' or 'tamacheq' are not written languages; only a few consonant signs ('tifinars') allow limited written communication. Despite contacts with other written languages such as French or Arabic, the maternal transmission of Tuareg languages has been fully preserved. Bourgeot (1995) ascribes a great importance to this socio-cultural link between confederations, and calls the Tuareg 'Kel Tamacheq'; they call themselves 'Imazighen' (*les hommes libres/* 'free men').

TOWARDS DESTABILISATION OF THE TUAREG WORLD

The first part of this discussion, a description of the Tuareg ecosystem, is necessary in order to understand the culture's evolution over the last 100 years, and why its disintegration is attributable to our impact on it.

The curtain rose on the Tuareg drama in 1902, at Tit, a Hoggar village, where northern tribes met a party of French soldiers under Lieutenant Costenet (an ambitious young man sent to the Sahara to undertake 'pacification', meaning colonisation) (Archives de Vincennes). The Tuareg lost not only the battle, but also face and, without realising it, took the first step towards the destruction of their complex ecosystem (just as happened around Timbuktu to the southern tribes).

We shall now describe broadly the history of the Tuareg world's degradation, not chronologically, but by analysing the main factors responsible for it: territory, slavery, sedentarisation and education.

Territory

Three main activities traditionally allowed the Tuareg a good control of territory:

(a) Pastoral nomadism

We have already mentioned that feeding animals such as camels or cattle in the desert is a type of climatic gamble in which quality of information and speed of movement play a major role. Tuareg herders wandered in any part of the desert territory, which they know perfectly. The French army could only try to emulate this command of territory using camel platoons and medical back-up. At the same time, in compensation, the Tuaregs' desert adaptations and social culture were preserved, including slavery. Despite a few armed conflicts, the situation remained stable until the years 1950–1960.

(b) Caravanning

The main trans-Saharan trails were controlled by Tuareg caravans carrying salt or taking camels to more salted southerly pastures; and by groups of warriors protecting caravans travelling from Maghreb or Sudan.

(c) Rezzou

These fast attacks took highly adventurous and unpredictable routes.

Between 1950 and 1960 the most important factors of Tuareg territorial disorganisation appeared. These factors are political, economic and climatic.

After the French pacification, the Tuareg lost armed control of trans-Saharan markets and caravanning; this loss was confirmed in 1950 by the emergence of oil-related activities and nuclear research in the Sahara, creating much motorised traffic between Algiers and Mali-Niger. A sense of exclusion first appeared mostly in the northern tribes, but was addressed in 1951 by the French project for an autonomous regional Saharan entity. This project became more focused in 1957, with the birth of the OCRS (Common Organization of Sahara Region). However, the southern Tuareg were excluded from this because they were supposed to share the territories of the young African states created in 1960 (Mali, Niger, High Volta) and in 1962 (Algeria).

Suddenly the Tuareg confederations were divided and trapped as minorities in new black states hostile to them because of the slavery that the 'veiled people' had imposed on them for centuries (millennia?).

The establishment of formal boundaries was immediately rejected by the Tuareg, who tried to free themselves from this control of their global pastoral strategies.

The departure of the French army from Occidental Africa was badly received, as were the following: the systematic disarming of Tuareg in Algeria (1963–64 confiscation of swords and guns); the census of tribes, families and individuals in all tribes, including slaves, with the compulsory assumption of patronymic names and identity cards; and to cap it all, the requirement that the Tuareg abandon nomadism and adopt a sedentary existence.

Throughout this period there was no reaction to these innovations except from the Kel Adagh Ifoghas, which led to their horrific massacre by the Malian army near the town of Kidal, and military control being maintained until 1987 (Bourgeot, 1995).

The definitive territorial changes of the Tuareg ecosystem happened with the droughts in 1970/1973 and 1984/1986. In a few months, the traditional pastoral migrations were totally modified. Northern Tuareg could not move from their mountains where some poor pasture still existed at high altitude. Mali and Niger Tuareg had to move more south and compete for water with Sahelian half-nomadic herders such as the Fulani. Drought and competition killed off cattle very quickly; over all Tuareg territory 80 per cent of livestock died. Families or tribes migrated to nearby towns such as Gao, Menaka, Timbuktu (Mali), Tahoua, Arlit, Agades (Niger), Diapaga (Burkina Faso), Inguezzam, Tamanrasset, Insalah (Algeria), Ghat and Bilma (Libya), where they turned into an unhappy population needing assistance and health care.

Shortly after the first drought, Algeria organised efficient nutritional aid at Tamanrasset and at some small border points. Then a monthly gift of food was distributed to all people, Tuareg or not, in the areas affected by drought. However, at the end of the second drought, Algeria decided to get rid of these foreign nomadic populations and transported many tribes or families to her southern borders.

The Tuareg could not accept this new exclusion, and staged an armed rebellion which has persisted now for ten years. These days, the Tuareg have better food and living conditions. They have agreed to meetings and talks on the position of pastoral nomads in the modern Sahara, but their political views and their understanding of democracy demonstrate a very slow and limited evolution. At a recent meeting, in answer to a fundamental question concerning their future, they said 'we want to be granted a large autonomous region to develop an armed pastoral society'

(Bourgeot, 1995). As we understand it, even though the Tuareg do not lay claim to the whole Sahara, they claim for their future the ghost of their past!

Let us now study some other aspects of the Tuareg ecosystem's degradation in a development parallel with that of their territorial problems.

Slavery

A high-ranking, 'noble' Tuareg possesses weapons, has dependent tribes, food stocks and slaves as workers.

During the first 50 years of colonisation, France was 'in love' with the 'Blue men', fascinated by their cultural and biological adaptation to desert, and preserved everything of their way of life, including slavery. Tuareg and French officers discussed honour, courage, martial arts, gallantry, poetry – but not slavery! In 1950, the development of oil and nuclear projects in Sahara needed workers; also the new trans-Saharan route, run by hundreds of lorries, needed many drivers and mechanical technicians trained for the difficult conditions.

Having observed this modernisation, nomadic populations such as Chaanba or Tuareg have answered its needs and discovered the monetary value of physical work. The Tuareg dependants and slaves were the quickest to understand the economic independence brought by money. In the 1970s, more opportunities for work began to appear in large building projects for the new nations, at the edges of the desert, and especially in Algeria; again, nomadic workers were welcomed and many migrated to these countries for months on end. In a similar quest for freedom, slave women escaped to the Saharan villages and small towns, usually working for farmers who protected them.

This loss of people has seriously affected Tuareg pastoral organisation for many years, and more so since the end of droughts. Also many dependent tribes have refused to pay tribute to 'noble' Tuaregs, and many of these high-ranking nomads now accept lowly jobs as guards, door-keepers, or (preferably) as guides, translators, etc.

In conclusion, while the Tuareg hierarchy may still exist in isolated aspects, it no longer operates in economic terms, and the feudal structure is vanishing from day to day.

Sedentarisation

The collapse of pastoral nomadism, territory control, slavery, and the appearance of wages and money have pushed the Tuareg to new types of settlements and ways of life; three of these types of habitat can be easily discerned:

(a) Agricultural villages, marketplaces, wells in the vicinity of which houses or huts have been authorised. In Algeria some villages have doubled their construction, providing electricity and running water (a shock for newcomers).
(b) Tents, huts or shanty houses (often equipped with TV) on the periphery of more important towns.
(c) Camping close to any inhabited place, where women and children live permanently, where access by car is easy, where there is privacy and protection.

The argument for sedentarisation made by African governments was that it would promote good health through improved hygiene and nutrition and through medical care. However, after 30 years this argument needs to be reviewed. For example, the Algerian nutritional assistance has actually created problems. Remember that the main nutriments for the Tuareg – milk, butter and dates – have been replaced in central Sahara by condensed milk low in lipids and vitamin C, by oil of low quality and by ordinary white sugar (potassium-free). Remember also that the loss of vitamin C and potassium is crucial for resistance to infections and for thermo-regulatory metabolism. Even the available cereals such as flour, noodles and bread lack their germs, the most important nutritional component. Water in towns and villages is not usually under regular and proper controls. An investigation in Tamanrasset (Grima-Badoc, 1979) revealed an incredible level of amoeba in water and in inhabitants alike.

As a consequence, a pathology of poverty and distress is developing in areas where nomadic populations were known as the healthiest in Africa.

Education

Education appears as a 'black spot' in the evolution of the Tuareg. In 1946 'nomadic schools' were organised in Aïr (Niger) and in Hoggar (Algeria).

Teachers followed the same programmes and methods as in the schools of (sedentary) Saharan villages and towns. The argument for nomadic schools, as explained to the Tuareg, was to promote French education and education in French among 'noble' tribes before it reached dependants, slaves and sedentary people – very democratic! (Gast, 1990).

Following the disappearance of nomadic schools, their prediction was quickly fulfilled: after 1962 Algeria promoted sedentary schools in French and in Arabic languages (the baccalaureate degree is now taught in Tamanrasset). In the Hoggar and Tassili villages some boarding-schools were built, which accept many nomadic boys (the education of nomadic Tuareg girls is still a problem).

Such education contributes to the disintegration of the Tuareg world. Typically, three groups are created: first, highly talented young intellectuals, often obliged to emigrate; second, people who are well educated (sometimes in two or three languages) and well informed about modern structures; and third, a large proportion of poorly educated nomadic or sedentary people and especially women who turn to agricultural or pastoral life.

Conclusions

In brief, the dynamic of the Tuaregs' ecosystem has suffered for a hundred years, not so much from the harsh conditions of the Sahara, but from the effects

---

BOX 14.1   THE ISSEQQAMAREN, A TUAREG TRIBE WITH A CHANCE OF SURVIVAL (WILAYA DE TAMENGHEST, ALGERIA)

The Isseqqamaren Tuareg make up a large tribe (tausit) of about one thousand individuals divided into 11 family groups (taqabilt) related through maternal links (see figure below). Some data prove their extension up to Tidikelt, to Ahnet and Immidir areas north of the Ahaggar; they relocated recently on the northern slope of this mountain and in the western part of the Tassili n'Ajjer; in both places they occupy large dry valleys. Therefore their entire habitat lies in a hyperarid area corresponding to the 50 mm per year mean rainfalls very irregular in time and place. Due probably to this history and localisation, the Isseqqamaren developed a unique combination of hyperadaptation and flexibility, the main factors of which we shall shortly mention.

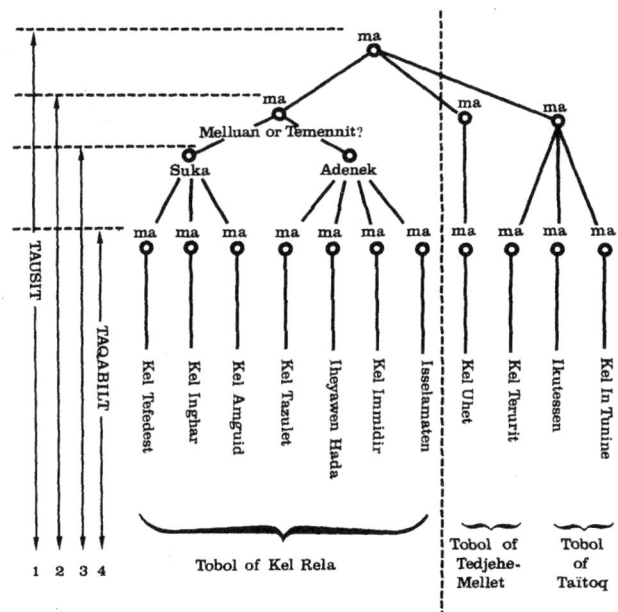

Parental relationships of Tuareg Isseqqamaren following maternal link
Source: after Lefèvre-Witier, 1982

Their hyperarid conditions for herding make Tuareg Isseqqamaren assume an extreme dispersion of animals and populations and as a consequence to occupy a very large territory equivalent to half of France (see figure below). Each taqabilt of 30 to 100 people controls some part of the territory and sometimes splits into herding groups of three or four people or one family only; in this nomadisation for the search of pastures, camping equipment is limited to one or two bags and a few litres of water, providing very fast moves. Let us stress also two social characteristics of this tribe: first, slavery is very limited and, second, their links of dependence towards 'noble' Tuareg are weak and their annual tribute very light.

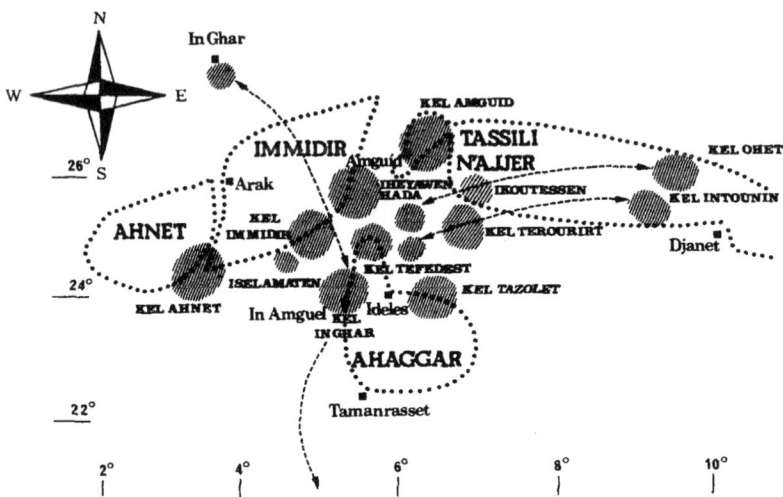

Territory and localisation of subtribal groups (taqabilt) of Tuareg Isseqqamaren in the north of Hoggar (Ahaggar) mountain and in the west of Tassili n'Ajjer
Source: after Lefèvre-Witier, 1982

Usually Tuareg Isseqqamaren herd an equivalent number of camels and goats. Their large flocks of goats provide them with good food. Camels were traditionally sold or exchanged with other Tuareg tribes but recently they entered the meat market in large towns of the central Sahara at a mean price of 20,000 FF per animal. Let us remember that the Isseqqamaren resisted the two recent droughts better than any other tribe may be because of their exceptional adaptation to hyper-arid conditions. These factors certainly play an important role in their easy modernisation. Another argument is their early participation in the process of sedentarisation as co-founders of small villages, in Ahaggar and Tassili n'Ajjer, such as Idelès, Tazerouk or Amguid where they have developed, since the last century, an agricultural production, where they have built some houses and where their children have been able to attend schools. Some of these young Isseqqamaren are now responsible for the local administration or economy.

As a whole Isseqqamaren appear as a sub-ecosystem of the Tuareg population; they did not really suffer the main problems analysed in our section on 'Tuareg ecosystem destabilisation'. Tuareg Isseqqamaren maintained efficient herding and accepted modernisation as a slow process and as a guarantee of their survival.

on its evolution of the recent history of foreign colonisation, and the poverty and financial ambition of nearby populations.

Tuaregs are well aware of the degradation of their complex system. In response, many of them try to preserve their identity, staying in what they call 'our earth, our country', that is the central Sahara.

A rebellion is underway at the borders of this territory of peaceful pastoralism. Arms for survival have become four-wheel drive cars, machine-guns; there are also political talks, media information, and above all a burning hope for new autonomy amidst a pseudo-republican, black and white environment under a fragile and changeable approximation of democracy. Tuareg cannot understand this new world, and to a greater or lesser extent are afraid of it. The power of economic games played in Africa, the strategies developed for energy production, the many ethnic wars, the aggressive spread of Islam, all contribute to the distress of the Tuareg.

ECOSYSTEM OF MIXTECOS INDIANS: EXCLUSION OR DISAPPEARANCE?

For purposes of comparison with the above description of the evolution and virtual disappearance of a nomadic pastoral ecosystem, we shall now consider a second example of sedentary food-gatherers and farmers.

This second example comes from studies of Mexican Indians. As we know, many autochtonous ethnic groups in this continent have lost their habitats, socio-economic organisation, parts of their culture and even (through massacres and epidemics) their lives. The causes of this have been colonisation wars and, more recently, technological innovations.

A very few of these different ecosystems have been maintained in more or less artificial conditions such as the 'reserves' in North America, in isolated areas as with the Nambikwara in Paraguay, among some tribes in Amazonia, or in very controlled situations as with the Wayampis in French Guyana.

In Mexico, since the Spanish colonisation, many Indians have been integrated at different levels of society in terms of power, economy, religion and genetic admixture. Each Indian group still identifiable in Mexico requires a special study to determine its traditional ecosystem and its evolution.

Our presentation concerns Mixtecos Indians of the Mixteca Alta in a high area (5,000 to 7,500 feet) in the Sierra Madre, State of Oaxaca. Western Mixteca Alta is on the Pacific slope, eastern in the altiplano of Tlaxiaco (Figure 14.7).

From 1983 to 1988 a French–Mexican interdisciplinary research project examined some problems of a network of villages of coffee growers around the communes of Santa Maria Yucuiti and Santiago Nuyoo. San Pedro Yosotato, part of Yucuiti, is situated on the very crest of the Sierra and faces the Pacific Ocean. The main theme of this bio-ecolog-

Figure 14.7 Map of the Mixtecas, Oaxaca Estate, Mexico

Figure 14.8 Coffee drying in Yosotato, Oaxaca Estate, Mexico

Table 14.1 Characteristics of marginal populations for food, lodging and education, Mexico

Characteristics	% of total population
Consumption of vegetal proteins only	38.3
Cooking of food on soil	40.6
Sleeping at the place where they cook	73.6
Living in only one room	31.7
Children of 10 years old who do not read or write	22.8
Children of 10 years old with only three years at school	33.8
No running water in the house	33.5
No electricity in the house	49.8
Defecation on soil of the village	68.9
Waste pollution on soil of the village	54.6

Source: after Instituto Mexicano de Seguro Social, 1983

ical study was to observe the relationships between health, nutrition and migrations to cities (Proyecto Franco-Mexicano, 1983–1988).

A few years ago the Mixtecos drew attention to their nutritional status. For Indians of the Mixteca Alta, bad nutrition was linked with extreme poverty and lack of communication systems (bad trails and roads, no phones, no TV, absence of health control and of visits by the authorities responsible, etc.). This type of complaint, more often heard in Mexico in rural and Indian communities, is now known as one of 'social deprivation' (Table 14.1).

The objectives of the research project were to assess the complaint and the deprivations the Mixtecos suffered, and to measure these using biological parameters. The idea was to evaluate the health and nutritional conditions and to estimate their compatibility with a survival of Mixtecos in their traditional ecosystem.

Bad health is in evidence in the Mixteca Alta partly because of humid tropical conditions combined with high-altitude cold and the increasing number of shelter roofs. Leishmaniosis, leprosy and Chagas' disease exist with low frequencies; the main problems are respiratory infections, the frequencies of which are at the record level of Mexico. Intestinal parasitosis, particularly with *ancylostoma duodenale*, affect the total community as a true faecal risk, with consequences of infection, anaemia, vitamin deficiencies and growth retardation by food spoliation (Lefèvre-Witier *et al.*, 1993). Surprisingly the nutrition

has been observed not to be disastrous either in quantity or in quality, thanks to low but regular and well-balanced food consumption, at least among the Mixtecan coffee growers' families (see Box 14.2).

The coffee plantations have been established in Yucuiti and Nuyoo for 50 years, and have been a co-operative enterprise for ten. Their income complements agricultural production and traditional foraging in the forest; more precisely, these few pesos in each village solved the usual problems of annual irregularity in production and of lack of protein.

In the altiplano Mixteco, climatic conditions are more severe and soils very poor. Farm production in some places has become very low and people survive thanks to a new activity: the weaving of a vegetable (palma) into hats; the salary for this very destructive practice is one US dollar per day for each worker. Apart from the market town of Tlaxiaco, poverty is total in this area and feelings of 'social deprivation' and individual depression are very strong. The Mixtecos react by running away to more exciting, better-equipped places such as Mexican towns in full expansion, as they were before the oil crash.

In Mexico city, for example, Mixtecos have organised their own settlements in poor places such as Nezzahalcoyotl, where their living conditions are not much better than those they have left. Some villages have lost around 50 per cent of their population; they are becoming half-rural, half-urban communities

BOX 14.2 PROBLEMS OF NUTRITION OF THE MIXTECOS INDIANS (MIXTECA ALTA, OAXACA ESTATE, MEXICO)

In this case study we shall present more information on the nutritional status of the Indian community of Yosotato (see Figure 14.7) to complete the analysis on the risk of migration to urban settlements.

Traditional food resources and the farming calendar

In Yosotato people cultivate small plots and kitchen gardens (0.25 to 3 hectares) and yields are not very high. Traditional subsistence crops (beans, squash, several varieties of maize) as well as other food plants are raised and there are numerous fruit trees. Gathering of wild plants plays an important part in subsistence, most being weeds collected in the source of agricultural work and consumed as greens. Most families raise pigs and poultry, but only a few have goats, sheep or cows. Hunting and insect gathering occur from time to time. Seasonality is well marked in the subsistence pattern and scarcity is very severe in some months, particularly February–March and August–September, the 'months of hunger'. In these periods the food shortage is supplemented with insects and larvae, with edible plants collected in the forest, and with mushrooms and wild seeds.

Coffee production

An important change appeared in 1930 with the introduction of coffee cultivated as a cash crop. Coffee is harvested in the dry season from October to March, so that from its sale eggs and meat can be purchased for two or three months avoiding the spring break in protein consumption. Despite this improvement a problem still exists during the rainy season when impassable trails make it impossible to buy food from outside. Only rich families can buy some staples such as maize or beans during the dry season and keep their own production for the rainy season.

A consequence of coffee production is the apparition of economic differences in the community deriving from inegality in land tenure. In 1986 many families harvested about 1 tonne of coffee per year and produced staples for 3 to 6 months, they ate meat once a week and eggs three times a week during the dry season. Some families who had extra incomes from trade or salaries in the co-operative (the local coffee enterprise) produced over 2 tonnes of coffee and enough staples for the whole year; they ate eggs and meat more frequently and often included festive foods in their daily diet. By contrast some families produced less than 0.5 tonnes of coffee and had to work for the richer ones; they ate eggs and meat less often (see table below).

Levels of wealth as defined by combining items of production and consumption

	Status 1	Status 2	Status 3
Coffee production: tonnes of coffee sold to the co-operative market	2	1.5	0.5
Staple production: comparative production of staple food	+++	+++	+
Extra income: comparative importance of extra income such as salaries	+++	+	+
Meat consumption: cannot be evaluated in terms of weight but by frequency (no. of times eaten per week)	1–5	1	0.5
Eggs: no. of times eaten per week	3–7	3	3
Festive food: consumption is frequent and abundant for people of status 1, infrequent for others	+++	+	+

Source: after Lefèvre-Witier *et al.*, 1993
Note: + = little money, production or consumption
+++ = much money, production or consumption

The faecal risk

The improvement of diet through coffee production is still limited by the epidemiological load of intestinal parasites and transmissible diseases. The table below shows evidence for the intestinal parasite invasions in Yosotato and Tlaxiaco which generate an incredible food spoliation and a true risk for infections of the intestinal tract leading to death, especially during the rainy season when transportation of patients is impossible.

→

Parasites found in stools in Yosotato and Tlaxiaco						
	1985 Tlaxiaco schoolchildren			1986 Yosotato children		
	Number		%	Number		%
Parasites	tested	positive	positive	tested	positive	positive
Entamoeba histlytica	112	24	24.5	292	43	15.0
Ascaris	112	23	20.5	292	156	53.0
Trichocephales	112	4	3.5	292	3	1.0
Giardia/Lamblia	112	8	7.0	292	32	11.0
Ancylostomoides	112	0	—	292	5	1.5
Entamoeba coli	112	0	—	292	100	34.0
Endolimax nana	112	0	—	292	64	22.0

Source: after Lefèvre-Witier, 1993

Conclusions

Nutrition appears to be better in Yosotato than in many other parts of rural Mexico. However, the food production and resources are inextricably linked to other different factors such as:

(a) the fluctuations of the coffee market and the diseases of the coffee plantations which can ruin one or more crops;
(b) health conditions, with a morbidity and a mortality maintained by faecal risk and respiratory diseases in places where the lack of medical control and sanitary education are total;
(c) the bad communication system and equipment of the village.

These main factors generate the feeling of social deprivation observed in Yosotato, as in all Mixteca Alta, and increase the risk of exclusion and migration out of the rural areas.

with a strong cultural identity: traditional festivals, weddings, economic customs and help in the community are all preserved between the two parts of the village.

Figure 14.9 Sunset on the Pacific Ocean, a view from Yosotan

The ecosystem of Mixteca Alta is at risk in all its aspects. Ever since the Spanish invasion, however, the Mixtecos have demonstrated great unity, an impressive ability to manage their territory, and strong resistance to Spanish power; they also avoid miscegenation.

Perhaps, like many Indian groups, the Mixtecos will now melt into the large Mexican community. The destruction of their human and economical resources, if not of their original culture, will proceed slowly and painfully, through episodic development projects or through rural rebellion as we have seen over the last three years in Yucatan and Guatemala.

GENERAL CONCLUSIONS

We shall briefly summarise the problems that emerge from the study of these two ecosystems and their evolution.

In view of current world economic imbalance, we must consider not only the new divisions appearing in European society (the 'dual society'), but also the complexities of socio-economic levels in developing countries, which can conceal the disappearance of marginal ecosystems.

To understand the structures and evolution of any marginal ecosystem clearly, we need a precise awareness of their complexity and specificity no such dynamic can be explained as a simple process merely because it concerns small populations with archaic modes of production. Not only are these human groups as complex as any others, they also show a type of desynchronised evolution vindicating ways of life and socio-economic mechanisms often passed over and forgotten by our more advanced societies. Furthermore, their mental representations and their communication systems are totally ignored by modern decision-makers; it seems that sometimes only biocultural anthropology can explain their particular ways of thinking, their behaviour, their hopes and projects.

However, in the context of world poverty and problems of development, there is neither the time nor the money to identify and study the degradation of many 'marginal ecosystems', nor, when the situation is identified, is there the capability of redressing it. A 'marginal ecosystem' is not recoverable once it loses its autonomy, its auto-organisation dynamics, corresponding to a loss of choices and strategies. As Bernus (1990) said of pastoral nomads: 'in the loss of choices, nomads have to accept the reduction of their values and their genius and the loss of the pastoral knowledge which formed the basis of their civilisation.'

Given this prospect, it is not only necessary to manage correctly the disappearance of 'marginal ecosystems', to facilitate its slowing-down, but also to try to save some elements whose adaptive value is irreplaceable; above all, to save those people (in this case Tuaregs or Mixtecos) who have survived a long period of misery and suffering, but who, thanks to their culture, their resistance and their individual adaptablity, could be shown to be very efficient in their new societies.

There is no doubt that to save the Tuaregs and Mixtecos and many others, ills such as humiliation, torture and murder, which lead to depression or rebellion, should be avoided. These people need nutritional assistance, aid agencies, solidarity petitions, etc. But what they need even more than these is an international committee, the aim of which (following the Rio world conference recommendations) (1992), would be to identify and take special care of 'marginal ecosystems' as specific and often disappearing entities. To conclude, such a committee should both conduct specific research and promote practical competence in anthropology and human ecology.

REFERENCES

Archives de Vincennes, fonds Maroc, section Algérie, AB, carton 29, Paris-Vincennes, France.

Bernus, E. (1990) 'Le nomadisme pastoral en question', in A. Bourgeot and H. Guillaume (eds) *Identité et sociétés nomades: Symboles, normes et transformations*, Etudes Rurales, 120, pp. 41–52 (in French).

Bourgeot, A. (1995) *Les Sociétés touarègues – nomadisme, identité, résistances*, Paris, France: Karthala (in French).

Capot-Rey, R. (1953) *Le Sahara français*, Paris, France: Presses Universitaires de France (in French).

Dop, M.C., Turc, R., Maiza, E., Keddari, M., Rochiccioli, P., Sevin, A. and Lefèvre-Witier, Ph. (1984) 'A cross sectional study of growth of Algerian children from the Tell and the Ahaggar (Sahara)', in J. Borms, R. Hauspie, A. Sand, Ch. Susanne and M. Hebbelinck (eds) *Human Growth and Development*, New York and London: Plenum Press.

Dubief, J. (1959) *Le Climat du Sahara*, T.1 and 2, Alger: Institut de Recherches Sahariennes (in French).

Gast, M. (1968) 'Alimentation des Populations de l'Ahaggar', in *Memoires du C.R.A.P.E.*, VIII, Paris, France: Arts et Métiers Graphiques (in French).

Gast, M. (1990) 'L'école nomade au Hoggar; une drôle d'histoire', *Revue du Monde Méditerranéen et Musulman* 57, 3 (in French).

Grima-Badoc, F. (1979) 'Résultats d'une enquête parasitaire dans la daïra de Tamanrasset, Algérie', unpublished Thesis of Medicine in the frame of the Franco-Algerian project 'A.T.P.2290-CNRS-Biologie des populations sahariennes', Ph. Lefèvre-Witier, co-ordinator (in French).

Instituto Mexicano de Seguro Social (IMSS) (1983) *Diagnostico de Salud en las zonas marginadas rurales de Mexico*, Mexico D.F.: IMSS (in Spanish).

Lambert, G. (1968) *L'adaptation physiologique et psychologique de l'Homme aux conditions de vie désertiques*, Paris, France: Hermann (in French).

Lefèvre-Witier, Ph. (1982) 'Ecology and biological structure of pastoral Isseqqamaren Tuareg', in *Current Development in Anthropological Genetics*, Vol. 2, Ch. 4, New York, USA: Plenum Press, pp. 93–124.

Lefèvre-Witier, Ph. (1996) *Idelès du Hoggar*, Paris, France: CNRS Editions (in French).

Lefèvre-Witier, Ph., Katz, E., Serrano, C. and Vargas, L.A. (1993) 'Health and nutrition in Mixtec indians: factors influencing the decision to migrate to urban centres', in L.M. Schell, M.T. Smith and A. Billsborough (eds) *Urban Ecology and Health in the Third World*, symposium 32 of the Society for the Study of Human Ecology, Ch. 11, Cambridge, UK: Cambridge University Press, pp. 163–171.

Proyecto Franco-Mexicano (1983–1988) *Biologia humana y desarollo en la Mixteca Alta – Estado de Oaxaca*, Synthetic Report, Ph. Lefèvre-Witier, co-ordinator (in Spanish).

Wilms, A. (1980) 'Die Dialectele Differenzierung des Berberischen', in *Africa und Ubersee*, Supplement 31, Berlin, Germany: Dietrich Reimer (in German).

SUGGESTED READING

Butterworth, D. (1975) *Tilantongo, communidad mixteca en transicion*, Mexico: Instituto Nacional Indigenista (in Spanish).

Cabot-Briggs, L. (1967) *Tribes of the Sahara*, Cambridge, Mass., USA: Harvard University Press.

Claudot-Hawad, H. and Hawad (1996) 'Voix solitaires sous l'horizon confisqué', *Ethnies* 20–21 (in French).

'Dix études sur l'organisation sociale chez les Touaregs' (1976) *Revue de l'Occident musulman et de la Méditerranée* 21 (in French).

Gast, M. (1979) 'Pastoralisme, nomade et pouvoirs: la société des Kel Ahaggar', in *Pastoral Production and Societies*, Cambridge, UK: Cambridge University Press and Paris, France: M.S.H., pp. 201–220 (in French).

Gast, M. (1981) 'La société traditionelle des Kel Ahaggar face aux problèmes contemporains', in *Islam, société en communauté, anthropologie du Maghreb*, Cahiers du C.R.E.S.M., No. 12, Paris, France: CNRS, pp. 107–139.

Instituto Mexicano del Seguro Social (IMSS) (1984) *Round Table on the Extension of the Social Protection to Marginal Groups of the Rural Areas* – Memoire, Mexico D.F.: IMSS.

Keenan, J. (1977) *The Tuareg, People of the Ahaggar*, London, UK: Allen Lane.

Spittler, G. (1993) *Les Touaregs face aux sécheresse et aux famines: les Kel Ewey de l'Aïr*, Paris, France: Karthala.

SELF-ASSESSMENT QUESTIONS

1   For the definition of a marginal human ecosystem, two main characteristics should be taken into account:
    (a)   isolation;
    (b)   fragility;
    (c)   adaptation;
    (d)   complexity.
2   Mean rainfall range per year in the Tuareg area is:
    (a)   50 mm to 100 mm;
    (b)   100 mm to 200 mm;
    (c)   more than 200 mm.
3   The main factor for Tuareg ecosystem destabilisation is:
    (a)   creation of national boundaries;
    (b)   loss of servile workers;
    (c)   droughts.
4   What is the actual status of Tuareg evolution?
    (a)   Integration to African surrounding nations.
    (b)   Revendication for their capacity as nomadic herders.
    (c)   Armed rebellion.
    (d)   Talks to obtain a regional autonomy.
5   The exodus to cities of rural and Indian populations of Mexico is due to:
    (a)   urbanisation;
    (b)   nutritional problems;
    (c)   soil erosion;
    (d)   social deprivation.
    Please choose the most important of these factors and explain in a few words why it is a key for rural and Indian populations' sustainability.
6   Is the future of a disappearing human ecosystem
    (a)   easily restorable;
    (b)   restorable under some conditions;
    (c)   impossible to restore.

GLOSSARY

Acrotelm: the active layer of a peat bog, often of the order of 10 cm in depth. The plant litter is little humified. The peat is saturated only periodically and conditions are generally aerobic.

Agenda 21: international agreement completed at the UNCED conference in 1992, without legal, but with moral and inspirational value. It addresses international problems concerning the environment and development from a sustainable point of view. It can be seen as a framework on which national environmental plans can be founded.

Agro-biome: a distinct vegetation community that is characterised by having similar agricultural qualities, for example, the height of the crops.

Agro-ecosystem: an ecosystem in which management is applied to enable crop and/or animal products to be harvested.

Agroforestry: forestry geared along food production lines.

Agro-industry: agriculture developed along industrial lines; industry connected with agriculture.

Altiplano: a plateau in high altitude.

Amazigh, Imazighen: self-designation as 'free' man (men) of all Berber-speaking people.

Arbovirus: arthropod-borne virus. Viruses that have an arthropod as intermediate host, which act as their vector.

Autotroph: organism that produces its own food.

Barrier island: generally narrow (*c.* 1 km) elongated island (tens of kilometres) parallel to the coast formed from unconsolidated wave-lain and wind-blown sand. Excellent examples are found offshore of the eastern USA from New Jersey to Texas.

Biodiversity (biological diversity): an ecological concept composed of five key components: species diversity (number of species), genetic diversity (variety of genes within a given population), habitat diversity (variety of biotic and abiotic elements), successional diversity (spectre of plant communities between early and late ecological succession), landscape diversity (landscape as a mosaic of forests in different succession stages).

Biodiversity convention: convention adopted at UNCED in Rio de Janeiro 1992, which is a legally binding instrument for over one hundred and sixty countries that have ratified it, in order to conserve the world's biological diversity.

Biogenetic diversity: existence of wide genetic variation within a species.

Biological oxygen demand (BOD): quantity of oxygen expressed in mg/litre required for the destruction or deterioration of organic matter in a body of water with the assistance of micro-organisms, which develop there over a period of five days at a constant temperature of 20°C.

Biological pest control: method of regulating plant and animal pests by natural predators.

Biomass: weight of living organisms in a particular place.

Biome: a major ecological community type, characterised by distinctive vegetation.

Boundary zone: the edge zone where two ecological systems of different maturity meet. A zone of intensive exchange of matter and energy.

Brown-air city: refers to urban areas mostly localised in warm, dry, sunny climates where the major sources of pollution are cars and electrical power plants, emitting NO_x, CO_x and hydrocarbons. Air pollution in these cities is characterised by the presence of ozone, formaldehyde and peroxy acetylnitrate causing a brownish-orange smog.

Bryophyte: a division of the non-flowering plants consisting of mosses and liverworts.

C3 grasses: grasses that fix CO_2 solely by means of the enzyme ribulose bisphosphate carboxylase and initially produce a three-carbon compound. Most frequent in forests and extratropical vegetation.

C4 grasses: grasses that fix CO_2 additionally by means of a more efficient enzyme phosphoenolpyruvate carboxylase and initially produce a four-carbon compound in the cells of the normal leaf tissue. This four-carbon compound (malate or aspartate) forms the substrate for subsequent carboxylation in larger, specialised cells around the conductive tissue. These plants use more energy for fixation of one molecule of CO_2 than the C3 plants. Occur mainly in tropical and subtropical regions.

Carrying capacity: level of biomass and/or yield that can be sustained in the long term in relation to the productivity of a particular ecosystem or region with regard to the diversity, structure and trophodynamics of populations and communities within that system.

Catchment area: area delivering inflowing water to a lake or other water body.

Chagas' disease: a chronic infection through the parasite 'trypanosoma cruzi' with severe cardiac failure, transmitted by some insects of the 'rheduvidae' type living in the walls of houses.

Char: a rapidly changing island forming from flood water fine deposition in the coastal zone of Bangladesh. Such islands prove to be fertile ground for landless people who accept the risk of staying in a naturally hazardous zone in order to stake out newly created land that emerges post storms and floods.

Chemical oxygen demand (COD): quantity of oxygen in mg/litre that is consumed by oxidisable matter present in the water.

Climax: the last stage or sere in a vegetational succession, a state of dynamic equilibrium that will last as long as the environmental conditions remain stable.

Climax community: the end product in any succession; relatively stable community, able to maintain itself over longer periods of time and to regenerate and replace itself without marked further change.

Closed coast (crenellate): an irregular coastline (in plan view) along which multiple reversals of longshore sediment transport are likely. Irregularities are often caused by variably resistant rock types outcropping along the coast.

Coastal engineering: civil engineering designed to support human occupation in the coastal zone. In recent decades the term has become identified with the ethos of building permanent structures, e.g. groins and sea walls which defend people living at the shoreline.

Coastal system: a collection of natural environments containing variable energy (tides, waves and currents) and minerals- and organic-mass (sediments) found at the land–ocean edge.

Community: all the organisms that live in a particular habitat and affect one another as part of the food web or through their various influences on the physical environment.

Compensation principle: see 'No net loss' principle.

Competition: use of resource by one individual that depletes the amount of that resource available to other individuals, either of the same (intraspecific) or of other species (interspecific).

Convention to Combat Desertification: convention, adopted in 1994, that sets the guidelines for combating desertification and drought in countries that are particularly vulnerable to this problem, recognising that the most important cause of desertification lies in bad human practices (deforestation, bad irrigation) originating land degradation.

Corridor: a route that allows movement of individuals or groups from one region or place to another.

Cut-and-burn (slash-and-burn): traditional method of forest clearance used by local people to clear land for agriculture, grazing and settlement.

Desertification: the diminution or destruction of the biological potential of the land, leading ultimately to desert-like conditions. It is an aspect of the widespread deterioration of ecosystems and has diminished the biological potential of the land, i.e. plant and animal production, for multiple-use purposes at a time when increased productivity is needed to support growing populations in quest of development.

Diuresis: quantity of urine per day.

Down-drift: the direction in which beach sediment is being transported. It is usually taken to mean the net or long-term direction of sediment movement as reversals of sediment transport are usual on a short-term scale.

Ecological diversity: the existence of a wide variety of communities within a particular environment.

Ecophysiology: ecology of the functions and vital processes of living organisms.

Ecosystem: an open ecological system where biotic and physical elements of the environment interact, characterised by the circulation of nutrients and energy transfer. Biogeosystem.

Ecosystem inventory: a systematic listing of ecosystems with emphasis on biodiversity at ecosystem level.

Ecosystem management: an ecological approach to resource use that harmonises biological and environmental considerations with economic and social needs.

Ecotone: a narrowly defined transition zone that exists between two or more merging communities or landscapes. It usually arises naturally (e.g. at woodland–grassland interfaces).

El Niño: infusions of warm surface waters originating in the mid-Pacific into areas off Ecuador and Peru occurring irregularly every 3–10 years for 6–18 months, during which nutrients are prevented from rising to the surface from the colder layers of water below, causing disruption of the food chain and significant reductions in marine populations.

Environmental impact assessment (EIA): an instrument used to aid and improve the decision-making process. The objective is to determine the potential environmental, social and health effects of a proposed development project. It attempts to assess these effects in a form that permits a logical and rational decision to be made. Attempts can be made to reduce or mitigate any potential adverse impacts.

Environmental wetland management: efforts to minimise the cumulative adverse effects of wetland use and alteration for society as a whole, by making optimal choices in wetland use and conservation.

Epilimnion: the surface water.

Epiphyte: plant that lives on the surface of another plant, used solely as a support.

Eutrophication: accelerated ageing of lakes caused by excessive quantities of nutrients, such as phosphates, increasing the rate of biomass production.

Eutrophy: rich in nutrients.

Exclusive economic zone: a zone of 200 natural miles off the shore of coastal states as defined in the Law of the Sea, allowing a state to exploit this zone exclusively.

Fadama: broad floodplain in the dry Savanna zone of West Africa, often with no clearly defined river channel. Of major importance as cultivation areas.

Flow of species: a permanent or periodic movement of individuals or groups of a given species from one home range to another one.

Gap analysis: an assessment of the protection status of biodiversity in a specified region which looks for gaps in the representation of species or ecosystems in protected areas.

Grey-air city: refers to urban areas mostly localised in cold, moist climates in which the major air pollutants are sulphur oxides and particulates, causing a greyish haze over the city.

Habitat: an environment defined by specific abiotic and biotic factors, in which a organism, a group of organisms or a species lives.

Habitus: general appearance, form or conformation of a plant or animal.

Harratins: people of the Sahara mostly devoted to agriculture in small oases.

Healthy city: an urban area characterised by a clean, safe, high quality physical environment and a sustainable ecosystem, a strong, supportive and participatory community, provision of basic needs, access to a wide variety of experiences and resources and a high health status.

Hepatitis: liver infection.

Heterotroph: organism that does not produce its own food.

Hypolimnion: the water at depth.

Ideles: name of a village of the mountain Ahaggar or Hoggar in central Algerian Sahara.

ILO Convention No. 169: the most comprehensive and up-to-date international instrument on the conditions of life and work of indigenous and tribal people, drafted as a treaty which is legally binding once it has been ratified. Adopted in 1989, it has been ratified by Norway, Mexico, Bolivia, Peru, Colombia, Costa Rica, Paraguay, Honduras, Denmark and Guatemala, as of June 1996.

Indeterminacy (principle of): individual reaches on any river system attempt to adjust their form in such a way as to provide just the energy distribution within the channel required to transport the load upstream. This happens according to adjustments that are impossible to determine.

Integrated river basin management: a management concept directed at making the best use of resources within a river basin, whilst satisfying the needs of different groups depending upon the resources of the river and its surrounding catchment area or flood plain.

Interdependence of wetlands (principle of): implies that wetlands cannot be considered independently or as an isolated ecosystem. They should be managed on a wider ecosystem basis.

Intertidal: coastal land between the low and high water tide marks.

Jokulhlaup: scoured stream side and mudflow.

Kel: the people of a precise geographical location in the Sahara.

Landscape: ecological system that is a spatial and temporal expression of interacting ecosystems and their (natural and/or social) environment; it is characterised by specific structure, functioning and developmental trends.

Landscape, cultural: landscape marked by impacts caused by disturbances brought about by humans.

Landscape, natural: (present) landscape showing no traces of human interference.

Landscape, primeval: (ancient) landscape in which human impacts were not significantly different from those of any other species, due to low population density of hunters and/or gatherers and/or due to their inability to introduce significant changes of ecosystems.

Landscape ecology: a study of structure, function and change of landscapes.

Leishmaniosis: parasitic disease due to different types of 'leishmania' and transmitted by a small fly called 'phlebotomus'.

Leprosy: disease caused by *Mycobacterium leprae*, a microbe similar to that of tuberculosis: usually it causes important mutilations and disability. Hanseniasis.

Lichen: a complex 'plant' made up of an alga and a fungus growing in a symbiotic relationship.

Liveable city: an urban area in which it is pleasant to live and to raise children, characterised by a quiet and friendly environment, areas with a neighbourhood identity, green spaces and urban spaces adapted to the daily needs of people.

M'rabtins or Merabtines: people of northern Sahara, mostly sedentary.

Ma: linear transmission by mother.

Maghreb: western and northern part of the African continent corresponding to the coastal parts of Morocco, Algeria and Tunisia.

Management plan: the formulated detailed method or procedure which a managing board or governing body uses to undertake, manage and/or organise different activities; designed or schemed by the organisation.

Mangrove: tree or shrub commonly growing in the intertidal zone along tropical sheltered coasts, or a community of such plants.

Mega-city: very big city, also megalopolis or metropolis.

Mesotrophy: intermediate nutrition level as compared to oligotrophy and eutrophy.

Metropolis: urbanised region with millions of people, powerful international settlement whose economic and political importance exceeds that of any city that existed in the centuries since the first human settlement.

Moorland: colloquial term for upland vegetation, especially in Britain and Ireland, consisting of coarse grasses and heathers.

Mudflat: area of mud submerged for part of every tidal cycle. Often supporting a high biomass of infauna as well as being important feeding areas for waders and wildfowl.

Mulching: a layer of organic or inorganic material spread on to the ground to protect the underlying soil from raindrop impact, frost or evaporation.

Mutualism: interaction among species that is beneficial for the individuals of each species.

National park: areas established for the protection and preservation of superlative scenery, flora and fauna of national significance which the general public may enjoy and benefit from.

Natural monument: a monument that has developed naturally in nature and capable of economic exploitation.

Nature reserve: a tract of land managed so as to preserve its flora, fauna and its physical features.

Niche: living space.

'No net loss' principle: if a wetland area is lost by a development, the developer should make the best effort to create or restore the functions of an equivalent area of wetland (compensation principle).

Nomadism: the practice of moving from place to place to find pasture.

Oligotrophy: low in nutrition.

Open coast: a regular coast in plan view, along which there is a consistent direction of net longshore sediment transport.

Orography: the study of the Earth's relief.

Pastoralism: the practice of tending animals, esp. sheep or cattle.

Peat: a mass of partly decomposed vegetation. Usually black or dark brown in colour, it occurs in waterlogged conditions which prevent biological breakdown of the plant litter.

Phylum: major division of animal or plant kingdom, containing species having common characteristics and assumed common ancestry.

Physiography: describing the physical geography.

Phytoplankton: tiny floating aquatic plants that form the first link in the marine food chain.

Podsol: a soil typical of the coniferous forest belt of, for example, Northern Europe. It has very clear horizons, including a strongly 'bleached' horizon below the litter and humus layers, where most minerals, organic and clay particles have been washed out of the soil. These are deposited in the lower B horizons, sometimes including a hard 'iron-pan'.

Pollination: process by which a pollen grain moves from the anther sac (the male reproductive organ) of a flower to the stigma (the receptive part of the female reproductive organ) of a flower.

Pollinator: an animal, mostly insects, capable of disseminating pollens among flowers.

'Polluter pays' principle: suggests that the polluter should bear the cost of preventing and controlling pollution. The intent is to force polluters to internalise all of the environmental costs of their activities so that these are fully reflected in the costs of the goods and services they provide.

Precautionary principle: holds that where there are threats of serious or irreversible damage, lack of full scientific certainty should not be used as a reason for postponing cost-effective measures to prevent environmental degradation.

Predator: organism feeding on other organisms.

Protected area: an area, set aside by law for a particular purpose, for example, forestry, wildlife management and/or historical site. Usually, it excludes all other normal public uses.

Radiation stress: as waves move onshore they contain an energy related to the momentum imparted by the moving wave's presence. This energy is not dissipated by the breaking of the wave, rather it is transmitted to a wave-generated current in the surf zone (observed as a longshore current). Radiation stress is the technical name for this momentum element.

Rain forest: evergreen, hygrophilous in character, at least 30 metres high, but usually much taller, rich in thick-stemmed lianes and in woody as well as herbaceous epiphytes.

Ramsar Convention: global intergovernmental treaty providing a framework for international co-operation for the conservation and wise use of wetlands. It was adopted in Ramsar, Iran in 1971.

Reclamation: the process or processes of change that create fertile and productive agricultural land from a previously unproductive state, by using techniques such as drainage, irrigation, deep ploughing, lining and fertilisation.

Resourcism: the traditional form of resource management that revolves around human needs. It promotes, through sustained ecosystem yields, maximum goods production and services provision for human markets. It tends to neglect the limits to ecosystem exploitation.

Rezzou: a group of warriors performing a fast ride to capture camels or slaves.

Rotation: a planned sequence of cropping for a particular area of land designed to produce a balanced demand on soil resources. Usually includes a fallow or rest period.

Rural areas: those areas where many agricultural activities are performed by farming communities.

Salinisation: accumulation of salts in the soil, often due to the excess application of irrigation water in areas where evapotranspiration rates are high and drainage poor.

Savanna: grassland with scattered shrubs and trees; temperate, subtropical or tropical area intermediate between desert and forest; any environment that is dominated by perennial grasses and herbaceous vegetation and supports shrub and tree cover in a heterogeneous pattern.

Seascape: a term suggesting that many landscape ecology concepts are probably useful in understanding oceans and their bottoms particularly in protected area management.

Sediment budget: the balance between sediment entering and leaving a unit of coastline. A budget is expressed as a net term when outputs are subtracted from inputs. A positive beach budget refers to sediment outputs less than inputs and implies a growing beach volume. A deficit beach sediment budget indicates a shrinking beach volume.

Sickle-cell anaemia: hereditary disease caused by anomalous red blood-cells shaped like a sickle.

Sink: a physical space in which sediment resides for a length of time regarded as long compared to the time-scale of the processes involved and where sediment, once in the sink, is not readily removed. Sand dunes can be seen as a sink for beach sediment, as can offshore shoals when supplied from the beach by offshore-moving currents, for example, rip current. Sinks can be permanent or transient depending on the relative time-scale of processes involved. A coastal cell by definition has a sink at the down-drift end of the cell.

Species capacity: limits of any biochemical, cellular or physiological mechanism permitting survival in a given species; example for Man, sodium in blood should maintain between 136 and 150 milli-equivalent/litre.

Succession: the sequential change in communities (vegetation) either in response to environmental changes or induced by the intrinsic properties of the plants themselves. Classically, the term refers to the colonisation of a new physical environment by a series of vegetation communities until a final equilibrium state, the climax, is achieved.

Surf zone: as waves move onshore they are readily deformed in height and spacing as they respond to a diminishing depth of water (refraction). In very shallow water a wave breaks and this defines the seaward edge of the surf zone. On wide, very low angle beaches, smaller waves can propagate after the initial breaking across the beach before final breaking occurs. This defines the landward limit of the surf zone. Within the surf zone much of the longshore directed sand transport is undertaken by wave-generated longshore currents. The surf zone is the area where longshore transport on sand beaches is dominant.

Sustainable agriculture: systems of farming that maintain the resources upon which production depends.

Swash zone: once a wave has achieved final breaking the energy of the wave form is translated into a kinetic energy of a volume of turbulent water rushing up a beach. This uprush is known as swash. The returning water draining back to sea is the backwash. The swash zone is the area between final breaking and the limit of this uprush. When waves break at an angle to the shoreline the resulting swash carries sediment longshore. The swash zone is the area where longshore transport on gravel beaches is dominant.

Tamahaq, tamacheq: the Berber language of Tuareg, 'tamahaq' in the north, 'tamacheq' in the south.

Tamezlayt: tribute paid annually by dependent Tuareg tribes to 'noble' ones.

Tifinars: Berber consonant signs inherited from punic writing.

Timber line: the upper limit of trees of forest stature and density on a mountain.

Troposphere: the lower layer of the atmosphere, reaching around 16 km above the earth at the equator, 11 km at 50°N and S and 9° at the poles.

Trypanosomiasis: parasitic diseases caused by protozoans of the genus *Trypanosoma*, for example, the sleeping sickness of Africa and Chagas' disease of tropical America.

Tuareg, targui: Berber-speaking nomadic herders of central Sahara.

Urban areas: integral centres of production and communication in a highly interdependent global network.

Wadi Rhir and Wadi Mya: dry valleys of northern Algerian Sahara where the towns of Tuggurt and Wargla and many smaller oases are located.

Water balance: summarises the inputs, outputs and storage of water for any given catchment.

Wave-sediment cell: the cell shown to be the fundamental basis for all human intervention in the coastal zone. The interaction of breaking waves generating longshore sediment transport potentials and sediment availability for longshore transport defines the basic energy and mass which are elements of the coastal cell. The source of coastal sediment, the transport corridor and the deposition sink are the three elements that define a wave-sediment coastal cell in a down-drift direction. The spatial link between these three units defines the extent of the coastal cell alongshore. Generally a cell identifies an absence of beaches at the start of the cell and then progressive development of beaches in a down-drift direction. Cells can be hierarchically grouped with a consistent direction but spatially variable longshore transport in open coasts. Closed coasts show a high potential for reversals in cell transport directions.

Wetland: place where dryland meets, or is inundated by, water. Land where an excess of water is the dominant factor determining the nature of soil development and the types of animals and plant communities living at the soil surface.

Wetland management: the integrated and simultaneous control over multiple features of wetlands, such as river water level, soil water level, fish stock, wildlife habitat, etc.

Wilderness: an area where the influence of the modern human is absent or minimised.

Wildlife management area: an area of communal land that contains natural resources where the people have use rights and the mandate to protect the natural resources contained therein, and from which they are allowed to retain a significant proportion of revenue.

Wildlife zone: an area set aside by the management plan or land-use plan for the purpose of managing wildlife and all other uses that are compatible with the permissible uses and objectives of each protected area.

Wise use of wetlands (principle of): sustainable utilisation of wetlands for the benefit of humankind in a way compatible with the maintenance of natural properties of the ecosystem.

World Heritage Convention: adopted in 1972, this convention promotes the international co-operation for the protection of the world cultural and natural heritage, composed of sites that are considered to have universal value due to their historic, artistic, scientific or aesthetic characteristics.

Zooplankton: tiny marine animals that feed on phytoplankton and are in turn fed upon by larger marine creatures.

ANSWERS TO SELF-ASSESSMENT QUESTIONS

CHAPTER 1

1 (d)
2 (d)
3 (a), (d), (f) and (g) should by definition appear in coastal cells as the main elements of the cell are the source area, the transport corridor and the sink area, with cell boundaries defined by positions of nil longshore beach transport.
4 A purist might say that all cited activities should be banned. However, socio-economic reasons may prevail and from a cost–benefit point of view not all of the choices are equally deleterious to the coast.
5 Match persons (a) and (g) with reason 1, persons (b), (c) and (d) with reason 2 and persons (e) and (f) with reason 3.
6 Items (a), (c) and (e) should not appear in an ecologically-led management programme as they are items that implicitly require human valuation above ecological valuation.
7 The UK and Bangladesh have reactive coastal management programmes, while the USA has elements of a proactive perspective.

CHAPTER 2

1 (c), 2 (c), 3 (e).

CHAPTER 3

1 (d), 2 (c), 3 (b), 4 (b), 5 (b)*, 6 (c), 7 (a) and (c), 8 (e), 9 (d).

* Increased slope may increase velocity. But increased velocity can also be achieved by increased depth and/or decreased roughness and/or increased slope. More usually, all the dependent channel morphology variables (width, depth, slope and roughness) react to changes in streamflow or sediment supplied from upstream and no one type of change can be predetermined. Remember also that slope decreases in the downstream direction while mean stream velocity in relation to streamflow increases or stays constant.

CHAPTER 4

1 (b), 2 (c), 3 (b), 4 (c), 5 (b), 6 (a), 7 (a), 8 (b) and (c), 9 (b).

CHAPTER 5

1 (a) (iii), (b) (ii), (c) (iii).
2 and 3 See subsection on Upland Reclamation, pp. 91–94.

4 Your diagram should include: drainage and its effects on plant communities, on the peat itself and on river regimes; fertiliser applications and their effects on plants, rivers, fauna; planting patterns and their effects on flora and fauna, on drainage, on scenery; effects of the size of area planted on fauna especially birds.

5 See section on Recreation, pp. 100–102 and 104, and associated reading.

CHAPTER 6

1 (c), 2 (e), 3 (d), 4 (c), 5 (a), 6 (e), 7 (a), 8 (e), 9 (d), 10 (a).

CHAPTER 7

1 (c) The others include underlying factors such as overpopulation, triggers such as climatic change and local factors such as urbanisation. Deforestation is the only widespread activity that leads directly to desertification.

2 (d).

3 There is no one correct answer to this question, the reader must ensure that the actions that are chosen apply to as many areas as possible, initiate other actions, are practically feasible and, most important, will markedly improve the environment and the lives of the people who live in threatened areas.

4 (b).

5 (b).

6 (d) Although the other definitions appear feasible, desertification can occur in areas at some distance from a desert margin (a) and the others refer to drought (b) and famine (c) — both of which need not result from or cause desertification.

CHAPTER 8

1 (b), 2 (c), 3 (a), 4 (d), 5 (a), 6 (d).

CHAPTER 9

1 (b), 2 (c), 3 (c), 4 (b), 5 (c), 6 (b), 7 (a), 8 (a), 9 (b), 10 (c).

CHAPTER 10

1 (a), (b), (c); 2 (a), (b), (c); 3 (b), (c); 4 none; 5 (a), (b), (d); 6 (b), (d).

CHAPTER 11

1 (b), 2 (a), 3 (b), 4 (a), 5 (a).

CHAPTER 12

1 (c), 2 (b), 3 (c), 4 (a), 5 (c), 6(b), 7(a).

CHAPTER 13

1 (b), 2 (d), 3 (d), 4 (c), 5 (c), 6 (a).

CHAPTER 14

1 (b) and (c).

2 (a) and (b) because 90 per cent of Tuareg herding activities are performed under the limit of 200 mm per year.

3 (a).
4 (d).
5 (d) because the feeling of social deprivation corresponds to a lack of information, of equipment, of education, and of development with bad consequences for the health of populations.
6 (c).

Index to *Environmental Management in Practice*, volumes 1–3

NIMBY syndrome **1** 163, 437; **2** 31, 34, 44
NIMTO syndrome **2** 44
nitrates **1** 109; **2** 20, 60
Nitrates Directive **1** 410
nitrogen **2** 19–21; biological fixation **2** 173–4; conversion in soil **2** *21*; cycle **2** 20–1
nitrogen dioxide, monitoring **2** 89
nitrogen fixation, savannas **3** 114
nitrogen oxides **2** 75, 236, 263; acid rain **2** 221; aircraft **2** 273, 274; transport **2** *263*
nitrous emissions **2** 118, 119, 265
NMHC *see* non-methane hydrocarbons
no effect concentration (NOC) **1** 173
no observed adverse effect level (NOAEL) **1** 112; *see also* zero....
no observed effect concentration (NOEC) **1** 173, 180–1
NOAEL *see* no observed adverse effect level
NOC *see* no effect concentration
NOEC *see* no observed effect concentration
noise dose formula **2** 103, *104*
noise pollution **2** 96–109; control **2** 104–7; emissions **2** 101–2, 104; generation factors **2** 102; health effects **2** 102–4; immission **2** 104–5; legislation **1** 412; **2** 107–9; sources **2** 101–2; tourism **2** 284; transmission **2** 104–5; urban environments **3** 206; well-being effects **2** 104
noise pollution level (NPL) **2** 101
noise rating (NR) **2** 100, 101
noisiness **2** 100–1
nomadic people, Tuareg **3** 252–61
non-carcinogens: drinking water **1** 116; guideline values **1** 112
non-fossil fuels **2** 126, 128–32, 242–3
non-governmental organisations (NGOs) **1** 372, 464, 473
non-malignant carcinogens **1** 115–16
non-methane hydrocarbons (NMHC) **2** 90
non-point pollution **2** 58
non-renewable resources **1** 220
non-wood pulp **2** 246
Nordhaus, W.D. **2** 136–7
normal forest concept **2** 223–4
normative contextualisation **1** 32–3
normative observers **1** 30
normative statements *see* values
norms *see* values
North America *see* United States
North American Free Trade Agreement (NAFTA) **1** 341
North Sea **2** 63; **3** 24–5
Northern Ireland: ESA **2** 176; growing season **3** *88*; Lough Neagh **3** 54–5, *54*; mountain regions **3** 91–2, *92*, 95; overgrazing **3** 95; peat extraction **3** 97–9; river management **3** 49; upland areas **3** 95
Not In My Back Yard *see* NIMBY syndrome
NPL *see* noise pollution level
NPV *see* Net Present Value
NR *see* noise rating
NRA *see* National Rivers Authority

NSR *see* New Source Review
nuclear power **1** 12–13; **2** 129, 131–2, 242
Nuisance Act **1** 438
nutrients **2** 18–22
nutrition, Mixtecos Indians **3** 263–4

objectives *see* values
oceans: atmospheric energy exchange **2** 120; forcing **3** 9; *see also* fisheries; sea water
ODSs *see* ozone-depleting substances
Odum's scheme **3** 232, 235
OECD framework **1** *96*, 135, 136, 180
Office of Management and Budget (OMB) **1** 384
Official Journal of the Communities (OJ) **1** 407–8
oil **1** 4; Amazonia **3** 161–2; pollution **2** 274–7; *see also* seed-oil
OJ *see* Official Journal of the Communities
oligotrophic lakes **3** 52, *53*
open access fisheries **2** 206
open-cycle economic systems **2** 230, 231
optimal global paths **2** 140
optimum yield (OY) **2** 207
option generation **2** 248, 249
Orangi Pilot Project (OPP) **3** 210
organic farming **2** 172, 174, 175, 191
organic manures **2** 173, 174
organic waste: lakes **3** 55; rivers **3** 45, *46*
organisational audits **1** 260
organisational change **3** 4
organismal endpoints **1** 177, 178
overcultivation **3** 139–40
overgrazing **3** 92–3, 95, 105, 140
overlapping consensus **1** 55–7, *55*
oxygen **2** 219–20
oxygen depletion, lakes **3** 55
oxygen sag curves **3** 45, *46*
OY *see* optimum yield
ozone **2** 75, 119, 266; depletion **1** 73–4, 366; **2** 273; monitoring **2** 89, 90; *see also* Vienna Convention on the Protection of the Ozone Layer (1985)
ozone-depleting substances (ODSs) **1** 352–3, 362

PACD *see* Plan of Action to Combat Desertification
PAHs *see* polycyclic aromatic hydrocarbons
Pakistan, irrigation **3** 142
palustrine wetlands **3** *66*, 67–8, *67*
Pan-European Biological and Landscape Diversity Strategy **3** 231
paper **2** 246
Papua New Guinea **2** 186
pararosniline method, SO[2] **2** 89
Paris, water **3** 206
Paris Summit **1** 405, 417
park and ride **2** 265–6
partial problem analysis **1** 31
pastoral nomadism **3** 257–8